Internet of Things

Technology, Communications and Computing

Series Editors

The series Internet of Things - Technologies, Communications and Computing publishes new developments and advances in the various areas of the different facets of the Internet of Things. The intent is to cover technology (smart devices, wireless sensors, systems), communications (networks and protocols) and computing (theory, middleware and applications) of the Internet of Things, as embedded in the fields of engineering, computer science, life sciences, as well as the methodologies behind them. The series contains monographs, lecture notes and edited volumes in the Internet of Things research and development area, spanning the areas of wireless sensor networks, autonomic networking, network protocol, agent-based computing, artificial intelligence, self organizing systems, multi-sensor data fusion, smart objects, and hybrid intelligent systems.

Indexing: *Internet of Things* is covered by Scopus and Ei-Compendex **

Rohit Sharma • Gwanggil Jeon • Yan Zhang
Editors

Data Analytics for Internet of Things Infrastructure

Springer

Editors
Rohit Sharma
SRM Institute of Science and Technology,
Delhi NCR Campus
Ghaziabad, Uttar Pradesh, India

Gwanggil Jeon
Incheon National University
Incheon, Korea (Republic of)

Yan Zhang
Xi'an Shiyou University
Xi'An, China

ISSN 2199-1073 ISSN 2199-1081 (electronic)
Internet of Things
ISBN 978-3-031-33807-6 ISBN 978-3-031-33808-3 (eBook)
https://doi.org/10.1007/978-3-031-33808-3

This Springer imprint is published by the registered company Springer Nature Switzerland AG
The registered company address is: Gewerbestrasse 11, 6330 Cham, Switzerland

Paper in this product is recyclable.

Preface

The edited book aims to bring together leading academic scientists, researchers, and scholars to exchange and share their experiences and results on all aspects of IoT and big data analytics. It also provides a premier interdisciplinary platform for researchers, practitioners, and educators to present and discuss the most recent innovations, trends, and concerns as well as practical challenges encountered and solutions adopted in IoT and data analytics.

Ghaziabad, Uttar Pradesh, India — Rohit Sharma
Incheon, Korea (Republic of) — Gwanggil Jeon
Xi'An, China — Yan Zhang

Organization of the Book

The book is organized into 17 chapters. A brief description of each of the chapters is as follows:

Chapter "Big Data in Cloud Today: A Comprehensive Survey" enunciates big data's importance, characteristics, and classification with relevant examples. It also presents the tools and techniques used for processing of big data. In addition, the concept, working, characteristics, and key features of cloud computing are discussed. Ultimately, this chapter correlates the technologies – big data and cloud computing – in the today's scenario with a case study.

Chapter "Cloud of Things Platform for a Water Meter Network" analyzes and modifies the IoT architecture for designing an autonomous and distributed IoT module. The module enabled M2M (machine-to-machine) communication to reduce the problem and add scalability to IoT devices. The designed module can be built on IoT devices to directly connect with the network through Ethernet or Wi-Fi and provide an interface to the user (cloud server, person, device) to communicate with each other.

Chapter "Online Newspaper Development within the Internet of Things Environment: The Role of Computer-Mediated Communication" includes a social approach to developing online newspapers within IOT environment and discusses how CMC plays a significant role in online journalism. CMC in journalism also enables the readers to put their thoughts and views in comments under the provided section of online newspapers. Communication within IOT environmental online newspapers has three inherent characteristics.

Chapter "FATS (Fuzzy Authentication to Provide Trust-Based Security) in VANET to Mitigate Black Hole Attack", In this chapter, Fuzzy Authentication is used to provide trust-based Security to find a Black Hole node and block the node from communication. FATS system uses Trust factor as the significant parameter to estimate a genuine node from a malicious node. The simulation is carried out in MATLAB for parameter estimation and implemented using Network Simulator 2.28 software.

Chapter "AI-Based Chatbot Agents as Drivers of Purchase Intentions: An Interdisciplinary Study" aims to determine the impact of chatbots on customers' purchase intentions. An empirical study was carried out on the impact of chatbot agent's informational support, emotional credibility, and trust on purchasing intentions. The data was collected through an online survey from 223 Delhi-NCR customers who use chatbots while making online purchases. PLS-SEM was used to analyze the data that was collected. The results of structural equation modelling (SEM) showed a significant impact of informational support, emotional credibility, and trust of chatbots on purchase intentions of customers.

Chapter "An Intelligent Model for Identifying Fluctuations in the Stock Market and Predicting Investment Policies with Guaranteed Returns" analyzes the current and market tendencies' shortcomings and constructs a time-series version for mitigating most of them by imposing more efficient algorithms. An expert machine is proposed to predict the marketplace risk's uncertainty and the assured return amount. Fuzzy Inference is deployed to deal with the uncertainty and make a prediction for the same.

Chapter "Sandwiched Metasurface Antenna for Small Spacecrafts in IoT Infrastructure" the main targets of this study is the use of an optimized sandwiched metasurface for increasing return loss and peak gain of the proposed antenna at the X-band. Moreover, the sandwiched metasurface is used for minimizing levels of generated back lobes and so interferences with electronic components inside the spacecraft box.

Chapter "Development of Laser-Beam Cutting-Edge Technology and IOT-Based Race Car Lapse Time Computational System" proposes a system that uses a laser beam transmitter and receiver to detect laps crossing. The ATMEGA 328 controller continuously triggers the laser transceiver. This system provides the solution for the time delay in lapse calculation for racing. The detection of the vehicle is based on beam cutting.

Chapter "A Study of Cloud-Based Solution for Data Analytics" studies the provisioning and usage of the cloud-based architecture of AWS and GCP for building a data analytics platform. With the introduction of big data technologies, cloud computing, and different types of data analytics technique, it now became easier to combine real world data and data generated from scientific experiments to extract meaningful insights and use them in real-world scenarios.

Chapter "An Intelligent Model for Optimizing Sparsity Problem Toward Movie Recommendation Paradigm Using Machine Learning" proposes a system that uses a laser beam transmitter and receiver to detect laps crossing. The ATMEGA 328 controller continuously triggers the laser transceiver. This system provides the solution for time delay in lapse calculation for racing. The detection of the vehicle is based on beam cutting.

Chapter "Techniques to Identify Image Objects Under Adverse Environmental Conditions: A Systematic Literature Review" references all the research articles published between 2011 and 2022 in various IEEE Xplore, ScienceDirect conferences, and journals for the systematic review on identifying different objects from images and videos taken in adverse environmental conditions. We used different tags and keywords to search papers about the topic under study.

Chapter "Technology-Enhanced Teaching and Learning During the COVID-19 Pandemic Using IoT Infrastructure" reviews learning during the COVID-19 pandemic-related studies and perspectives on technology-assisted language teaching and learning to make recommendations for administrators, teachers, and learners. It first reviews current perspectives on technology-assisted teaching and learning. Then, it critically examines second language acquisition theories aligning with computer-mediated communication.

Chapter "The Symbiotic Relation of IoT and AI for Applications in Various Domains: Trends and Future Directions" aims to showcase the current trend of IoT with artificial intelligence and future directions. Many domains have been analyzed and showcased in the tabular format where the methodology advantages and future scope with respect with AI-assisted IoT technologies are identified.

Chapter "Text Summarization for Big Data Analytics: A Comprehensive Review of GPT 2 and BERT Approaches" looks at extensive data methodologies and methods such as Bidirectional Encoder Representations (BERT) and Generative Pre-trained Transformer 2 (GPT 2) transformers for multi-document summarization. In transformers, the BERT and GPT 2 models in-text summarization give very close results in terms of accuracy, and they need to be compared to give a model that performs better.

Chapter "Leveraging Secured E-Voting Using Decentralized Blockchain Technology" provides detailed information on the blockchain concept, features, and types. The chapter also discusses a few major consensus algorithms frequently utilized in different blockchain networks. Further, the major challenges faced by existing voting methodologies are addressed and the chapter explains how integrating the e-voting systems with blockchain technology can help overcome most of these challenges.

Chapter "Multilayer Security and Privacy Provision in Internet of Things Networks: Challenges and Future Trends" demonstrates security challenges and countermeasures. Moreover, the chapter also gives the existing security models and solutions. The chapter concludes with future trends in IoT.

Chapter "A Methodology for the Development of Soft Sensors with Kafka-ML" presents a methodology to support soft sensor development based on Kafka-ML, an open-source framework to manage ML pipelines. Kafka-ML will allow researchers to develop, train, and validate ML models, and visualize real-time predictions using streaming data.

Key Features

- This book will help generate interest in new trends in low-voltage circuit design for IoT applications.
- This book addresses the challenges in designing low power architectures for IoT applications.

- This book provides a comparative analysis of different techniques used for designing low power architectures.
- This book contains an analysis of different low power architectures.
- This book provides a practical understanding of the uses of semantic technology in designing, so the readers can improve their strengths in making better decisions.

About the Book

This book focuses on the use of the Internet of Things (IoT) and big data in business intelligence, data management, Hadoop, machine learning, cloud, smart cities, etc. IoT and big data emerged from the early 2000's data boom, driven forward by many of the early Internet and technology companies. The Internet of Things (IoT) is an interconnection of several devices, networks, technologies, and human resources to achieve a common goal. A variety of IoT-based applications being used in different sectors have succeeded in providing huge benefits to users. The generation of big data by IoT has ruptured the existing data processing capacity of IoT and recommends adopting data analytics to strengthen solutions. The success of IoT depends upon the influential association of big data analytics. New technologies like search engines, mobile devices, and industrial machines provided as much data as companies could handle—and the scale continues to grow. In a study by IDC, the Market Intelligence firm estimated that global data production would grow 10x between 2015 and 2020. So, the proposed book will cover all the aspects discussed in the field above.

Contents

About the Editors

Rohit Sharma is currently an Associate Professor in the Department of Electronics and Communication Engineering, SRM Institute of Science and Technology, Delhi NCR Campus Ghaziabad, India. He is an active member of ISTE, IEEE, ICS, IAENG, and IACSIT. He is an editorial board member and reviewer of more than 22 international journals and conferences, including the topmost journal *IEEE Access* and *IEEE Internet of Things Journal*. He serves as a book editor for 24 different titles.

Gwanggil Jeon received the B.S., M.S., and Ph.D. (summa cum laude) degrees from the Department of Electronics and Computer Engineering, Hanyang University, Seoul, Korea, in 2003, 2005, and 2008, respectively. He was with the Department of Electronics and Computer Engineering, Hanyang University, from 2008 to 2009. He was with the School of Information Technology and Engineering, University of Ottawa, Ottawa, ON, Canada, as a Post-Doctoral Fellow, from 2009 to 2011. He was with the Graduate School of Science and Technology, Niigata University, Niigata, Japan, as an Assistant Professor, from 2011 to 2012. He is currently an Assistant Professor with the Department of Embedded Systems Engineering, Incheon National University, Incheon, Korea. His current research interests include image processing, particularly image compression, motion estimation, demosaicking, and image enhancement, and computational intelligence, such as fuzzy and rough sets theories. Dr. Jeon was a recipient of the IEEE Chester Sall Award in 2007 and the ETRI Journal Paper Award in 2008.

Yan Zhang received the Ph.D. degree from the School of Physics and Optoelectronic Engineering, Xi'dian University, China. The main research directions are optical communication, quantum communication, nonlinear optics, and device reliability.

Big Data in Cloud Today: A Comprehensive Survey

K. Umapathy, S. Omkumar, S. Chandramohan, D. Muthukumaran, and Wasana Boonsong

1 Introduction

Big data refers to a great collection of data which cannot be stored or processed by any means of data storage or processing units at traditional level. It is produced at a larger scale and employed by various companies for processing and analyzing in view of improving their business activities [1]. Moreover, it is also meant for uncovering the insights relevant to those companies. Big data is a blend of organized and unorganized data collected by various organizations, meant for mining and used for applications such as predictive modeling, projects relevant of machine learning, and data analytics. As a typical example, the application of amazon – "Elastic Map Reduce" – illustrates how the power of cloud elastic computers is employed for the processing of big data [2]. Cloud is a blend of servers with each server having a unique function. It is not a physical entity, but a very wide network that comprises remote servers all over the world. These servers are integrated together in order to operate as a single entity. They are meant to store and manage data, execute applications, and provide services such as video streaming, electronic e-mail, and software for productivity. Data or information can be accessed anywhere from online rather than retrieving from the local computing system. Hence, cloud computing can be referred to as utility computing or on-demand computing [3].

K. Umapathy (✉) · S. Omkumar · S. Chandramohan · D. Muthukumaran
Department of ECE, SCSVMV Deemed University, Kanchipuram, India
e-mail: chandramohan@kanchiuniv.ac.in

W. Boonsong
Department of EE, Rajamangala University of Technology, Songkhla, Thailand
e-mail: wasana.b@rmutsv.ac.th

R. Sharma et al. (eds.), *Data Analytics for Internet of Things Infrastructure*, Internet of Things, https://doi.org/10.1007/978-3-031-33808-3_1

Fig. 1 V's of big data

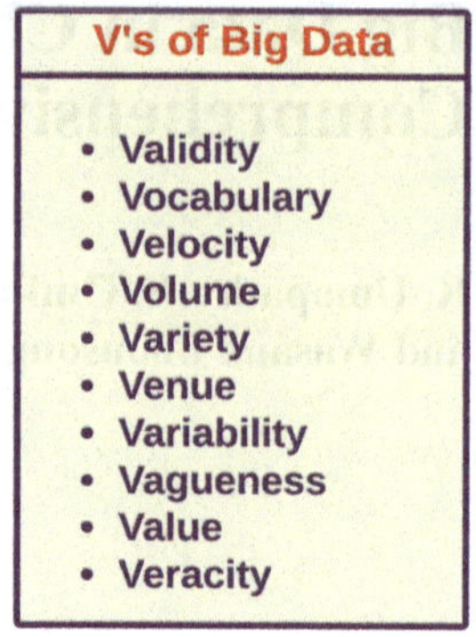

2 Characteristics of Big Data

Data management architectures in various companies include systems to process and manage big data integrated with relevant tools for analytics. Figure 1 shows the various types of V's connected with big data [4]. Big data is usually characterized by the following three V's:

- Large *volume* of data in many environments
- Wide *variety* of data types stored in big data systems
- *Velocity* at which data is produced, collected, and processed

The above three characteristics were resolved in the year 2001 by Doug Lancy, an analyst who belonged to a consulting company – Meta Group Incorporation. Recently, in addition to the existing V's, other V's were included to explain big data [5].

Figure 2 shows the six V's of big data. There is no specific volume to equate with big data but applications generally involve terabytes, petabytes, and even exabytes of data generated and collected with respect to time. Big data is a blend of organized and unorganized data collected by various institutions and is subject to mining for information [6].

- Organized data such as documents and transactions linked with finance
- Unorganized data such as text, records, and multimedia files
- Semi-structured data such as web logs and data streaming

Hence big data is voluminous and more complex sets of data especially from new sources of data. The voluminous of data makes it very difficult for processing using suitable traditional software. But big data can be employed to sort out problems in business which would have been not possible in the past earlier [7]. Figure 3 shows the five key characteristics of big data.

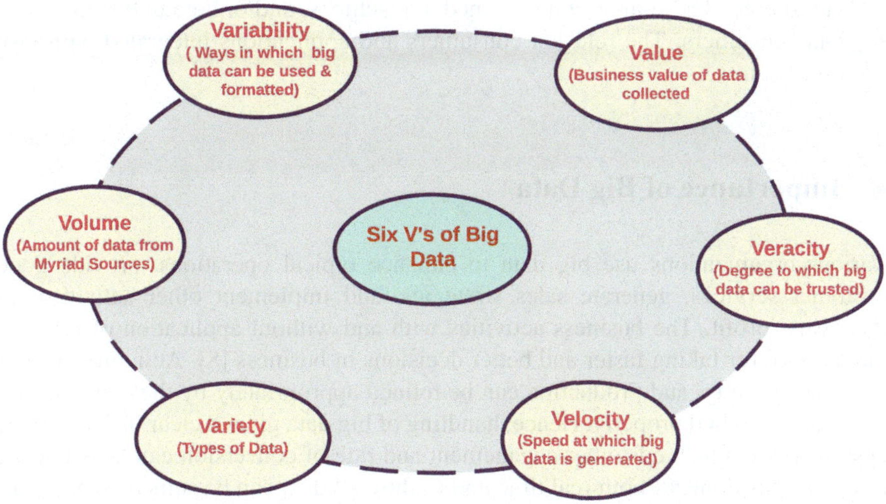

Fig. 2 Six V's of big data

Fig. 3 Five characteristics of big data

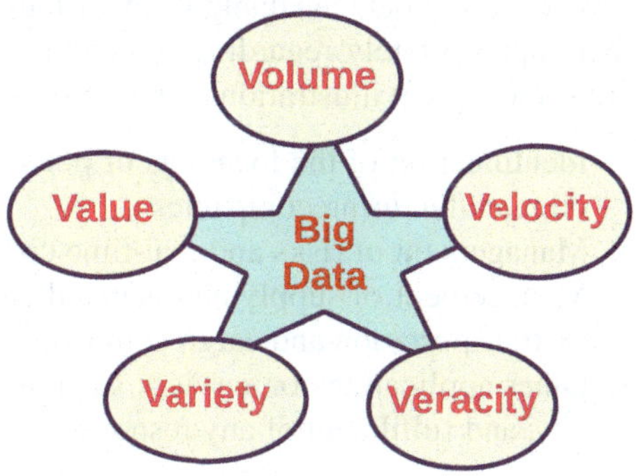

3 Classification of Big Data

Big data is classified into the following types:

- Analysis type: The type in which data is analyzed either in real time or processed in batch. Typical examples for real time and batch process are fraud detection in banking and strategic decisions in business, respectively.
- Processing methodology: The business requirements have to decide the methodology for processing of data – either predictive or ad-hoc in nature.
- Data frequency: Indicates the frequency of data being ingested and arrived. The nature of data shall be either continuous or time dependent.
- Data type: Indicates the stream of data – transactions, history, or real time.
- Data format: The type of format used for data –structured, unstructured, or semi-structured. The format type decides the type of data for storage and processing.

- Data source: The source of data – media, machines, and/or human beings.
- Data consumers: The relevant consumers and applications integrated with processed data

4 Importance of Big Data

Various organizations use big data to enhance typical operations, provide good customer services, generate sales strategies, and implement other activities for increasing profit. The business activities with and without application of big data matter a lot for taking faster and better decisions in business [8]. Activities such as sales, advertising, and production can be refined appropriately by the companies if big data is handled properly. Hence, handling of big data gives a clear insight for the customers by which customer engagement and rate of conversion can be enhanced easily. In this context, both real time and traditional data can be subject to analysis. Medical professionals also use employ big data for identifying health complications and risk factors, making a diagnosis, and establishing the medical history of patients. Moreover, the data available from various sources of media and internet will guide them appropriately regarding infections of diseases and pandemic. The following are some typical illustrations for employment of big data by certain organizations.

- Identification of the locations of potential drilling and operations of pipeline by oil manufacturing companies
- Management of risks and real-time computation of data by financial companies
- Management of supply and demand chains with appropriated routes of delivery by transportation and cargo companies
- Other applications or services such as prevention of crime, initiatives for smart city, and fulfilment of any response on demand.

The following are the list of activities connected with business ranging from consumer experience to statistical analytics:

- Development of products
- Maintenance of predictive services
- Experience of the customers
- Machine learning
- Efficiency in operations
- Innovative activities

Big data provides an opportunity for all business models, which includes the following vital actions:

- Integrate
- Manage
- Analyze

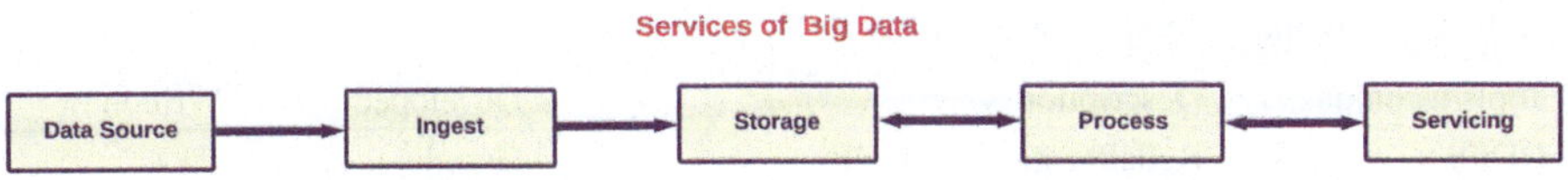

Fig. 4 Services of big data

Fig. 5 Examples of big data

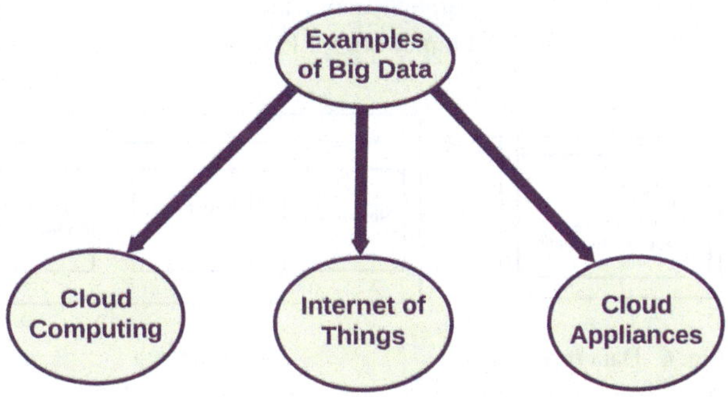

5 Examples for Big Data

Figure 4 shows the services of big data – data sources, ingest, storage, processing, and servicing. Big data comes from different forms of sources. The typical illustrations are processing systems, databases of consumers, files, electronic mails, medical histories, etc. [9]. It also comprises data generated from log files of server and from sensors connected with manufacturing and industrial equipments and IoT devices. Figure 5 shows the various examples for big data usage.

Apart from the availability of internal data from the concerned systems, the environments of big data also include external information from consumers, open markets, traffic conditions, weather conditions, information connected with geographical locations, research, etc. The different forms of big data are images, videos, audios, and various applications that represent streaming to data subject to processing and collection on a regular arrangement [10].

6 Tools and Techniques

The big data applications include various tools and methodologies for processing and analysis of the data. Table 1 shows few relevant tools and techniques [11].

Table 1 shows that Hadoop Map Reduce is a technique for distributed processing framework, which is meant for preparing applications that handle huge volume of data with hardware resources in an efficient way with minimum number of errors. The task of Map Reduce work is to decompose the given information into small pieces which are handled by it in parallel manner. The values resulting from these mapping by the above framework are provided as inputs for the reduction of tasks.

Table 1 Tools and techniques used for Big Data

Tools/techniques	Description	Developed by	Written in
HDFS	Reliable storage of data	Google	Java
Map reduce	Framework for distributed processing	Google	Java
YARN	Framework for resource management	Apache	Java
Spark	Parallelism of data	Berkeley	Scala
Pig	Framework for Pig Latin	Yahoo	Java

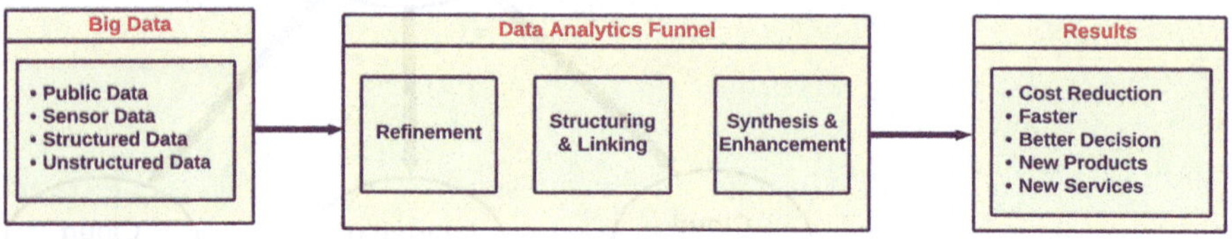

Fig. 6 Data processing

There is storage in the file system for preserving both the inputs and outputs. Scheduling of tasks, tracking, and executing of unsuccessful tasks are managed by the framework [12].

HDFS is another technique that covers all the nodes in the form of cluster for the purpose of storing data. It integrates all file systems of the network connected with local nodes to construct a larger system of file. This tool enhances the reliability by duplicating the data across various relevant sources for overcoming failures in nodes.

7 Big Data Analytics and Its Benefits

Big data analytics is one among the analytic techniques at advanced level for diversifying great volume of big data in terms of sets, which involve both organized and unorganized forms of data from various sources in different sizes ranging from terabytes to zettabytes [13]. The following are the advantages of Big Data Analytics:

- Speeder and comparatively better decision-making
- Economical and efficient in operation
- Enhanced driving of data with respect to market (Fig. 6)

Fig. 7 Concept of cloud computing

8 Cloud Computing

A cloud can belong to either private or public party. A cloud in public provides services to the concerned with the help of internet. A cloud in private is an owned data center of a person, which provides hosted services to certain people with specific settings of permissions. However, the objective of cloud computing is to give easy means of access to specific resources of computing and services of IT [14, 21]. Figure 7 shows the concept of cloud computing.

9 Working of Cloud Computing

The concept of front end comprises client device usage, type of browsing software, and applications connected with cloud software. The concept of back end comprises databases and server computers that operate as repository for storage of data used by the front end. A server at the center manages the communication between the front and back ends of cloud computing. This server depends on certain protocols to implement the data exchange [15, 22]. It employs both software and middleware to establish connection between various devices and servers in the cloud. Figure 8 shows the features and characteristics of cloud.

Cloud computing mainly depends on the virtual concept and technologies of automation. By the concept of virtualization, customers can access services of cloud

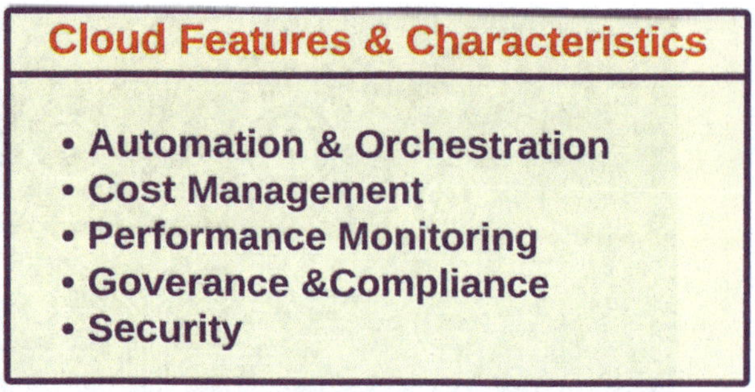

Fig. 8 Cloud features and characteristics

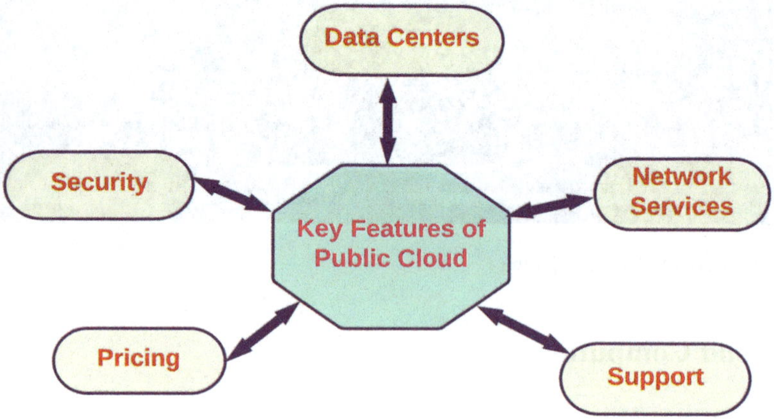

Fig. 9 Key features of a public cloud

and utilize them appropriately. The customers can access the relevant resources, integrate services, and employ workloads without any sort of disturbance from the IT staff of cloud provider [16, 23]. Figure 9 shows the key features of a public cloud. The following are the characteristics and benefits of cloud computing.

- Reliability
- Flexibility
- Reduction of cost
- Global deployment in shorter time
- Payment based on usage
- Resilience in workload
- Provision of self services
- Broad Access in Network
- Pooling of available resources
- Management of cost toward essential services
- Mobility in data and job

Fig. 10 Service categories of cloud computing

Disadvantages of Cloud Computing

The following are the demerits of cloud computing:

- No Security in the cloud
- Unpredictability in cost factor
- No expertise in specific domain
- Mismatch of governance in IT
- Non-compliance with laws in the industry
- Cloud management is hard to implement as every cloud is unique
- No optimum performance in cloud
- Hard to construct a private cloud
- Migration in cloud

Cloud Computing Versus Traditional Web Hosting

The characteristics of a cloud are totally different from that of traditional web hosting for the following reasons:

1. Huge volume of power computing can be accessed by the consumers on the basis of demand. Hence, it is sold at that particular minute or hour of demand.
2. It is totally flexible – service is provided to the users as much as they want with respect to capacity and time interval.
3. Management of all services will be done by the provider completely except for a computing system and internet connectivity. The interest in cloud computing increased due to better innovations in computing and virtualization.

Cloud Service Providers

Figure 10 shows the service categories of cloud computing. There is no shortage of providers in the market of cloud computing. Figure 11 shows the major service providers in this field. Following are the companies who are experts the industry:

- Amazon Web Services (AWS)
- Google Cloud Projects (GCP)
- Microsoft Azure

Fig. 11 Cloud service providers

The list of other companies who excel in cloud computing are as follows:

- Apple
- Citrix
- IBM
- Salesforce
- Alibaba
- Oracle
- VMware
- SAP
- Joyent
- Rackspace

Factors Influencing Companies to Become Cloud Computing Providers

- *Money factor:* Most of the large datacenters used to buy hardware, bandwidth of network, and computing power for one-fifth of the cost provided to middle-level datacenter [17]. Moreover, the money spent for the development of software and deployment can be extended to any number of computers. Hence, there is a possibility for a large company to make a huge profit with less investment spent as such for that of a small company.
- *Existing investment*: If the services of a cloud computing are concatenated to the existing network, then it will provide a stream of revenue system with less cost and investments. The web techniques prepared by Amazon are well applicable for the internal operations of Amazon.
- *Defend a franchise*: Since the server and applications at traditional level are able to embrace the technique of cloud computing, vendors having interest in building franchise will be encouraged to provide the choice of cloud. As a typical example, Microsoft Azure gives a migrating way for the existing users to cope with a cloud environment [18, 24].
- *Attack an incumbent*: If a company has sufficient resources and proper datacenter, it may be able to build a path for deployment of cloud leading to automation with respect to features in scalability and balancing of load.
- *Customer relationships:* The relationship with the customers can be build by providing service in the name of branded cloud computing which leads to totally free pathway for both the companies and the customers.

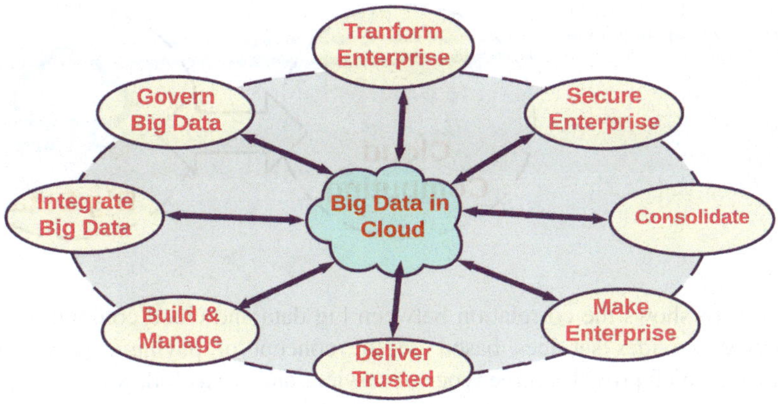

Fig. 12 Big data in the cloud

- *Become a platform:* If the application like plug-in facility is introduced by a company, then that company will become a good provider in cloud.

Typical Successful Example

The web service provided by Amazon known as Elastic Compute Cloud (EC2) markets ISA "slices" at a frequency of 1 GHz for 10 cents per hour, and new additions can be done within a time period of 2 minutes. Amazon's Scalable Storage Service (S3) costs about $0.12 to $0.15 per gigabyte per month to send data in and out of AWS over the Internet [19, 25, 26]. If many instances are multiplexed with a single box, then lot of profit can be earned without any interference among the customers. Three factors are essential for an organization to become a provider in cloud computing – minimum investments on data centre, software infrastructure, and operational cost (Fig. 12).

Big Data and Cloud Computing Relationship

Both the techniques – big data and cloud computing – have their own unique advantages. Many companies across the globe are planning to integrate these two methodologies for earning huge profit in their business. Their objective is to improve the income of the company, thereby reducing the cost of investment in parallel. Cloud and big data help them in the management of software and decision-making, respectively. Big data is concerned with collection, storage, and processing of huge data either in organized or in unorganized manner. The five aspects of big data are already described under Sect. 2 And are as follows:

- Volume – amount of data
- Variety – types of data
- Velocity – rate at which data flows
- Value – depends upon the content of data
- Veracity – confidentiality present in data

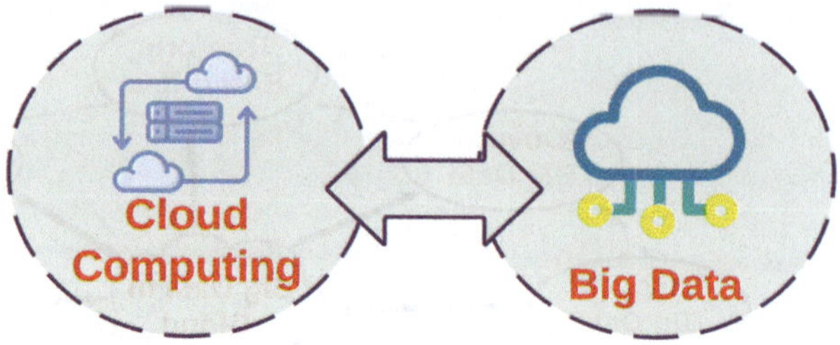

Fig. 13 Correlation between big data and cloud computing

Figure 13 shows the correlation between big data and cloud computing. Cloud computing provides services based on the concept of paying capacity of the customers, which provides three types of services and are as follows.

- *Infrastructure as a Service (IAAS)*

This service provides the infrastructure on an entire basis in parallel with tasks related with maintenance.

- *Platform as a Service (PAAS)*

This service provides a set of resources like storage of objects, execution time, process of queuing, databases, etc. The consumer is totally responsible for the configuration and implementation of connected tasks.

- *Software as a Service (SAAS)*

This service provides a list of essential settings against the platform and infrastructure.

Figure 14 shows the architecture of big data, cloud computing, and data analytics. Hence from the above explanation, it is seen that cloud can be thought of "As-a-Service" indication by which challenges are abstracted by means of flexible services [20, 27, 28]. Big data also satisfies the same requirement but the end customers abstract the processing of huge data in a distributed manner. The following are the various advantages of big data analysis with respect to cloud:

- Improved analysis: The analysis of big data has got better with the integration of cloud. Hence most of the organizations show interest to analyze big data with respect to cloud. Additionally, cloud helps to concatenate the data from various types of sources.
- Simplified infrastructure: The analysis of big data is a good task on the part of infrastructure since the data arrives in huge level with changing speed. Since the infrastructure provided by the cloud is flexible, it is very easier to handle any sort of workloads with appropriate scaling.
- Lowering the cost: These two techniques reduce weightage of ownership by providing appropriate values to the companies. Cloud enables customers to process big data without the need for larger resources. Hence these techniques help the companies to reduce the cost factor for their objectives and add values to them.

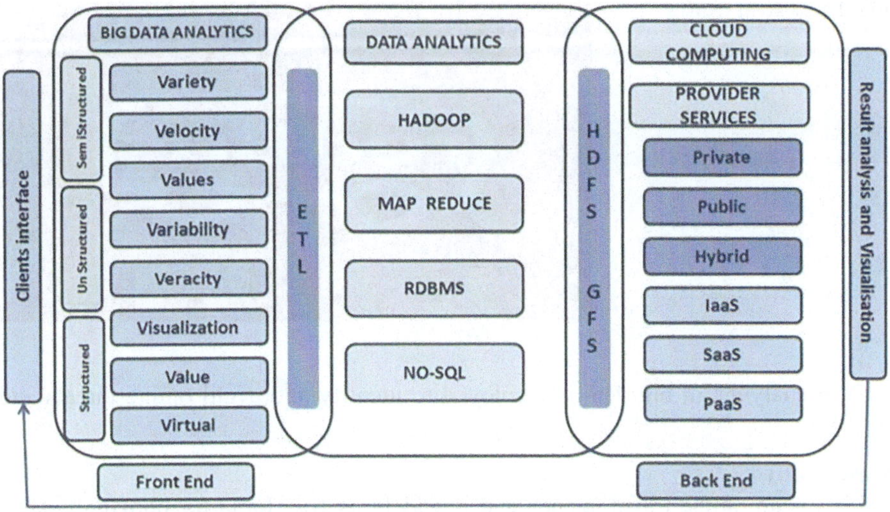

Fig. 14 Architecture of big data, cloud computing, and data analytics

- Security and privacy: The security of data is a major concern while connecting with data of an enterprise. The security will become a bothering factor if any application is executed in the platform of cloud because of its open access. Solutions of big data like hadoop employ services of a third party and its relevant infrastructure. These days, solutions for private cloud are available with flexibility and extendibility. In addition, processing of data is done in a distributed and scalable manner.

Generally, there will be a common location called cloud storage server used to store and process the data of cloud. The service agreement has to be signed by the service provider then after identifying the cloud server for gaining the confidence. If it is needed, the level of security will also be leveraged by the service provider. Following are the rules connected with the agreement of service for data protection:

- Big data protection from threats
- Maintenance of storage and data

The following issues are covered by the factor of security in big data with the aid of cloud computing:

- Data
- Capacity
- Scalability
- Security
- Privacy
- Storage and growth of data

Fig. 15 Logo of redBus

The analytics in big data is employed to identify and avoid online threats and hackers.

Case Study
redBus is an online travel agency and is widely used in India for booking tickets. In order to enhance the booking system and to increase its sales, redBus decided to employ infrastructure of Google data. One such application is Google BigQuery. This application is meant for handling huge amounts of booking and data within a short span of time. By this application, consistent streaming of data is available with respect to location of customers, information about seats, and booking. These data will be centralized in a common place. This data shall be provided to BigQuery for answering any sort of queries within seconds. Figure 15 shows the logo of redBus.

This complete arrangement of big data integrated with cloud helps redBus application to identify the glitches in a speedy manner and reduce the marketing losses, thereby enhancing customer service.

10 Conclusion

The integration of big data and cloud computing is a perfect choice to improve the efficiencies of organizations. Even though certain constraints like storage of data and accession in open environment do persist, they are negligible when compared to net benefits provided by this blend of big data and cloud computing. Management of data in cloud-based environment will help companies to integrate both master and big data in all respects. This chapter clearly presents the importance, characteristics, and classification of big data with appropriate examples. It also explains the tools and techniques used for the processing of big data. Moreover, the concept, working, characteristics, and key features of cloud computing are discussed in a detailed manner. Ultimately, this chapter correlates both the technologies – big data and cloud computing – in today's scenario with a case study. Both the techniques significantly modified the path, in which companies handle their data and promote it for their business activities. Additionally, these techniques improve the success rate of a business by enhancing the big data analysis and the process of decision-making. The future is going to be brighter if there is appropriate development in these technologies.

References

1. Jagani, N., Jagani, P., & Shah, S. (2021). Big data in cloud computing: A literature review. *International Journal of Engineering Applied Sciences and Technology, 5*(11), 185–191. ISSN No: 2455-2143.
2. Berisha, B., Meziu, E., & Shabani, I. (2022). Big data analytics in cloud computing: An overview. *Journal of Cloud Computing, 11*(24), 34–45.
3. Begam, S. S., J. V., Selvachandran, G., Ngan, T. T., & Sharma, R. (2020). Similarity measure of lattice ordered multi-fuzzy soft sets based on set theoretic approach and its application in decision making. *Mathematics, 8*, 1255.
4. Thanh, V., Rohit, S., Raghvendra, K., Le Hoang, S., Thai, P. B., Dieu, T. B., Ishaani, P., Manash, S., & Tuong, L. (2020). Crime rate detection using social media of different crime locations and Twitter part-of-speech tagger with brown clustering. *Journal of Intelligent & Fuzzy Systems, 38*(4), 4287–4299.
5. Nguyen, P. T., Ha, D. H., Avand, M., Jaafari, A., Nguyen, H. D., Al-Ansari, N., Van Phong, T., Sharma, R., Kumar, R., Le, H. V., Ho, L. S., Prakash, I., & Pham, B. T. (2020). Soft computing ensemble models based on logistic regression for groundwater potential mapping. *Applied Sciences, 10*, 2469.
6. Jha, S., et al. (2019). Deep learning approach for software maintainability metrics prediction. *IEEE Access, 7*, 61840–61855.
7. Sharma, R., Kumar, R., Sharma, D. K., Son, L. H., Priyadarshini, I., Pham, B. T., Bui, D. T., & Rai, S. (2019). Inferring air pollution from air quality index by different geographical areas: Case study in India. *Air Quality, Atmosphere and Health, 12*, 1347–1357.
8. Sharma, R., Kumar, R., Singh, P. K., Raboaca, M. S., & Felseghi, R.-A. (2020). A systematic study on the analysis of the emission of CO, CO2 and HC for four-wheelers and its impact on the sustainable ecosystem. *Sustainability, 12*, 6707.
9. Dansana, D., Kumar, R., Das Adhikari, J., Mohapatra, M., Sharma, R., Priyadarshini, I., & Le, D.-N. (2020). Global forecasting confirmed and fatal cases of COVID-19 outbreak using autoregressive integrated moving average model. *Frontiers in Public Health, 8*, 580327. https://doi.org/10.3389/fpubh.2020.580327
10. Malik, P. K., Sharma, R., Singh, R., Gehlot, A., Satapathy, S. C., Alnumay, W. S., Pelusi, D., Ghosh, U., & Nayak, J. (2021). Industrial internet of things and its applications in industry 4.0: State of the art. *Computer Communications, 166*, 125–139. https://doi.org/10.1016/j.comcom.2020.11.016. ISSN 0140-3664.
11. Sharma, R., Kumar, R., Satapathy, S. C., Al-Ansari, N., Singh, K. K., Mahapatra, R. P., Agarwal, A. K., Le, H. V., & Pham, B. T. (2020). Analysis of water pollution using different physicochemical parameters: A study of Yamuna River. *Frontiers in Environmental Science, 8*, 581591. https://doi.org/10.3389/fenvs.2020.581591
12. Dansana, D., Kumar, R., Parida, A., Sharma, R., Adhikari, J. D., et al. (2021). Using susceptible-exposed-infectious-recovered model to forecast coronavirus outbreak. *Computers, Materials & Continua, 67*(2), 1595–1612.
13. Hashem, I. A. T., et al. (2015). The rise of "big data" on cloud computing: Review and open research issues. *Information Systems, 47*, 98–115.
14. Savaglio, C., & Fortino, G. (2021). A simulation-driven methodology for IoT data mining based on edge computing. *ACM Transactions on Internet Technology, 21*(2), 30., 22 pages. https://doi.org/10.1145/3402444

15. Xiao, W., Miao, Y., Fortino, G., Wu, D., Chen, M., & Hwang, K. (2022). Collaborative cloud-edge service cognition framework for DNN configuration toward smart IIoT. *IEEE Transactions on Industrial Informatics, 18*(10), 7038–7047. https://doi.org/10.1109/TII.2021.3105399
16. Erhan, L., Ndubuaku, M. U., Mauro, M. D., Song, W., Chen, M., Fortino, G., Bagdasar, O., & Liotta, A. (2020). *Smart anomaly detection in sensor systems: A multi-perspective review*. arXiv: Learning.
17. Fortino, G., Messina, F., Rosaci, D., & Sarné, G. M. L. (2020). Using blockchain in a reputation-based model for grouping agents in the internet of things. *IEEE Transactions on Engineering Management, 67*(4), 1231–1243. https://doi.org/10.1109/TEM.2019.2918162
18. Golam Morshed, M., & Yuan, L. (2017). Big data in cloud computing: An analysis of issues and challenges. *International Journal of Advanced Studies in Computer Science and Engineering, 6*(4), 345–350.
19. Arya, S. (2016). Big data with cloud computing. *International Journal of Computer Science and Information Technology Research, 4*(4), 34–38.
20. Sandhu, A. K. (2016). Big data with cloud computing: Discussions and challenges. *Big Data Mining and Analytics, 5*(1), 378–380.
21. Umapathy, K., Mangayarkarasi, T., Subitha, D., & Sivagami, A. (2021). Android application and SMS alert based garbage monitoring and navigation system. *Journal of Physics: Conference Series, 1964*(6). https://doi.org/10.1088/1742-6596/1964/6/062064
22. Mangayarkarasi, T., Umapathy, K., Sivagami, A., & Subitha, D. (2021). An IoT based safe assembly point alert system. *Journal of Physics: Conference Series, 1964*(7). https://doi.org/10.1088/1742-6596/1964/7/072013
23. Chandramohan, S., & Senthilkumaran, M. (2020). Cluster based scheduling approach for SDN with edge computing. *International Journal of Advance Science and Technology, 29*(10S), 7463–7468.
24. Muthukumaran, D., & Chandramohan, S. (2020). Testing and measurement criteria for the internet of things (IoT). *International Journal of Engineering Research & Technology (IJERT), Special Issue*. ISSN: 2278-0181.
25. Umapathy, K., Sai Swaroop, V., Viswam, P., & Balaswami Sairaja, T. (2020). Counterfeit bank note detecting system. *International Journal of Scientific & Technology Research (IJSTR), 9*(3), 1033–1035. ISSN: 2277-8616.
26. Umapathy, K., Sridevi, T., Navyasri, M., & Anuragh, R. (2020). Real time intruder surveillance system. *International Journal of Scientific & Technology Research (IJSTR)*., ISSN: 2277-8616, *9*(3), 5833–5837.
27. Umapathy, K., Sree Sai Sindhu, M., Nithyasri, S., & Nandhini, P. (2022). Automatic engine locking system for drunken drivers. In *AIP conference proceedings*., ISSN: 1551 7616 (Vol. 2519, pp. 050019-1–050019-6).
28. Umapathy, K. (2022). Wireless technique based vehicle speed control system. In *AIP conference proceedings* (Vol. 2519, pp. 050022-1–050022-5). ISSN: 1551 7616.

Cloud of Things Platform for a Water Meter Network

Biswaranjan Bhola, Raghvendra Kumar, and Ahmed Alkhayyat

1 Introduction

With the steep rise in the population on Earth, there is an exponential decline in the availability of essential commodities such as electricity, water, and food. For the proper management and maintenance of essential things, different types of devices have been designed. In order to maintain all the devices properly and gather information and facts to generate knowledge so that the devices are interconnected with each other is called the Internet of Things (IoT). The implementation of systems such as traffic management on the roads, medical services, streetlight systems, and educational systems should be managed in a systematic manner. Some examples of contributions of the IoT system to the technical world are described in the following. Home appliances are monitored by the IoT system to avoid the abuse of electricity [1]. When people are absent from home, this technology can be used to monitor and provide security to the home [2]. It can also be used for monitoring and controlling the office [3]. In medical services, the IoT is used to provide patients with a better service as well as to warn the management regarding fire and leakage of different gases, which are harmful to patients [4]. Using the IoT, traffic systems and street lights can be controlled efficiently in localized urban areas [5–7]. It can also be applied for monitoring water distribution in a centralized manner in townships [8]. In agriculture, the system is in use for providing video surveillance over the land and for operating the machinery remotely [9].

B. Bhola · R. Kumar (✉)
Department of Computer Science and Engineering, GIET University, Kharling, Gunupur, Odisha, India
e-mail: biswaranjan.bhola@giet.edu; raghvendra@giet.edu

A. Alkhayyat
College of Technical Engineering, The Islamic University, Najaf, Iraq

R. Sharma et al. (eds.), *Data Analytics for Internet of Things Infrastructure*,
Internet of Things, https://doi.org/10.1007/978-3-031-33808-3_2

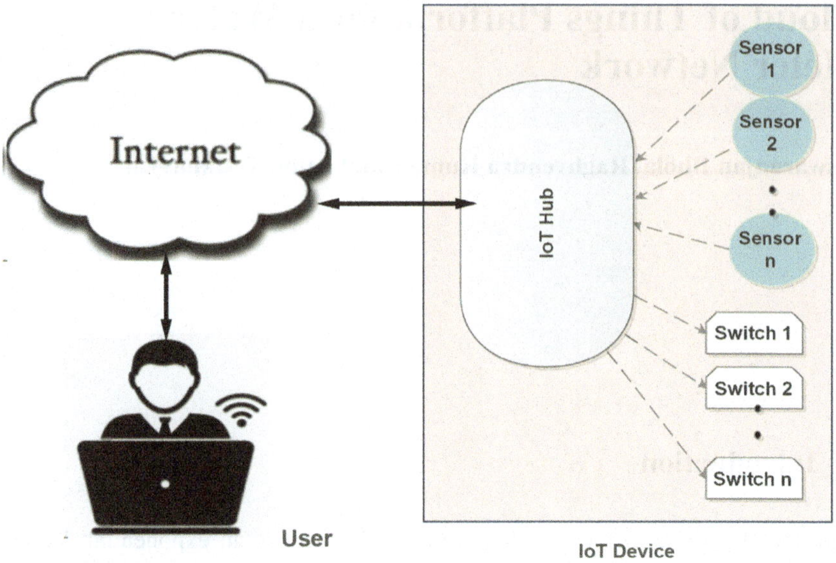

Fig. 1 Traditional internet of things (IoT) architecture

To develop the IoT, different types of communication technologies are used, such as the Global System for Mobile Communications (GSM), Bluetooth, Zigbee, etc. [10]. Here, we analyze the architecture of the traditional IoT system. Figure 1 describes the architecture of the IoT system. In this system, the main components of the IoT architecture are:

- IoT Hub
- Sensors
- Switch
- Smart Device
- Connectivity

IoT Hub
The IoT hub is an interface between the Internet and the Controlling and Monitoring Components (CMCs). The user monitors the CMCs using the Internet and an IoT hub. The IoT hub controls and monitors CMC components such as sensors and switches. Moreover, it uses a separate network to communicate with sensors and switches present within a CMC.

Sensors
Using sensors, the IoT hub detects the location of a particular place. For example, a gas sensor is used to identify the leakage of gas.

Switch
Using a switch, the IoT hub switches the electrical and electronics components on or off.

Smart Device
Using a smart device (Android-based mobile phone, computer) end users communicate with the IoT hub through the internet to monitor and control the device remotely.

Connectivity
The Internet is the public network through which a user's smart device can communicate with the IoT hub present in any part of the world.

The discovery of the IoT has been exceptionally beneficial to human lives. For instance, the IoT facilitates Machine to Machine (M2M) communication. Owing to M2M, physical devices can stay connected. With the introduction of the IoT, physical devices are connected digitally within a wireless infrastructure. Without much human interaction, devices can communicate seamlessly, which leads to more automation. Generating a variety of information through IoT-enabled devices assists in making better decisions. Another major advantage of the IoT is in monitoring the environment. This monitoring capability helps to make our lives easier. For instance, air-quality monitoring, milk-density monitoring, gas-leakage monitoring, an IoT monitoring system can improve safety. Moreover, the development and innovation of the IoT play a key role in the continuously changing lifestyle of modern society. In other words, the amount of time saved by this technology is huge.

Although it has many advantages, the IoT still faces a variety of challenges. To utilize its fullest potential, it is essential to solve some grave issues. For instance, there is no standard of compatibility for monitoring the devices. The compatibility issue is obvious because devices from different manufacturing companies are being interconnected. The risk of losing privacy and security is always high in the IoT system. Moreover, existing IoT architecture is more complex. Although a lot of research is going on, it is still in its infancy. Hence, in this article we have proposed a new modified IoT architecture, which has certain advantages over the existing system. Some of the advantages of the proposed system are described below.

- Low maintenance cost
- Network independence
- Architecture independence
- Scalability

The prime contributions of this paper are highlighted as follows.

- With the idea of an IoT-hubless infrastructure, a novel architecture has been explored. By doing so, this work introduces the concept of a smart switch module. It first verifies the user, then it switches on the device and sends information to the user.
- This work also introduces another smart module called a smart sensor. The smart sensor reads the environmental information, and sends information to the smart switch and user for the necessary action.
- Further, the proposed architecture has been tested within a physical test-bed.

This article is further described in the following order. The next section discusses the methodology of components, followed by the working structure, results, and conclusion.

2 Related Work and Problem Motivation

This section focuses on the present success and challenges of the IoT. Moreover, we discuss a few challenges of the existing IoT system and the problem motivation for our research background.

2.1 Related Work

According to a survey, trillions of devices have converted to the IoT, which indicates the popularity of the IoT. However, certain limitations exist in the traditional system. In Yakoub et al. [11], the authors have stated that the increased requirement of the IoT formed a complex network structure, which is caused by the creation of various kinds of problems such as inter-operability, flexibility, scalability, energy efficiency, and security. For example, owing to the increased population, motor vehicles, and industries, the pollution of the cities is increasing day by day. To monitor the pollution, Kaur et al. proposed an IoT architecture named green IoT architecture [12]. To improve the efficiency of above-mentioned architecture needs a distributed platform.

Chien et al. [13] proposed an IoT architecture in which the sensors and actuators of devices are interconnected with each other using private networks and the Internet. Each device uses its own network, which is heterogeneous in nature, and every sensor also has a different technology. An improvised architecture is essential for handling interoperability and heterogeneity in a seamless way. In [14], the authors used a Software-Defined Network (SDN) to secure IoT devices. However, it does not meet the other inherent challenges. Table 1 discusses a few important forms of IoT architecture used in the literature. From the above literature survey, we observed that the IoT hub is an essential part of the architecture, which may lead to high maintenance cost, network, and architecture dependency. However, in the proposed architecture, the smart switch and the smart sensor have been introduced, which connect directly to the network. This architecture can pave the way for various challenges faced by the current IoT system. The methodology and experimentations are discussed in the following chapters. It is an integrational physical, computational, and electronic environment. Liu et al. describe the same concept in the form of a Cyber Physical System (CPS) [15]. An organization such as the European Telecommunications Standards InstituteETSI explains it as an M2M

Table 1 Existing IoT architecture

References	Year	Short analysis
[17]	2008	A four-layer architecture for the IoT system. The lower layer consists of an electronic controller to connect with the sensor and relay and contains the control information. The second layer contains the executable program. The third layer is the interface with higher-level languages and the task- and actor-oriented level is present in the fourth layer
[18]	2008	A prototype architecture defined by Tan et al. that in the physical architecture of an embedded system in lower layer the outer environment information is sensed by the sensor in the form of a electrical signal and the control system converts it into the high-level information that is required by the user or operational system
[19]	2015	To access the physical world through the IoT is dependent on the control network connected with sensors and actuators
[14]	2016	The IoT devices can be secured using advanced IoT security architecture, which supports Software-Defined Network (SDN) architecture
[20]	2017	To make the IoT device comparable and light weight it should be scalable. The proposed architecture solves the scalable problem created in the IoT device
[21]	2017	IoT resources can be properly managed using multiagent distributed architecture. But the shortcoming of this idea is that each agent is an independent module. If there is any dependency this type of architecture may not function properly
[12]	2018	To monitor the pollution, green IoT architecture has been proposed. To improve the efficiency of this type of architecture, it should be converted to a distributed platform, which can be solved using the proposed architecture
[22]	2018	Much research work has been done on architecture and the protocol of the IoT, but some problems remain
[13]	2019	Each IoT device uses its own network and Internet technology, which is heterogeneous is nature. Every sensor also uses different technology
[23]	2020	The shortcoming of this work is that it communicates with the user using a GSM module through SMS communication. Hence, the user only knows the device information by using SMS, and there is a maximum than can be communicated with the 24 sensors, because the controller has only 24 I/O ports
[24]	2020	A revised architecture is presented to add scalability and dynamism through the application layer

communication system [16]. Hence, in such a type of ambiguous scenario the real fact that is discussed by most technical reports is that physical objects around the world try to communicate with others using a common virtual platform to improve the performance and take decisions on a real-time basis or for the future called the IoT platform.

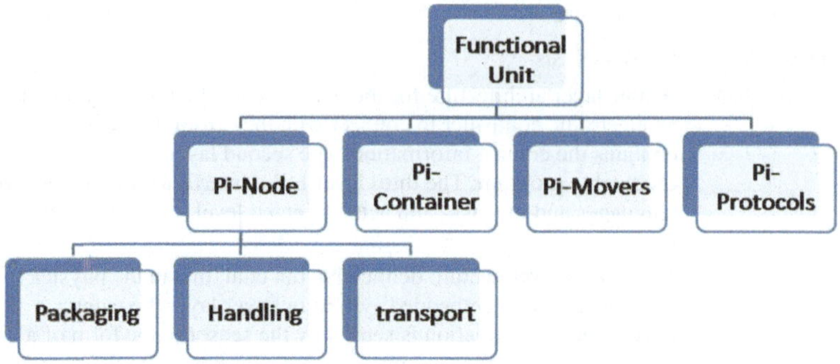

Fig. 2 Functional unit of the IoT

2.2 *IoT Paradigm*

The survey by Tran-Dang et al. [25] defines the services of the IoT into four categories:

1. Things Oriented (physical and virtual entities are the focus)
2. Internet Oriented (development of an IP-based network that enables the things)
3. Semantic Oriented (handles a huge amount of data in communication)
4. Service Oriented (developing the IoT intelligent service and application based on the above three perspectives)

and all the services are processed using four functional units explained in the Fig. 2.

In the above, all the pi-nodes are the end-to-end and intermediate components such as the control unit, edge server, cloud server, client smart devices, etc. pi-containers are the data packets, which are transmitted and the pi-movers are the technology that describes how pi-containers are transmitted from one pi-node to another pi-node. pi-protocols explain which protocols (http, tcp/ip, oli, etc.) are used for communication. One computing platform has been designed by Kondo et al. for dynamic module (pi-nodes) configuration for the IoT [26]. Ranganathan et al. explain the data communication in the IoT and an SDN switch, which can reduce the distributed denial of service attack. Using the proposed architecture and implemented module the above-mentioned things can be achieved and some of the problems defined below can be minimized.

2.3 *Problem Motivation*

In the present IoT system, to communicate between end users and sensors or switches to dissimilar networks (e.g., the Internet, an IoT hub network) two basic intermediate heterogeneous application layers are used. One is an Internet web server and the other an IoT hub. However, certain limitations persist in this type of architecture, which are listed below.

- High maintenance cost
- Network dependency
- Architecture dependent
- Unscalable

Hence, considering the above limitations, in this article, we propose a novel architecture, in which the switches and the sensors are directly connected to the Internet. In turn, the intermediate layer, i.e., the IoT hub, is not required. These are the main motivations for this work.

3 Proposed IoT Architecture

In this section, we describe the system model and functionality of the proposed scheme followed by analysis.

3.1 System Model

In the present IoT system Fig. 1, the switches (also called Relay) are used to switch the devices on and off. The sensors are used to read the current environmental information of a particular location. In this system we cannot directly access the switches or sensors. They can be accessed through the IoT hub. Hence, switches and sensors are tightly coupled with the IoT hub. Such tight coupling makes it difficult to add new sensors or switches to the IoT hub. The IoT hub will be reprogrammed to add new sensors or switches. Figure 3 describes the architecture of the proposed IoT system The proposed smart switch or smart sensor can be used to design any IoT application. Let, in an IoT system $\mathcal{I}$ there be n smart switches and n smart sensors. It can directly connect through the network, either using Wi-Fi or Ethernet. Each smart switch or sensor can be configured with one IP address, which is used to identify it in the network uniquely. Figure 4 describes the architecture of a smart switch, and Fig. 5 describes the architecture of a smart sensor.

3.2 Methodology

The traditional switch is a passive component in the IoT. It connects to the IoT hub. Only the IoT hub can change the state of the switch, which is on/off according to the user request or as the information is received by the sensor. It is also connected to the IoT hub. The IoT hub reads the information from the sensor, analyzes it, and

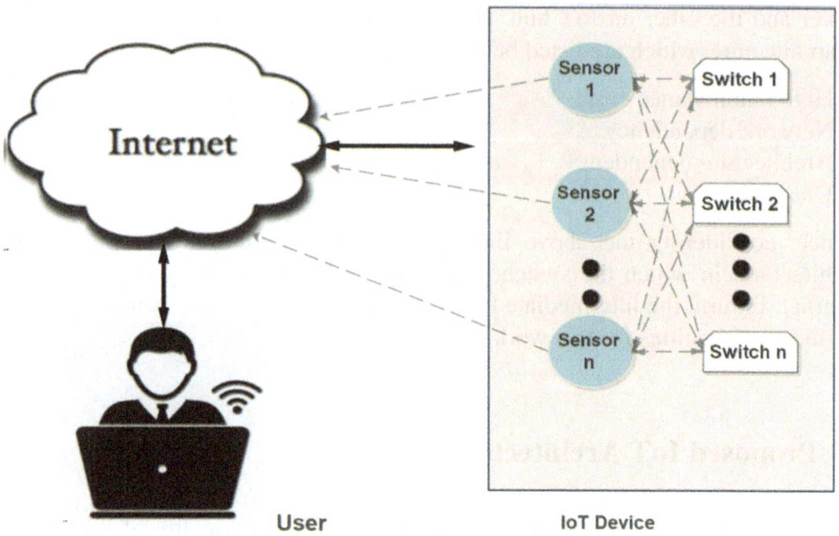

Fig. 3 Architecture of the proposed internet of things (IoT) system

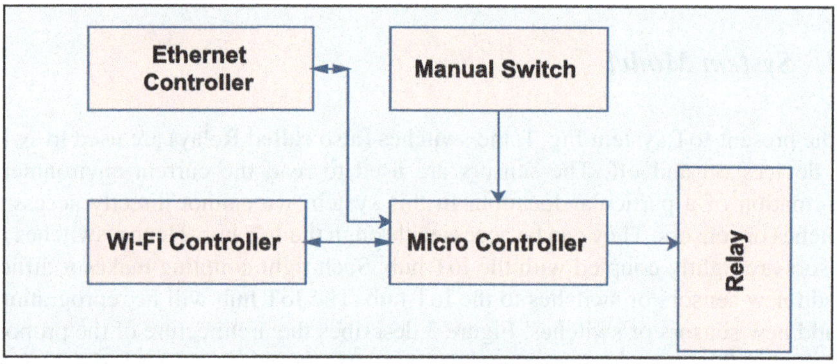

Fig. 4 Architecture of a smart switch

then sends the control information to the switch. The clear definition between the smart switch and smart sensor is defined below.

Definition 1 A smart switch directly receives the data packet from the user or sensor to switch on/off under its secure control and resends the confirmation or acknowledgment.

Definition 2 A smart sensor is an active component. It is directly connected to the network or Internet, analyzes the environment situation, and then transmits the information to the user as well as the smart switch for action.

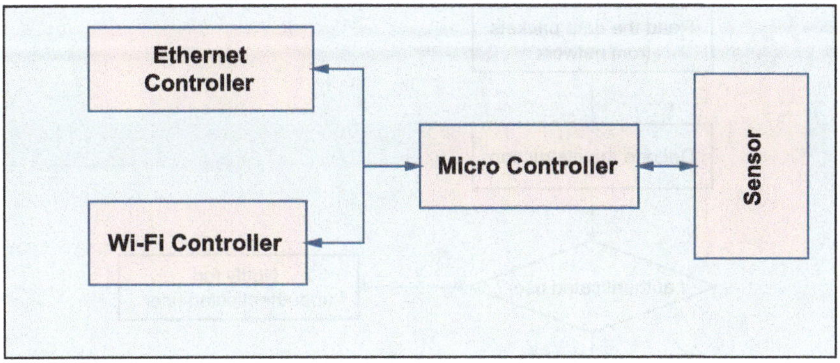

Fig. 5 Architecture of a smart sensor

Both the smart switch and the smart sensor are the active and intelligent modules. They can communicate with each other through networks or the Internet and can also take decisions if required. The basic function of the smart switch is that it first verifies the user, then switches the device on or off, and also sends the information to the user, sensor, or both, regarding the state of the switch. The step-by-step process of the switched-off mode processed by the smart switch is shown in Fig. 6.

The flowchart depicted in Fig. 6 describes when the switch OFF mode is processed by the smart switch module.

In a similar fashion, the switched ON mode is processed by the smart switch module. Worldwide, the authenticated sensor or user can access the smart switch to switch ON/OFF directly through the Internet.

The smart sensor reads the situation of the environment. For example, using a light sensor, we can find out the brightness of light present at a particular time or using a temperature sensor we can find out the temperature of the environment at a specific time. The smart sensor reads the environmental information and then sends it to the user and sends a control message to the particular switch to switch ON/OFF. The flowchart illustrated in Fig. 7, describes the smart sensor sending process. When the user wants to know about the situation of the environment from the smart sensor, it works like the previous process.

3.3 Development of Active Switch

The electronic name of the switch is Relay. It is used to make a device switch ON or OFF. There are five pins of Relay. Two pins, let us say A and B, are used to connect to the power source; one pin let us say C, is used as a Pole; and the remaining two pins, let us say D and E, are used as a Terminal, as shown in Fig. 8. Initially, the pole is connected with Normally Closed (NC), that is, pin E. When the current flow is available in pin A and pin B, the pole is shifted to Normally Opened (NO), that is,

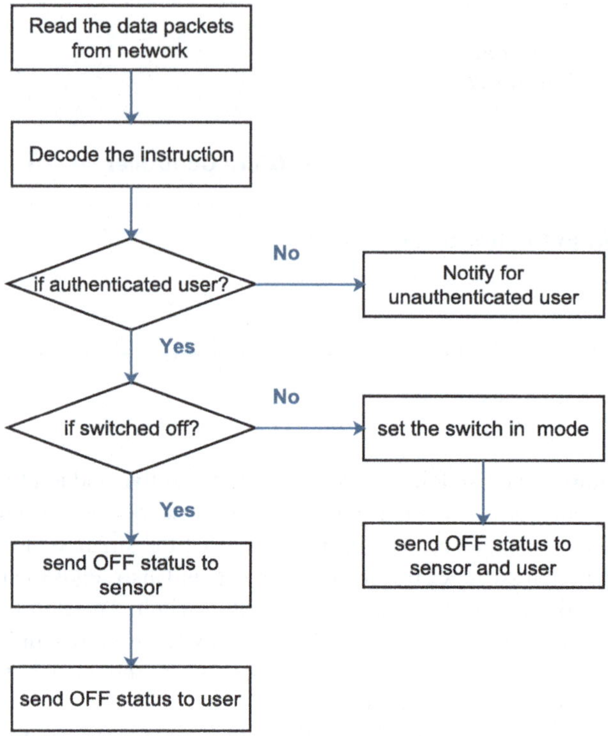

Fig. 6 Flowchart for switched OFF mode processed by the smart switch module

the pin D. Hence, pins D and E can be used for ON and OFF. Using Relay, transistor, diode, and resistor, one Relay module can be designed that has six pins. The pins numbered 1, 2, and 3 can be used as input pins and 4, 5, and 6, can be used as output pins. Using this module the relay can be triggered to ON or OFF by providing the signal in to pin 1 of the Relay module, as depicted in Fig. 9.

Active relay is a relay module that can directly connect with the Internet through an Network Interface Card (NIC). To convert a Relay module to Active Relay, the components are Ethernet, Wi-Fi gateway, Memory, Processor, Relay module. The interconnection structure is depicted in Fig. 10.

This module can be triggered directly using the Internet and can be monitored remotely to discover whether it is in ON/OFF state. Algorithm 1 describes the message passing technology between the user and the active switch. The user may be a person, device, or server that can directly communicate with the Active Switch using the httpRequest() and httpResponse() method.

Before sending data packets the active switch initializes all the parameters of the Ethernet controller/Wi-Fi controller, MAC address, and the Service Set Identifier (SSID). Further, it initializes the IP address automatically if the Dynamic Host Configuration Protocol (DHCP)-enabled router is there; otherwise, it initializes to

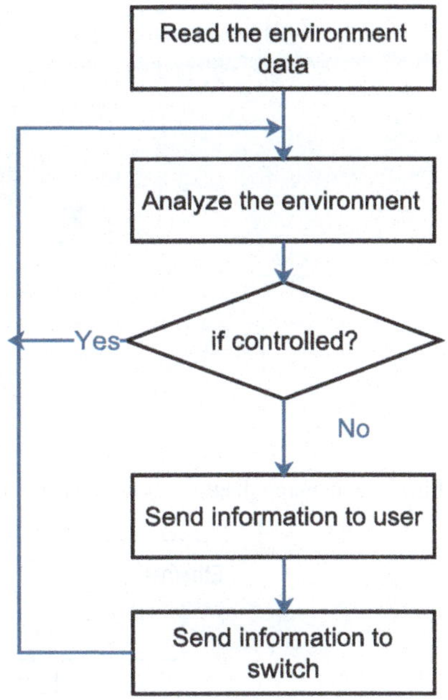

Fig. 7 The sensor sends the information to the switch and the user

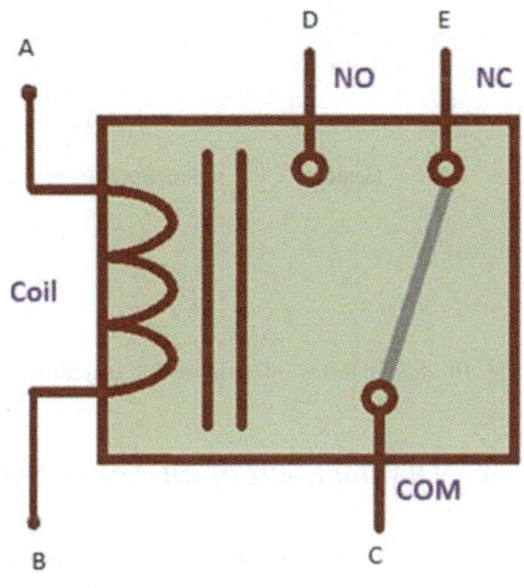

Fig. 8 Basic diagram of a Relay. COM = common, NC = normally closed, NO = normally open

the user-provided IP address. Next, it initializes the port as an output to operate the relay. After that, it reads the packet until the program receives a valid packet. Upon receiving an authenticated request, it sends the switch information in the form of a httpResponse() format. In the case of an unauthenticated request, it discards all the previous operations. Further, it starts to read the new packet, as discussed in step 1. The Algorithm 1 discusses the active switch interaction through the Internet.

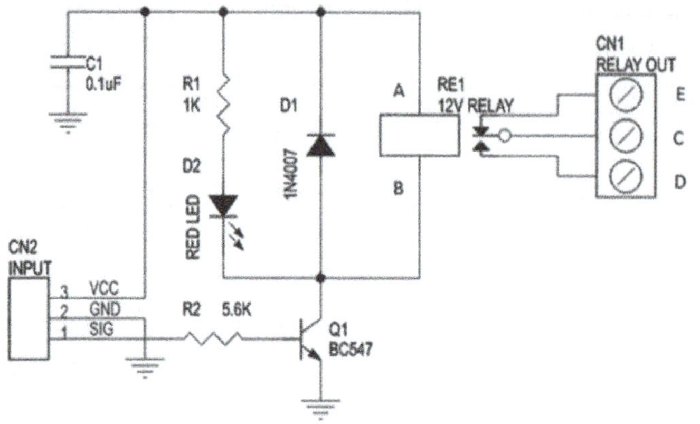

Fig. 9 Schematic diagram of a Relay module

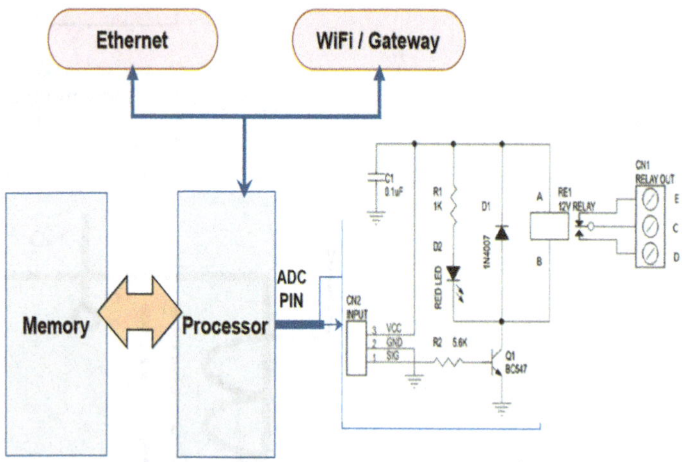

Fig. 10 Active Relay interconnection structure

3.4 *Development of an Active Sensor*

With the existing technology, an active sensor module can be designed with its own SSID, MAC Address, and IP address. That module can directly connect to the Internet using Wi-Fi/Ethernet. It also communicates the message to the user (person, server, device) using the httpRequest(), or httpResponse() method. The block diagram of an active sensor is shown in Fig. 11. The operation of the sensor is described using the Algorithm 2.

In Algorithm 2, the processor initializes all the parameters of the Ethernet controller/Wi-Fi controller, MAC, SSID, DHCP, etc., similar to Algorithm 1. Next,

Algorithm 1 Sending data packets by an active switch

Input: Ethernet Controller or Wi-Fi controller
Output: Get Details

```
    Initialization :
Initialize MAC/SSID
Initialize IP and Port
    Optimization Loop :
while 1 do
    Read TCP packets
    if Valid GET then
        if Valid REQ then
            Get Details
        else
            Unauthorized
        end if
    end if
end while
```

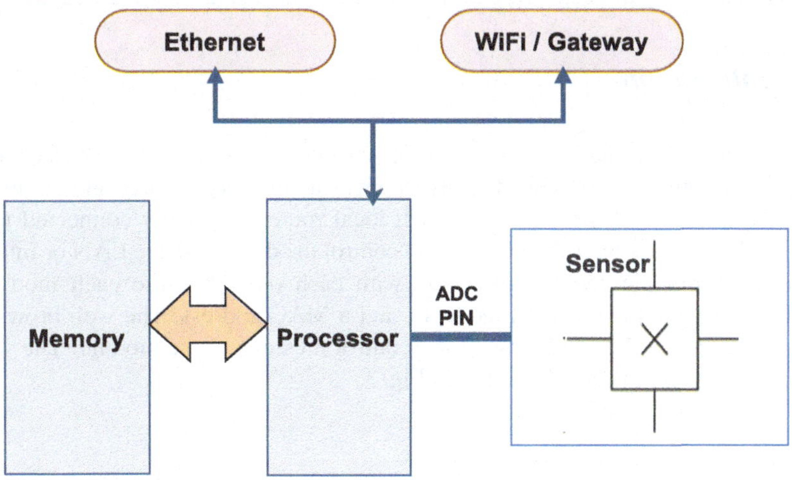

Fig. 11 Block diagram of an active sensor

it initializes the port as an input port and enables Analogue to Digital Converter (ADC) to read the information of the environment from the sensor. If the program finds that the read packet is valid, then it checks the request (REQ). If the request is authenticated, then it reads data from the sensor and sends it in the form of an httpResponse() format; otherwise, it discards all the previous operations and again reads the new packet.

Algorithm 2 Sending data packets by active sensor through the internet

```
Input: Ethernet Controller
Output: Send Status
  Initialization :
  Initialize MAC/SSID
  Initialize IP and Port
  Optimization Loop :
  while 1 do
      Read TCP packets
      if Valid GET then
          if Valid REQ then
              Read Environmental Status
              Send Status
          else
              Unauthorized
          end if
      end if
  end while
```

3.5 Integration

The integration of an active sensor and active switch is discussed below. Each active sensor and active switch can directly connect to the local router, either through an Ethernet port or Wi-Fi. Further, each local router is directly connected to the Internet. Users can remotely monitor and control the device via the LAN or Internet. Each active module can communicate with each other because each module is uniquely identified using an IP address, and a MAC address. The web browser is used to access the data and to send the control message to the module. The whole integration concept is demonstrated in Fig. 3.

4 Simulation and Result

In this section, the performance of the proposed architecture is evaluated by comparing the results with the existing architecture.

4.1 Setup Process

The proposed device (Fig. 12) is directly connected with the network through Ethernet (ENC28J60) or wireless module (ESP8266). The packets are processed using a microcontroller (Atmega32). The magnetic switch (Relay) or sensor is connected with the microcontroller. Through the web browser, the user can directly send instructions to the relay for the switch on or switch off instruction. The

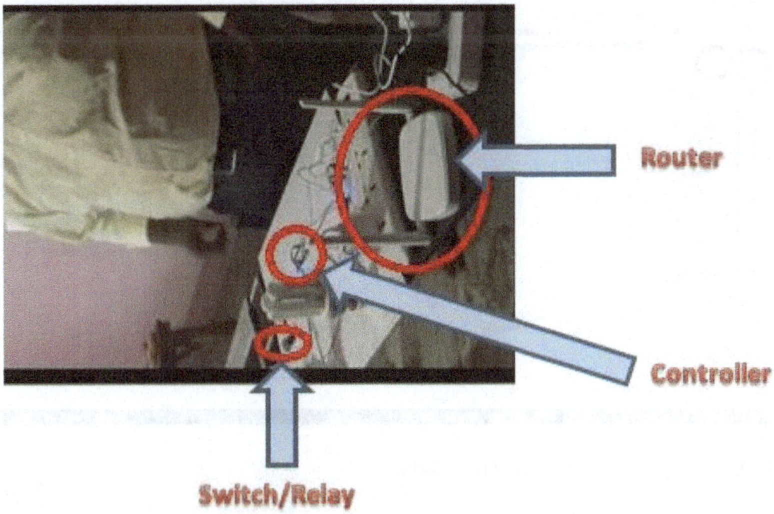

Fig. 12 Setup process and initialization

sensor information is also directly accessed through the web browser. The control information is transmitted in the form of an HTTP request or HTTP response using the GET method. It is described using the datasheet of ENC28J60 Ethernet controller.

To set up the proposed IoT architecture certain steps have to be followed. These steps are discussed below.

- Configure the IP address of the smart switch or smart sensor.
- Connect with the network using Wi-Fi or Ethernet.
- Access using any web browser.

This module can easily interface with any web application or stand-alone application. Moreover, owing to the absence of an IoT hub, it is purely platform independent, network independent, and provides a faster response.

Figure 12, shows the setup for the implementation of the proposed architecture. Figure 13 illustrates the initial process and Figs. 12, 13 show the remotely received stated details using a web browser. The Relay ON or OFF state has been experimented with in Figs. 14 and 15 respectively.

4.2 *Analysis*

One of the basic functions of IoT devices is to read all the information of the environment using the sensors and transmit it to the cloud for analysis and monitoring purposes. This section discusses and analyzes the shortest job first

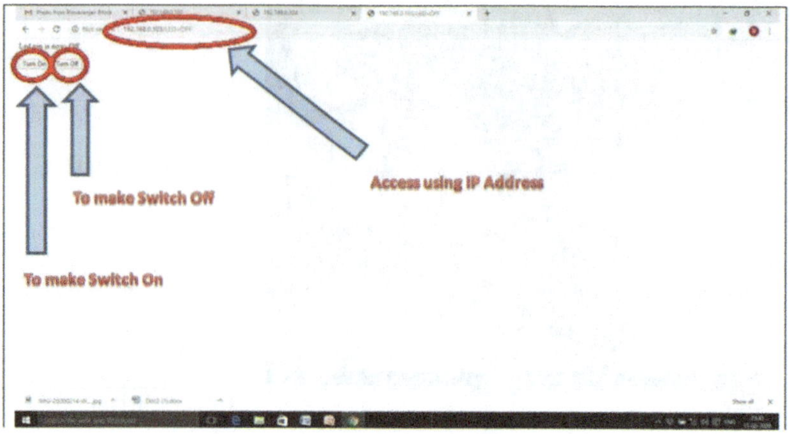

Fig. 13 Accessing the switch using an IP address

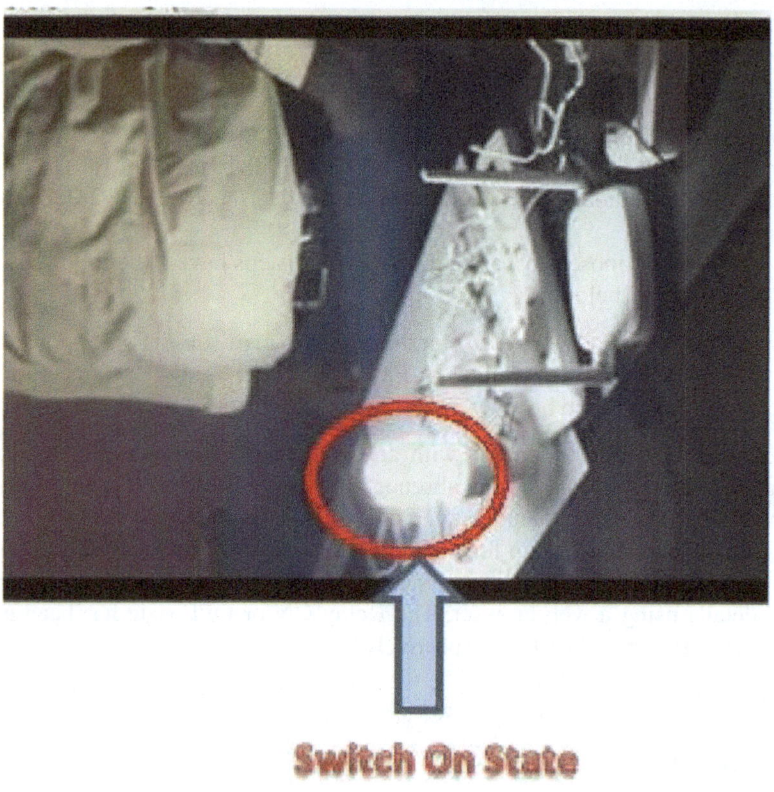

Fig. 14 Remotely operate Relay ON state

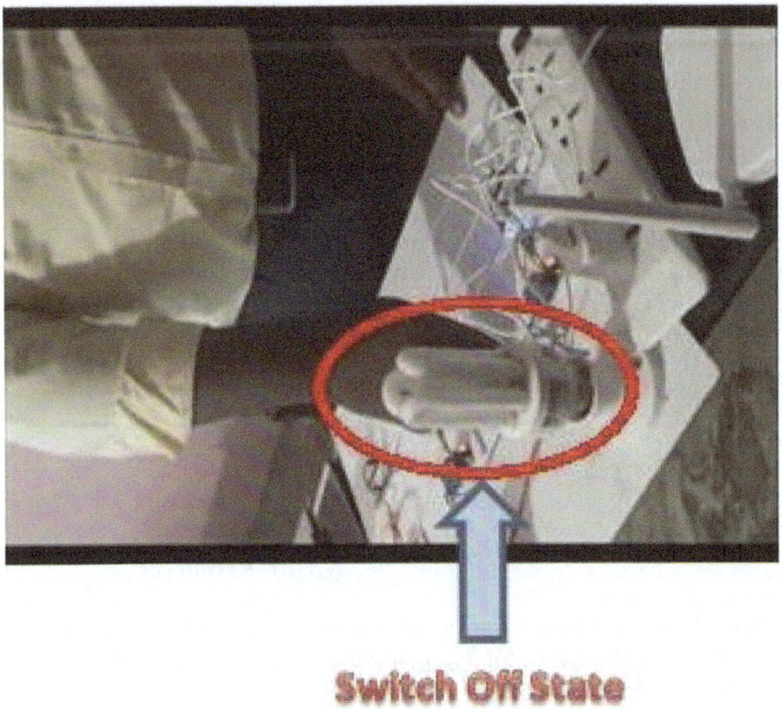

Fig. 15 Remotely operate the Relay OFF state

Fig. 16 Waiting time of different sensors

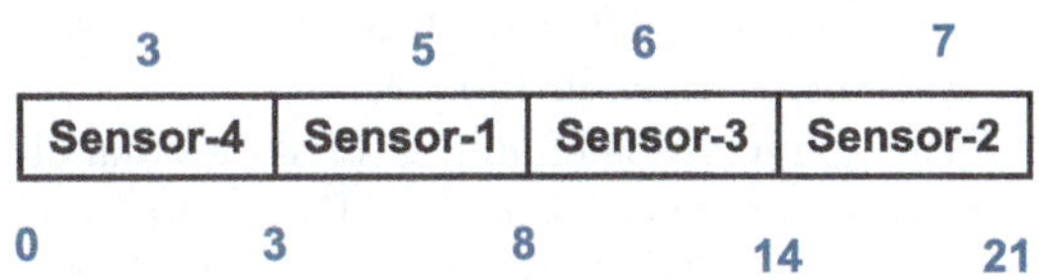

(SJF) scheduling algorithm on the proposed architecture. Further, it compares the waiting time, average waiting time, turnaround time, and average turnaround time required to upload the data to the cloud by traditional IoT devices with the proposed IoT module (Fig. 16). The analysis was done using a hypothetical IoT device. For experimentation, IoT device is connected with four sensors named Sensor1, Sensor2, Sensor3, and Sensor4. The time taken by these sensors to read the data from the environment is 5, 7, 6, and 3 units of time (ms) respectively. The basic function of the particular device is to repeatedly upload the sensor information onto the cloud. The time taken by the device, using the traditional sensor and switch, to upload one instance of data is given in Table 2.

If we observe the total waiting time it will be,

Total waiting time $= 0 + 3 + 8 + 14 = 25$ unit Here, the $waiting\ time = Service\ time – Arrival\ time$.

Table 2 Time taken by the devices to upload data using the existing architecture

Resources	Arrival time	Execution time	Service time	Waiting time (Service time – Arrival time)
Sensor 1	0	5	3	3
Sensor 2	0	7	14	14
Sensor 3	0	6	8	8
Sensor 4	0	3	0	0

Table 3 Time taken by the devices to upload data in the proposed architecture

Resources	Arrival time	Execution time	Service time	Waiting time
Sensor 1	0	5	0	0
Sensor 2	0	7	0	0
Sensor 3	0	6	0	0
Sensor 4	0	3	0	0

$$Average\ waiting\ time = \frac{Total\ waiting\ time}{No.\ of\ Resources} \tag{1}$$

The waiting times of different sensors are shown in Fig. 16. So, $Average\ waiting\ time = 25/4 = 6.25\ unit\ Total\ TurnAround\ Time = 3+8+14+21 = 46\ unit$

$Average\ Turnaround\ time = \frac{Total\ Turnaround\ time}{No\ of\ Sensors}$

$= \frac{46}{4}$

$= 11.5$ unit

To upload one instance of data, the time taken by the devices using the proposed architecture is given in Table 3.

The graph presented in Fig. 17 shows that although one sensor starts reading the data, other sensors have to wait. The waiting time is rather due to the IoT hub present in the system. The hub acts as the interface between the Internet and the CMC. However, using the proposed architecture, smart sensors can simultaneously read and send the data to the cloud. Hence, the overall waiting time has been reduced. In Fig. 18, we can observe that at 1 ms, all sensors have executed their task.

The advantages of this integrated architecture are listed below.

- Each module connects with the network through the Ethernet/Wi-Fi; hence, interfacing with the modules is very easy.
- It is scalable because each module contains its own program for operation
- The module can communicate using the browser for which it supports being platform independent.
- Different setup networks are not required to communicate with sensor and switch.
- There is no intermediate device such as an IoT hub between the user and the sensor; as a result, it can provide the user with a good-quality service.
- Interoperability is possible owing to the implementation of the httpRequest() and HTTPresponse() method for message communication.

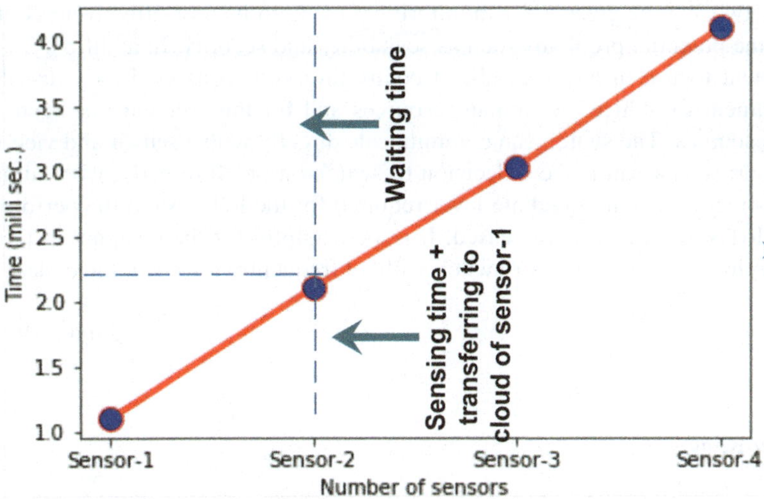

Fig. 17 Sensing time and waiting time using the existing architecture

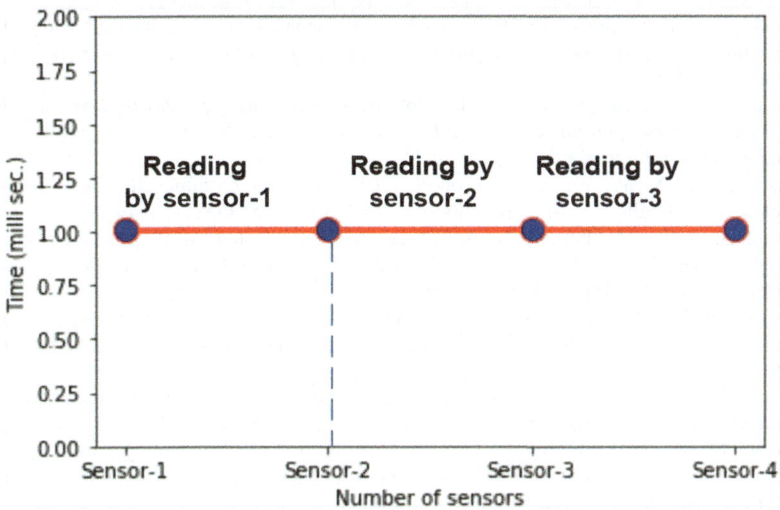

Fig. 18 Sensing and waiting time using the proposed architecture

5 Conclusion

Today, the Internet is the world's powerful communication system for interaction in any part of the world. This article is about a smart switch and smart sensor that help users of every IoT automation system such as smart home automation systems, smart traffic automation systems, smart street light automation systems, etc. The proposed architecture is platform independent and does not require an extra network to be created for monitoring. Owing to its cost-effectiveness, it may overcome potential problems such as scalability and security. In addition, it is very convenient to use in any centralized controlling system as well as a distributive environment to reduce the human resources and for the efficient management of other resources. The switch can communicate directly with a sensor and vice versa. The smart sensor can make a decision to switch on or off directly, without an IoT hub. As there is no intermediate layer required for the IoT system the performance of the IoT system can be increased. It is very helpful for the designer in terms of easy deployment of the sensor/switch without frequently rewriting the code.

mds
August 26, 2015

References

1. Motlagh, N. H., Khajavi, S. H., Jaribion, A., & Holmstrom, J. (2018). An IoT-based automation system for older homes: A use case for lighting system. In *2018 IEEE 11th Conference on Service-Oriented Computing and Applications (SOCA)* (pp. 1–6). IEEE.
2. Quadri, S. A. I., & Sathish, P. (2017). IoT based home automation and surveillance system. In *2017 International Conference on Intelligent Computing and Control Systems (ICICCS)* (pp. 861–866). IEEE.
3. Shaikh, S. A., & Kapare, A. S. (2017). Intelligent office area monitoring and control using internet of things. *International Journal of Advanced Research in Electrical, Electronics and Instrumentation Engineering, 6*, 4705–4711.
4. Baker, S. B., Xiang, W., & Atkinson, I. (2017). Internet of things for smart healthcare: Technologies, challenges, and opportunities. *IEEE Access, 5*, 26521–26544.
5. Javaid, S., Sufian, A., Pervaiz, S., & Tanveer, M. (2018). Smart traffic management system using internet of things. In *2018 20th International Conference on Advanced Communication Technology (ICACT)* (pp. 393–398). IEEE.
6. Dheena, P. F., Raj, G. S., Dutt, G., & Jinny, S. V. (2017). IOT based smart street light management system. In *2017 IEEE International Conference on Circuits and Systems (ICCS)* (pp. 368–371). IEEE.
7. Sahoo, K. S., et al. (2019). ESMLB: Efficient switch migration-based load balancing for multi-controller SDN in IoT. *IEEE Internet of Things Journal, 7*, 5852–5860.
8. Srihari, M. M. (2018). Intelligent water distribution and management system using internet of things. In *2018 International Conference on Inventive Research in Computing Applications (ICIRCA)* (pp. 785–789). IEEE.
9. Sushanth, G., & Sujatha, S. (2018). IOT based smart agriculture system. In *2018 International Conference on Wireless Communications, Signal Processing and Networking (WiSPNET)*. IEEE.
10. Vaidya, V. D., & Vishwakarma, P. (2018). A comparative analysis on smart home system to control, monitor and secure home, based on technologies like GSM, IOT, Bluetooth and PIC microcontroller with ZigBee modulation. In *2018 International Conference on Smart City and Emerging Technology (ICSCET)* (pp. 1–4). IEEE.
11. Yaqoob, I., Ahmed, E., Hashem, I. A. T., Ahmed, A. I. A., Gani, A., Imran, M., & Guizani, M. (2017). Internet of things architecture: Recent advances, taxonomy, requirements, and open challenges. *IEEE Wireless Communications, 24*(3), 10–16.
12. Kaur, G., Tomar, P., & Singh, P. (2018). Design of cloud-based green IoT architecture for smart cities. In *Internet of Things and Big Data Analytics Toward Next-Generation Intelligence* (pp. 315–333). Cham: Springer.

13. Chien, W. C., Lai, C. F., Hossain, M. S., & Muhammad, G. (2019). Heterogeneous space and terrestrial integrated networks for IoT: Architecture and challenges. *IEEE Network, 33*(1), 15–21.
14. Jerald, A. V., Rabara, S. A., & Bai, D. P. (2016). Secure IoT architecture for integrated smart services environment. In *2016 3rd International Conference on Computing for Sustainable Global Development (INDIACom)* (pp. 800–805). IEEE.
15. Liu, Y., et al. (2017). Review on cyber-physical systems. *IEEE/CAA Journal of Automatica Sinica, 4*(1), 27–40.
16. Oriwoh, E., & Conrad, M. (2015). Things' in the internet of things: Towards a definition. *International Journal of Internet of Things, 4*(1), 1–5.
17. Lee, E. A. (2008). Cyber physical systems: Design challenges. In *2008 11th IEEE International Symposium on Object and Component-Oriented Real-Time Distributed Computing (ISORC)*. IEEE.
18. Tan, Y., Goddard, S., & Pérez, L. C. (2008). A prototype architecture for cyber-physical systems. *ACM Sigbed Review, 5*(1), 1–2.
19. Lee, J., Bagheri, B., & Kao, H.-A. (2015). A cyber-physical systems architecture for industry 4.0-based manufacturing systems. *Manufacturing Letters, 3*, 18–23.
20. Ren, J., Guo, H., Xu, C., & Zhang, Y. (2017). Serving at the edge: A scalable IoT architecture based on transparent computing. *IEEE Network, 31*(5), 96–105.
21. Liu, X., Leon-Garcia, A., & Zhu, P. (2017). A distributed software-defined multi-agent architecture for unifying IoT applications. In *2017 8th IEEE Annual Information Technology, Electronics and Mobile Communication Conference (IEMCON)* (pp. 49–55). IEEE.
22. Kumar, R. P., & Smys, S. (2018). A novel report on architecture, protocols and applications in internet of things (IoT). In *2018 2nd International Conference on Inventive Systems and Control (ICISC)* (pp. 1156–1161). IEEE.
23. Krishnan, R. S., Julie, E. G., Robinson, Y. H., Raja, S., Kumar, R., & Thong, P. H. (2020). Fuzzy logic based smart irrigation system using internet of things. *Journal of Cleaner Production, 252*, 119902.
24. Gushev, M. (2020). Dew computing architecture for cyber-physical systems and IoT. *Internet of Things, 11*, 100186.
25. Tran-Dang, H., et al. (2020). Towards the internet of things for physical internet: Perspectives and challenges. *IEEE Internet of Things Journal, 7*, 4711–4736.
26. Kondo, T., et al. (2019). Development of the edge computing platform with dynamic modular configuration for an IoT platform. In *2019 IEEE Pacific Rim Conference on Communications, Computers and Signal Processing (PACRIM)*. IEEE.

Online Newspaper Development within the Internet of Things Environment: The Role of Computer-Mediated Communication

Pham Chien Thang and Ta Thi Nguyet Trang

1 Computer-Mediated Communication

In simple words, computer-mediated communication (CMC) is a type of communication in which more than one or two people interact or coordinate with each other using social software on different computers having a good internet network. CMC does not refer to how two computers interact; instead, it refers to how people connect with computers. While the word has typically been attributed to CMC (e.g., instant messaging, e-mail, discussion forums, internet forums, and social network services), it has also been extended to other forms of text-based contact like text messaging. CMC helps remove geographical barriers allowing effective communication between people over long distances [39]. Any type of document in written form could be edited, stored, copied, exchanged, or broadcasted. Written documents can be used to share and transfer data within no minutes. It is used for instant communication purposes, which are cheap and simple. Messages irrespective of sizes can be broadcasted, and additional operations, e.g., automated copying to a predefined distribution list, could be programmed. People over distance can review or edit the document if any issues are found. Concerning vertical hierarchy in social connections and organizations, software used for electronic communication remain blind [3]. When people use technology to communicate, their roles, ranks, and power aren't as clear as in face-to-face interactions. Hence, influential individuals with high status aren't as dominant; meanwhile, group members can participate more equally in digital communication. Because so much hierarchical dominance

P. C. Thang (✉)
Faculty of Journalism and Communication, TNU-University of Sciences, Thai Nguyen, Vietnam
e-mail: thangpc@tnus.edu.vn

T. T. N. Trang
Department of Economics and Management, TNU-International School, Thai Nguyen, Vietnam
e-mail: trangttn@tnu.edu.vn

R. Sharma et al. (eds.), *Data Analytics for Internet of Things Infrastructure*,
Internet of Things, https://doi.org/10.1007/978-3-031-33808-3_3

and power information is put down, social influence among CMC people becomes fairer and more impartial [39].

"Communication" involves the exchange of information; "mediation" refers to the fact that a technological medium is used for communication purposes; and "computer" refers to the fact that the means of mediation are technological, such as computers/internet, cell phones, video conferencing, and so on. Furthermore, it is beneficial to discern between the concepts of synchronous means (e.g., chats) and asynchronous means (e.g., blogs) of CMC and also to analyze these scenarios and technical aspects that may have an impact on language habits [45].

The mass media sector has undergone a massive transformation during the previous decades. New media, for example, computers, mobile phones, and other communications networks such as the Internet and multimedia technology, have sparked this revolution. These "new media" exhibit various communication configurations; their many forms reveal the link between interpersonal communication features. New media are distinguished from traditional mass media by their interactivity, demassification, selectivity, synchronization, immediacy, affordability, pro-democracy forums, marketing tools, and speed, among other characteristics. Interactivity, on the other hand, is widely regarded as the defining feature of new media. As a result, many experts in communication technology have tried to describe the qualities, aspects, and attributes of interactive communication within the area of new media and explore the level of interactivity that new media entails. No doubt, interactivity is a concept related to mass communication and a basic idea in CMC, but this is also a truth that there is still no agreement on how to define it [21]. Interactivity is a theoretical concept investigating the origins of enchantment, intrigue, and appeal in computer-mediated groups [2].

Language usage in CMC is a very significant subject. Language can be used to cover and exploit technology concerns such as short terms used in texting on mobile phone chats. A swift may be seen in observing language use in inventive language development. In addition, Wilbur [46] noted that CMC is actually text-based and hence belongs to the same field of study as linguistics. Moreover, communication on the internet is as genuine as communication in person. This remark was prompted by the frequent usage of the phrase "in real life," which refers to offline living and implies the existence of "an unreal life" [20].

2 Scholarly Information Related to CMC

Despite geographical distances, people in 1970s believed that usage of computers to transfer and share information to people could be of great benefit. At that time, such ideas were unique and revolutionary. But few people thought the idea could be dangerous as it is not appropriate to use computers for storing political and religious information and e-mails for longer than 1 month. People exerted great effort to establish the system but were unsuccessful. In 1990, when the internet was made available to the public, the concept of using CMC became successful

[27]. In contrast to non-mediated, face-to-face communication, earlier research focused on how technology dynamically modifies the performance and effectiveness of social interaction and team processes, addressing issues such as how people communicate and comprehend self-identity, form and maintain impersonations, develop and sustain relationships, make new connections, work collaboratively at a distance, and make intelligent decisions. As a result, past research's critical theories highlighted the lack of socio-contextual information as a character attribute of CMC that works against it [42]. The ability of CMC to make communication accessible to others was previously impracticable.

Some scholars believe we're moving into an era where communication is becoming more visible, which could be a significant area of study for social sciences. Modern CMC gives people many choices for displaying or attempting to access communication. Communication can now be seen by third parties, and even by those it wasn't intended for. This suggests big changes in how we express and understand intent through communication. Early CMC theorists understood that innovations that aided interpersonal communication were created to facilitate "socially transparent networks," hence visibility was the main consideration – visibility within those early research works linked to how other users were fully informed of an individual's existence, behavior, and engagement. Visibility, for example, was defined by Bregman and Haythornthwaite [47] as "the ways, strategies, and possibilities for representation; in interpretation, it largely emphasizes the presenters' concerns with the presentation of ownself" [39].

We consider CMC as a comprehensive framework encompassing multifunctional human-to-human social contact facilitated by ICT infrastructure. All kinds of social data transmission are included in this category, ranging from simple social attention (e.g., scrolling through the Facebook timeline) to intensive communication (e.g., a chat through Facebook messenger video call and social network sites, but not games). Computers facilitate CMC because they provide access to things centered on assisted socialization and user-generated content. Various programs are researched under a shared title since they often share a standard set of essential traits and functionalities [23].

3 CMC and Development of Online Newspapers

The number of print newspapers that also have an edition online has increased tremendously. According to the different surveys performed, it has been evident that increased number of online newspapers is being offered in the fax, audio tax, and voice services than printed newspapers [16]. However, it is pointed out by Morris and Ogan that mass communication is actually not so much attentive to the implications of CMC for internet publishing. In recent decades, however, the concept of communication technologies such as electronic mail, bulletin boards, and chat rooms has grown. This improvement in communication technologies has prompted numerous developments in CMC and related social research [31].

3.1 Advantages of Online Newspapers

Online journalism provides us with an easy, simple, fast and effective way of breaking the news. Synopsis of events as they occur can be provided to the people of the society within no time. In an online journalism, journalists can provide the people sitting on internet with accurate information which is also up to date [37]. Online newspapers in CMC help in various ways. They facilitate the reader's relationship with the newspaper. Readers can post their comments and opinions below. They can contribute to the authoring of the paper's sections. Let's take the example of a newspaper that is available in printed form and the publisher prints an online edition of this newspaper in which the online content differs from the printed content. It could be advantageous for the readers to access rare information that is not available in the printed edition. Sometimes, information available in the printed version could also be left out of the online version [28]. Some little newspaper articles might not ever reach the internet during a busy news day when information online has already been flooding, leaving online readers marginally less updated than print readers. The effectiveness of the newspaper available online is additionally enhanced by its different locations where they can be accessed easily. It is possible and easy to access the news form the mobile or the computer. In a second, people can have access to the news articles and the information available online. Most newspapers will provide free access to most of the information and the little pieces or summaries of the news stories. However, there is one thing about the printed newspaper; it can be read in more detail than the online edition. Simultaneously, an online newspaper is available and even extremely old copies can be accessed with a single click. If we wish to obtain an older edition of any newspaper, we must first obtain permission from the library [11].

3.2 Delivery of Online News and Information

The collapse of hard copy newspapers and the emergence of online news articles have sparked much debate among researchers and journalists who have looked into the powerful influence of the emerging innovations on the people and the newspaper industry. Despite uncertainties about how successful such initiatives will be in the future, market forces have driven newspaper publishers to embrace the latest platform and launch trials in online news providing. In the new internet marketplace, the established economic structure and process for hard copy newspapers, where the money or profit is made primarily from customers, newsagents, and advertisements, is now not relevant enough [5]. There may still be a market for Internet news (readers and advertisements), but the essential question is whether and how this market can be profitable. One characteristic that distinguishes online news as a viable (alternate solution) media delivery mechanism is its capacity to use technologies to ensure the same material as print newspapers to a broader public at a lower cost and in real time. E-newspapers are already a fact, fast expanding, and accessible to individuals all over the globe. E-newspapers are phenomena that are here to stay, based on their current rate of expansion and integration. Thousands of newspapers have invaded the world of internet publishing. This 'migration of information to the Internet' trend has already represented a pivotal shift in information exchange and communication from traditional formats [32].

The term "electronic transmission of news and information" relates to providing a starting point of sharing news and data digitally. This encompasses CD-ROM publication and information distribution via internet platforms, for example, the World Wide Web. According to one interpretation, an online newspaper is "a newspaper that needs the user to use an electronic gadget at some point for its transmission and reading" [17]. As a result, online newspapers have a substantial global reach and impact. However, the concern here is whether the new "online newspaper" could supplant the traditional "printed" one and how this would impact journalism. Some experts and skeptics say that, for the time being, online newspapers would not be able to take the place of printed newspapers. A number of publishers view their online newspapers as supplements rather than replacements for the printed editions. Furthermore, several observers have stated that online newspapers are auxiliary offerings that have less in common with the hard copy–printed newspapers. The publishers could not eliminate the print newspapers. The usage of online newspapers, and hence the expansion of the e-newspaper business, is further limited by local technology. Limited processing capacity and the screen and memory requirements of online newspapers mean that electronic news provision and layout actually would not be able to be user-friendly. Furthermore, online newspapers are also difficult to transport on laptop PCs. Despite these constraints, online newspapers have replaced their print counterparts significantly [35].

3.3 CMC Journalism Is a Better Option

Printed and online newspapers have content that covers the same topics and can satisfy the needs and requirements of the same reader and advertiser. Despite their differences, online and print newspapers both have the main common purpose of providing the readers with the information same in nature. Since Web and internet use in IOT environmental is expanding in terms of subscriber or user markets and the quantity of content, services, and programs available, CMC and rising online newspapers must receive assistance. When it comes to online and printed news, producers and consumers have different financial obligations. The traditional hard-copy newspaper requires expensive production equipment but no equipment for the reader.

In contrast, CMC and the online newspaper require less expensive production equipment and more complex receiving equipment on the consumer side, requiring greater upfront investment on the user's part. Some significant distinctions between online newspapers and printed newspapers may be observed due to the dynamic nature of technology, which is the backbone of the online newspaper production and distribution system, and the various formats in which they exist [44]. In addition, CMC and online newspapers can post more material, allowing for a significant depth of news coverage unrestricted by location or time and allowing readers to access back issues and instant online archives. Online newspapers are accessible at any time and from any location with a computer and Internet connection, whereas printed newspapers have established circulation times and require readers to travel to certain locations. On the other hand, a printed newspaper does not require the reader to purchase any particular equipment to read it. "The main distinguishing feature of what characterizes news is its immediacy." As a result, one of the key advantages of online newspapers is the elevated level of news update within seconds, which can occur at any time of day or night. The editor can provide the readers with up-to-date information and news by uploading or posting the latest stories by using the internet through its computer, which can ultimately aid in easy and simple CMC [25].

4 Interactivity with News Through CMC and Issues Within IoT Environment

Readers of newspapers and online newspaper employees may also communicate by exchanging views and thoughts using internet technologies. Sending e-mail-based communications to the media company and exchanging messages in an online group discussion are two essential options for CMC. Messages in text and voice chats are another form of interactivity that allows two or more people to exchange messages in real time. According to studies, readers are ready to let their ideas be known in public conversations and personal e-mail to specific individuals. This feature of the internet has caused people to expect a level of involvement from the internet that would be impossible to achieve anywhere else. Unlike other forms of news sources available online, Newspapers have always been slow to recognize that interactivity is a necessary condition for effective Web communication [4]. One of the most significant traits of the internet and CMC is interactivity. The features of interactivity in mass communication are the most critical elements. Between journalists and readers, these features have helped people shift from one-way communication to two-way interactive communications. For example, there is an opportunity for the readers of online newspapers to drop down their opinions, views, and thoughts in comments. In this way, readers can take part in discussions in online newspapers. Commenting on the online newspapers provides the readers with an advanced level of mass communication engagement. The power of mass communication has increased in public; hence, it is important to understand how the content features in online newspapers can help in the interactive behavior of readers, for example, the thought of sharing journalistic content. In other words, the importance of mass communication engagement of readers relies on the idea that if the level of engagement were higher, there would be more readership of online newspapers and CMC. According to past studies, the term interactivity is also controversial because the concept is used in different academic areas and the general public or society. Numerous scholars have attempted to define the term conceptually or theoretically, but the term has always been too "loose." Two types of comments in online newspapers can be posted. One type is "reactive comments" related to the content written in the news article, and the second type is "interactive comments" related to responding to the comments of others. Both the types aid in knowing what perception the newspaper content left on the readers [33].

4.1 CMC as a Tool for Organizations and Governments to Spread Information and News

CMC has become a crucial instrument for governments, agencies, and organizations to disseminate information and news to the country's general population and the global community in order to keep them abreast of global events. In recent years, reports have been involved reporting and online news sharing, which is computer mediated. The leading journalists from the organizations use their personal computers to collect and disperse news among people. The government uses its websites to spread important information to its people using internet, and so are the other agencies. Online newsrooms have been created for people's ease since it gives knowledge about news reporting. It has been made simple and easy for journalists to write and gather information more effectively [14]. The government uses CMC for its political purposes also. Online debates are being done by the government and politicians using CMC. Higher accessibility of the computer and internet has impacted the public positively also. A proper and separate space can be created for the government persons for CMC to engage and listen to public issues. The improved communication capability allows political communication to reach a broader public instead of only the inner circle of political insiders. New kinds of political communication based on new technology may make it easier for the public to participate in the political communication process. Further, than the simple sign language that politicians produced and employed in the past, enhanced participatory democracy via the digital age contributes to substantial shifts in communication methods and approaches. Certain emerging political groups with limited resources as well as those associated with minorities may be able to compete with the established political strength as a result of this [10].

4.2 *The Use of Social Media and Its Social Outcomes Concerning CMC*

In the twenty-first century, the internet and CMC have become the center of technology development. The inclusion of the internet has made computers more advanced than ever before. The combination is said to be a powerful technology for communication. It has become a great revolution in the field of technology and communication. The concept of CMC has given birth to social media, which has many benefits, including the authority for people to create their profiles and interact with other people. The emergence of social media is vital to make people across different locations engage and interact with one another using a communication network. The use of social media has also enabled people to share news, information, and updates in articles. These days, social media applications, for example, Instagram and Facebook, are being used as media channels to post the latest news.

On the other hand, ethics is one factor that should always be kept in mind while using CMC. Bigo Live is the example when technology has a deleterious impact if used uncontrolled [9]. Social media has been a major worry among security experts in the last two decades. Parents are deeply concerned about their children's online privacy and how to safeguard their personal information. However, the drawbacks of social networking go far beyond confidentiality. Numerous research works have linked social media to poor mental health in recent years. Multiple studies have demonstrated an association and causality, according to a recent Forbes article. According to the findings, persons who restrict their social media usage to 30 minutes each day feel substantially better than those who use it. People reported minimizing despair, stress, and sadness by limiting the amount of time they spend on social media [1]. Social networking sites have also become a very important part of everyone's life in today's era. Without the use of it, something in life seems missing. Also, it is not easy to find a person who does not connect to social media these days. CMC has become easy through social media, but the cons are also in great number if there are pros. Significantly, many students are interested in using social media for CMC purposes. No doubt, it gives the people chance for connectivity, collaboration, information, learning, and entertainment but the addiction to it could cause harm to the lives of people also. More and uncontrolled use of CMC via social media can cause addiction to the students. Gradually, the addiction may reach a level where the students get distracted from their studies. Students are then unable to focus on their studies since such students are always wasting their time scrolling their timelines and newsfeeds. No doubt, CMC helps us build friendships and good relations with people sitting on the other side of the computer, but if we see the broader picture, it could affect the relationships with a closed one. Cyberbullying is the other primary concern as well as the inappropriate content and health concerns [7].

4.3 Emerging Issues Related to the Use of CMC

Four main issues are rising day by day related to the use of CMC. The first issue revolves around the anonymity on the internet, which plays its role in increased misconduct of the news and other crimes. The second issue revolves around the reconfiguration of territorially and interest-based associations. The use of media artifacts and their social outcomes are the third main concern. Lastly, the mutual shaping of consumers and technologies also plays a role in the emergence of issues related to CMC.

Anonymity is described as "the state of being anonymous," whereas an interlocutor is "a person who participates in a dialogue or discourse." The new technology incorporates anonymity, a crucial aspect of CMC. Because much of the globe connect to the internet, the internet community is rapidly and constantly evolving. Freedom of expression and anonymity has always been significant societal challenges in the real world. Because more people learn online technology and are involved in anonymity in this socialist environment, these challenges are becoming incredibly influential. According to internet users, anonymity is essential for the safeguarding of freedom of expression. It is self-evident that the technology for internet anonymity is readily available. Anonymity, as shown in online activities, impacts the task and social components of CMC, such as information sharing and interpersonal communication [43].

Sociologists developed the deindividuation theory, which asserts that absorption in a group, most notably addressing lowered personal visibility, results in the loss of identity and even a sense of helplessness, based on this assumption of the crowd mentality in early anonymity studies. People lose their originality, integrity, and selfhood when they become anonymous and blend into the herd, leading to anti-social activities. As a result, the early description of anonymity is distinguished by emphasizing the loss of self through immersion in a shared belief and the resulting detrimental social implications. Deindividuation is characterized as a condition in which persons in groups are not viewed or treated as individuals.

Early deindividuation theory argued that anonymity in society may lead to a "loss of self" and to at least less self-awareness to describe how anonymity causes deindividuation. Whether "loss of self" identity occurs in anonymous conditions is the determining factor for distinguishing another class of deindividuation, scholars who argue that anonymity in the group typically does not result in a loss of identity, but rather encourages a switch from individual identity to social identity and tends to increase the significance of social constructs. In addition to preventing anonymity as a source of deindividuation, initial deindividuation theorists suggested that anonymity could negatively affect social behaviors, such as impulsive behavior and attraction to deviant groups, due to the loss of identity and self-control among crowd members. According to early CMC research, online communications are more likely to become involved in the aggressive expression of intense thoughts and feelings, known as flaming, due to deindividuation [43].

Social media platforms are platforms where anonymous are sitting in a large number hiding their identity and producing content for the other users so that they could have interaction through it. Several social applications aid in CMC. The rise of mobile applications that promise anonymity as the main feature has resulted in previously unstudied social configurations [6]. Due to increased anonymity on the Web, it has been challenging to identify people responsible for misconducting the information sitting behind their computer screens. Fake ID can be easily created and used to spread false information and sometimes panic people. Additionally, the anonymity of the Internet, which has rendered society defenseless, has opened the door to antisocial conduct [13]. Beyond all this, anonymity on social media has also played a role in promoting other crimes like hacking and harassment.

The crimes are responsible for damaging nearly around about billions of dollars per year as well as human dignity, faith, reputations, and moral destruction. Consequently, it is crucial that the information of those who perpetrate such crimes in the name of CMC be traced and reported to law enforcement. Free flow communication and increased anonymity allow criminals to walk free, which is the biggest issue regarding the CMC and internet. This must not have happened, and there should be a solution to solve this issue and enjoy the advantages of CMC [15]. Anonymity plays a big part in the freedom of speech; thus, it is brought up again. Anonymity can be misused, transforming free speech into hate speech or objectionable criminal material. Since anonymity is a strong cloak for whoever is behind it, it allows them to avoid taking blame or consequences for their actions.

On the other hand, anonymous communication is not required for freedom of expression. Messages sent via anonymous are sometimes uncertain about having such a significant impact on their own unless the receiver knows and believes the sender. Since, in case the communication is anonymous, it is not easy to establish a relationship with communicators [13]. Anonymity also plays a role in affecting the behavior of people. Online interactions with unknown people in CMC may have positive and negative effects. However, the number of positive effects is less than the negative ones. Adverse effects have been discussed above. One main positive impact of anonymity in CMC involves people revealing the personality traits they might hesitate in person. Being in a crowd can sometimes cause people to act rudely and viciously. Despite this positive effect, anonymity is problematic because it allows for accountability to be avoided. According to Moore et al. [24], the capacity to act openly while hiding an individual's identity might also hurt public discussion on the internet.

4.4 Reconfiguration of Territorially and Interest-Based Associations

The technology that enables readers to write online comments on newspaper websites anonymously provides readers with extraordinary chances to contribute, raising questions about the journalistic value of accessibility, self-regulatory practice, and competence formation. Even though technology aids journalists in providing quick information via CMC, they do not participate with the readers in the mutual shaping of news content. Scholars have suggested that online journalism has the power and ability to give readers a voice of their own [26]. At a global level, CMC also helps develop a new paradigm of human activity across the territorial borders providing feasibility. Electronic communications, no doubt, play havoc within different territorial boundaries. Since the territorial boundaries in a virtual world are separated via screens, codes, and passwords in an electronic environment, the new boundary develops a cyber space that helps legislation and legal institutions safeguard the territorial electronic environment. Nevertheless, lawmakers and law enforcers think that this new environment is very dangerous. Geographical borders are separated from the real world in the virtual world, providing the people with various advantages and disadvantages that are of great significance. Various and different sets of rules and regulations apply at the electronic territorial borders [12]. With respect to territorial concern, the main emerging issue in CMC is that people sitting behind the computer screen can access the personal information of people of different regions within the country and across the territorial borders. By just pushing a button, one can discover a person's origin, phone number, current location, and even driving instructions to his home. Hence, it has become complicated to protect people from unsafe access to personal data. Through CMC, people from different locations engage in relationships with anonymous and then due to blackmailing from them, they may try to harm themselves. This is one biggest disappointment from the internet and CMC [41]. When the internet and the computer were not so common in use, many people thought that engaging with different people via CMC would make people to become mechanical. It would also somehow be able to cause harm to humanization. The fear of people was challenged later on, and various different experiments suggested that people who communicate via CMC were ruder than that of those who communicate face to face [8].

4.5 Introduction of New Artifacts and Their Social Outcomes

The term "mass media" refers to a segment of the media designed and created to reach large groups of people such as a state's population. With the rise of regional broadcast stations, mass-circulation newspapers, and magazines in the 1920s, mass media became popular. Newspapers, television, and other forms of broadcasting are examples of mass media. Internet media is a relative newcomer to the mass media definition. Online media include personalized web pages, podcasts, and journals [36]. Web applications, forums, hashtags, and newsfeeds, among other social media tools, have revolutionized the way people interact, engage, and communicate. Furthermore, most of these innovations have found their way into agile cooperative approaches and modern software development tools, either as an add-on or as part of a broader set of tools that include text editors and issue trackers to IDEs and internet sites [40]. However, the issues related to such artifacts are rising day by day. A fundamental "infrastructure of involvement" that facilitates audience and numerous broadcast mechanisms can be found in social media applications. Their design encourages and fosters collaboration, often a by-product of individual actions. It also empowers people's participation and involvement in activities traditionally held by a small range of participants.

Many researchers acknowledge that software development procedures are more than just generating source code and that "articulation work" in a software engineering project must be encouraged. "Articulation" is defined by Gerson and Star as "all tasks required coordinating a specific activity, including scheduling subtasks, recovering from errors, and assembling resources." Discussions on design decisions, provision of bug-fixing duties to developers, and agreement on interfaces are all examples of articulation work. On the other hand, social media has changed how humans generate and curate information artifacts online in recent years [18]. For example, Wikipedia, a free encyclopedia created collaboratively by using wiki software, is one great example of several people coming and joining together to create web content using the technology and inventiveness of social media. Again, one major issue related to media artifacts is the possibility of misconducting information. False information and news articles on the Web can lead people to have false knowledge about a particular thing. Another social outcome that is of great concern is of ethical point of view. Wikipedia has now established the de-facto standard for encyclopedias, regardless of the lack of institutional systems to ensure the reliability and conciseness of content.

Moreover, social networking, especially the micro-blogging site Twitter, increasingly influences daily politics and affairs. Twitter, for example, was crucial during Egypt's 2011 revolt and in the aftermath of Japan's earthquake and tsunami in 2011. These instances demonstrate how vast groups of people may efficiently generate and curate web content without the use of formal rules and processes, utilizing a range of social media technologies such as Wikipedia articles, tagging, and weblog [38].

4.6 The Mutual Shaping of Consumers and Technologies

Journalism is rapidly transforming the way, most of it due to advances in information technology, but expecting that technological breakthroughs cause changes in news creation directly and discontinuously is a myopic view of dynamic processes. The journalism profession has adapted to technological advancements and helped shape communication technology in the past. Depending on the velocity of technical change in a specific medium in the twentieth century, news outlets could take years to figure out the best ways to adapt and shape innovations to meet their needs [29]. The present rate of electronic communication technology development and dispersion is becoming more challenging to manage. Media sources increasingly rely on digital platforms, which frequently have low entrance hurdles. This allows a slew of new competitiveness to emerge swiftly, posing a threat to news organizations' consumers and sponsors. This extremely fast-paced technology environment is not the main cause of falling viewership and profits. However, it does cause some companies, particularly the newspaper industry, to discuss sustainability rather than expansion. Many news outlets are modifying their production methods and people to adapt. A few news organizations have succeeded in introducing good services and tools, but achievement could be transitory, and also the social conventions ingrained in the environment of media companies or organizations would also be there, which many in the business would struggle to retain, even though if it is related to stifling innovation [30]. The mutual shaping of technology (MST), according to Boczkowski [3], includes "the orientation and velocity of technology adoption as well as the development of media artifact." He also discusses how mutual shaping of technology might be used to analyze technology change in the context of journalistic organizations.

The mutual shaping of the technology concept is designed to help us better understand how innovation from the communication perspective can impact and be influenced by a media outlet. Theoretical approaches to mass communication inquiry provide perspectives into the links between cultures, technologies, and communications. Many of the widely accepted theoretical asserts are predictable. They depict CMC developments as "floods of innovation" flowing over communities. They overlook or dismiss the cultural forces that determine the evolution of information technologies. Harold Innis [48] provides a framework for many types of research on the association between communication technology and society. Although his method is not directly linked to the transmission of MST, it addresses similar problems, such as technical inventions in communication in dynamic connections with cultures across time in the modern hemisphere. Innis, who examines diffusion processes through the prism of history, is more concerned with the social as well as cultural consequences of the new and latest communication inventiveness and technologies than with their adoption rates. Innis' reputation among media scholars is undoubtedly due to the influential role he gives mass communication innovation in social and cultural development. He allocates a significant influence to improvements in communication technology [19].

The mutual shaping of technology construct, as previously said, aims to maintain a positive attitude toward two competing paradigms: diffusion of innovations theory and social shaping of technology construct (SST). The diffusion of innovations theoretical model examines how news and information can be dispersed throughout a community or a society. Scholars of media innovation have emphasized the importance of taking a comprehensive perspective on the history and future of communication technology development. In retrospect, the internet was socially shaped as it emerged, and there is no reason to believe that future digital innovations will be any different. The SST approach studies the same coin from a different angle. It consists of studies on "social contexts of innovation...social and economic elements that may affect technology..." and has brought to light the significance of a vast array of stakeholders and those who are impacted [22]. The subject of concern is how technologies and artifacts are created, as well as what societies do with them once they have been adopted. According to SST, social factors determine how, not just if, technologies are adopted, and how the technology that is adopted shapes future generations of the innovation. Academic and industrial circles are interested in researching new business models for online journalism. Thousands of news agencies have been laid off or bought out recently, particularly in the newspaper industry [34].

5 Conclusion

CMC is a type of communication in which more than two people are involved, sitting behind their computer screens and communicating with each other. CMC has made everyone's life easy. Nowadays, CMC in journalism has benefitted journalists and readers worldwide. The online newspaper has not entirely but majorly replaced the printed newspaper. People can read news articles by just using the computer or tapping the mobile screens. Within the IoT environment, online journalism provides us with an easy, simple, fast, and effective way of breaking the news. A synopsis of events as they occur can be provided to the people of the society within no time. In online journalism, journalists can provide the people sitting on the internet with accurate information which is also up to date. The classic hard-copy newspaper necessitates costly production equipment but no equipment on the reader's end.

In contrast, CMC and the online newspaper require less expensive production equipment and more complex receiving equipment on the consumer side, requiring a more significant upfront investment on the user's part. The concept of CMC has given birth to social media, which has many benefits. CMC has also become easy through social media, but the cons are also in great numbers if there are pros. Some issues are also rising day by day related to the use of CMC within an IoT environment. Anonymity on the internet playing its role in increased misconducting of the news and other crimes is one major rising concern. MST and the use of media artifacts also have issues related to social points of view. If, on the one hand, computer-based communication provides benefits, then, on the other hand, it also raises issues for which we must seek answers.

References

1. Alobaidi, E. (2021). A study of outline to the advantages and disadvantages of social media. *Academicia: An International Multidisciplinary Research Journal, 11*(1), 1437–1444.
2. Ariel, Y., & Avidar, R. (2015). Information, interactivity, and social media. *Atlantic Journal of Communication, 23*(1), 19–30.
3. Boczkowski, P. (1999). Understanding the development of online newspapers: Using computer-mediated communication theorizing to study Internet publishing. *New Media & Society, 1*(1), 101–126.
4. Broekhuizen, T., & Hoffmann, A. (2012). Interactivity perceptions and online newspaper preference. *Journal of Interactive Advertising, 12*(2), 29–43.

5. Chandra, A., & Kaiser, U. (2015). Newspapers and magazines. In *Handbook of media economics* (Vol. 1, pp. 397–444). North-Holland.
6. Curlew, A. E. (2019). Undisciplined performativity: A sociological approach to anonymity. *Social Media + Society, 5*(1), 2056305119829843.
7. Darvishi, S. (2020). *The use of Slack as a social media in higher education: Students perceptions of advantages and disadvantage of Slack during learning process.*
8. DeAndrea, D. C., Tong, S. T., & Walther, J. B. (2010). Dark sides of computer-mediated communication. In *The dark side of close relationships II* (pp. 115–138). Routledge.
9. Delliana, A. S., & Wook, A. M. W. (2020, May). Bigo live: Ethical degradation in communication (A critical study of communication in a computer-mediated communication perspective). In *Tarumanagara international conference on the applications of social sciences and humanities (TICASH 2019)* (pp. 83–88). Atlantis Press.
10. DePaula, N., Dincelli, E., & Harrison, T. M. (2018). Toward a typology of government social media communication: Democratic goals, symbolic acts and self-presentation. *Government Information Quarterly, 35*(1), 98–108.
11. Everett, C. E. (2011). Transformation of newspapers in the technology era. *The Elon Journal of Undergraduate Research in Communications, 2*(2), 102.
12. Ford, R. T. (2012). Law and borders. *Alabama Law Review, 64*, 123.
13. Fortunati, L., & Edwards, A. (2020). Opening space for theoretical, methodological, and empirical issues in human-machine communication. *Human-Machine Communication, 1*, 7–18.
14. Garrison, B. (2020). *Computer-assisted reporting*. Routledge.
15. Hall, K. (2016). Selfies and self-writing: Cue card confessions as social media technologies of the self. *Television & New Media, 17*(3), 228–242.
16. Hong, S. (2012). Online news on Twitter: Newspapers' social media adoption and their online readership. *Information Economics and Policy, 24*(1), 69–74.
17. Hrdličková, Z. U. Z. A. N. A. (2018). Improving reading skills using on-line newspapers and magazines. *Lingua et vita, 14*, 75–84.
18. Huang, Q., Peng, W., & Ahn, S. (2021). When media become the mirror: A meta-analysis on media and body image. *Media Psychology, 24*(4), 437–489.
19. Kraidy, M. M., & Vilanova, J. (2019). 2 time, space, and communication. *The Media World of ISIS*, p. 33.
20. Locher, M. (2010). Introduction: Politeness and impoliteness in computer-mediated communication. *Journal of Politeness Research, 6*(1), 1–5. https://doi.org/10.1515/jplr.2010.001
21. Mahmoud, A. E., & Auter, P. J. (2009). The interactive nature of computer-mediated communication. *American Communication Journal, 11*(4), 19–21.
22. Mascia, M. B., & Mills, M. (2018). When conservation goes viral: The diffusion of innovative biodiversity conservation policies and practices. *Conservation Letters, 11*(3), e12442.
23. Meier, A., & Reinecke, L. (2021). Computer-mediated communication, social media, and mental health: A conceptual and empirical meta-review. *Communication Research, 48*(8), 1182–1209.
24. Moore, A., Fredheim, R., Wyss, D., & Beste, S. (2021). Deliberation and identity rules: The effect of anonymity, pseudonyms and real-name requirements on the cognitive complexity of online news comments. *Political Studies, 69*(1), 45–65.
25. Nah, S., Yamamoto, M., Chung, D. S., & Zuercher, R. (2015). Modeling the adoption and use of citizen journalism by online newspapers. *Journalism & Mass Communication Quarterly, 92*(2), 399–420.
26. Nielsen, C. E. (2014). Coproduction or cohabitation: Are anonymous online comments on newspaper websites shaping news content? *New Media & Society, 16*(3), 470–487.
27. Palme, J. (2010, October). Before the internet: Early experiences of computer mediated communication. In *IFIP conference on history of nordic computing* (pp. 271–277). Springer.
28. Panda, K. C., & Swain, D. K. (2011). *E-newspapers and e-news services in the electronic age: An appraisal.*

29. Poepsel, M. A. (2011). *The mutual shaping of technology in a news establishment: Social journalism and organizational change*. University of Missouri-Columbia.
30. Poepsel, M. A. (2018). Mutual shaping of a CMS for social journalism in a hierarchical News organization. *The Journal of Media Innovations, 4*(2), 55–70.
31. Rahmatov, F. (2013). *Social network uses of newspaper websites*.
32. Rieis, J., de Souza, F., de Melo, P. V., Prates, R., Kwak, H., & An, J. (2015). Breaking the news: First impressions matter on online news. In *Proceedings of the international AAAI conference on web and social media* (Vol. 9, No. 1, pp. 357–366).
33. Riskos, K., Dekoulou, P., & Tsourvakas, G. (2019). User interactivity in online newspapers: Exploring the relationship between content features and user response. *Newspaper Research Journal, 40*(2), 155–176.
34. Salaverría, R. (2019). Digital journalism: 25 years of research. Review article. *El profesional de la información (EPI), 28*(1), 1–26.
35. Sevincer, V., Biseth, H., & Vaagan, R. W. (2018). *Faktuell: Youths as journalists in online newspapers and magazines in Norway*.
36. Singh, R. (2009). Digital preservation of mass media artifacts: Technologies and challenges. *Journal of Digital Asset Management, 5*(4), 185–195.
37. Steensen, S. (2011). Online journalism and the promises of new technology: A critical review and look ahead. *Journalism Studies, 12*(3), 311–327.
38. Storey, M. A., Singer, L., Cleary, B., Figueira Filho, F., & Zagalsky, A. (2014). The (r)evolution of social media in software engineering. In *Future of software engineering proceedings* (pp. 100–116). New York: Association for Computing Machinery publisher.
39. Treem, J. W., Leonardi, P. M., & van den Hooff, B. (2020). Computer-mediated communication in the age of communication visibility. *Journal of Computer-Mediated Communication, 25*(1), 44–59.
40. Treude, C. (2012). *The role of social media artifacts in collaborative software development* [Doctoral dissertation].
41. Wakefield, M. A., & Rice, C. J. (2008). *The impact of cyber-communication on today's youth (ACAPCD-14)*. American Counseling Association.
42. Walther, J. B., Van Der Heide, B., Ramirez, A., Jr., Burgoon, J. K., & Peña, J. (2015). Interpersonal and hyperpersonal dimensions of computer-mediated communication. In *The handbook of the psychology of communication technology* (pp. 1–22). United States: Wiley-Blackwell publisher.
43. Wang, Z. (2017). Anonymity effects and implications in the virtual environment: From crowd to computer-mediated communication. *Social Networking, 7*(1), 45–62.
44. Xu, Y. (2017). Modeling the adoption of social media by newspaper organizations: An organizational ecology approach. *Telematics and Informatics, 34*(1), 151–163.
45. Yao, M. Z., & Ling, R. (2020). "What is computer-mediated communication?"—An introduction to the special issue. *Journal of Computer-Mediated Communication, 25*(1), 4–8.
46. Wilbur, S. P. (1996). An archaeology of cyberspaces: Virtuality, community, identity. In D. Porter (Ed.), *Internet culture*, (pp. 5–22). NewYork: Routledge.
47. Bregman, A., & Haythornthwaite, C. (2001). Radicals of presentation in persistent conversation. In *Proceedings of the 34th Annual Hawaii International Conference on System Sciences*. https://doi.org/10.1109/HICSS.2001.926499
48. Innis, H. A. (2022). *Empire and communications*. Toronto, Canada: University of Toronto Press.

FATS (Fuzzy Authentication to Provide Trust-Based Security) in VANET to Mitigate Black Hole Attack

M. Gayathri and C. Gomathy

1 Introduction

VANET is derived from the concept of the MANET Mobile networking. In mobile ad hoc networks, cell phones are used in communication, and in VANET, vehicles are used for communication [1]. The concept of distributed network computing is used in VANET. Smart transportation or intelligent transportation is only possible by VANET communication technology. VANET provides a wireless mode of communication between vehicles. This type of communication allows the vehicle node to connect to another unknown vehicle node and share needed information [2]. VANET users can share information about nearby fuel stations, hospitals, ATM centers, and so on. If a road is blocked by traffic due to an accident, VANET users can intimate the other user to take a diversion to another road which will reduce the road blockage. It can even intimate about the pre-crash warning, share the emergency information, and seek help from the other node user. All these applications provide a way to a smart transportation system. VANET communicates utilizing wireless communication by broadcasting messages. When a wireless mode of communication is used, secure mechanism plays a significant role. All messages being transmitted are broadcasted messages and the hackers may secretly listen to the conversation between vehicle users and steal the information. The hackers can hack the communication and inject false messages and forward them to the destination to divert the user. The attackers can even cause a car accident by interoperating a false message in the communication link [3]. Security mechanisms and services play a big part in vehicular communication. The onboard unit, application unit, roadside unit, and trusted authority make up the VANET architecture. Vehicles

M. Gayathri (✉) · C. Gomathy
Electronics and Communication Engineering, College of Engineering and Technology, SRM Institute of Science and Technology, Vadapalani campus, Chennai, Tamilnadu, India
e-mail: gm0717@srmist.edu.in; gomathyc@srmist.edu.in

R. Sharma et al. (eds.), *Data Analytics for Internet of Things Infrastructure*,
Internet of Things, https://doi.org/10.1007/978-3-031-33808-3_4

and roadside equipment can communicate, thanks to WAVE (Wireless Access in Vehicular Environment) [4]. Roadside units are positioned by the sides of the road to offer local connectivity to any passing vehicles. For communication, IEEE 802.11p radio technology-based DSRC-dedicated short-range communication protocols are employed [5]. Every car has onboard electronics. It comprises a GPS-based tracking gadget that transmits data to a roadside unit and other VANET node users. Vehicles are furnished with an event data recorder, an onboard unit, and sensors that give information to the other vehicle nodes [6, 7]. Communication via the VANET network is secured in large part by trusted authority. All of the vehicle's onboard identities are registered by a reliable authority, which then transmits that data to the roadside device. Before opening a communication channel with other nodes, it authenticates the vehicle, the user, and the user's identity. Together, these pieces of technology offer a communication link between vehicles as well as between vehicles and roadside equipment, infrastructure, pedestrians, and even train users. On the basis of trust models, many trust-based authentication systems are addressed in [8]. Trusted approach provides a highly secure communication. Section 2 discusses in brief about VANET architecture, Sect. 3 discusses the various attacks and threats generated in VANET, Sect. 4 explains the views on the prominent issues caused by black hole node, Sect. 5 concentrates on the fuzzy logic and its role in the proposed approach, Sect. 6 explains the proposed algorithm FATS, and Sect. 7 gives a brief note on implementation of algorithm in ns 2.28 software. Parameter estimation, framing of rules in MATLAB for detecting black hole attacks, and networking are done using ns 2.28 software [9].

2 VANET Architecture

Vehicular ad hoc networking is being introduced to provide intelligent transportation and minimize road accidents. Vehicle node consists of numerous sensors and actuator that enable the vehicle to sense the vehicle, pedestrian in its 360-degree range, and act based upon the road conditions. The major components used in VANET communication are vehicle node, onboard unit, road side unit, Trusted Authority required to communicate between vehicles, pedestrian, and other infrastructures. Onboard units are built inside the smart cars which are used in the communication these devices serve as a transceiver between the source and the destination [10]. The Certificate Authority is crucial in ensuring verified and secure communication between vehicles. In internet connected with all components of the VANET, it is very important to pay attention on providing security and privacy in the VANET networking. Hackers or attackers would try to steal the information and collapse or the smooth functioning of the network by inducing false message in the stream that would result in the crashing between vehicles, create traffic, can divert the user by modifying his/her location, and so on. Certificate authorities are third-party authority responsible for producing genuine certificate to the user based on the identity and behavior of the node. Figure 1 describes the architecture of the

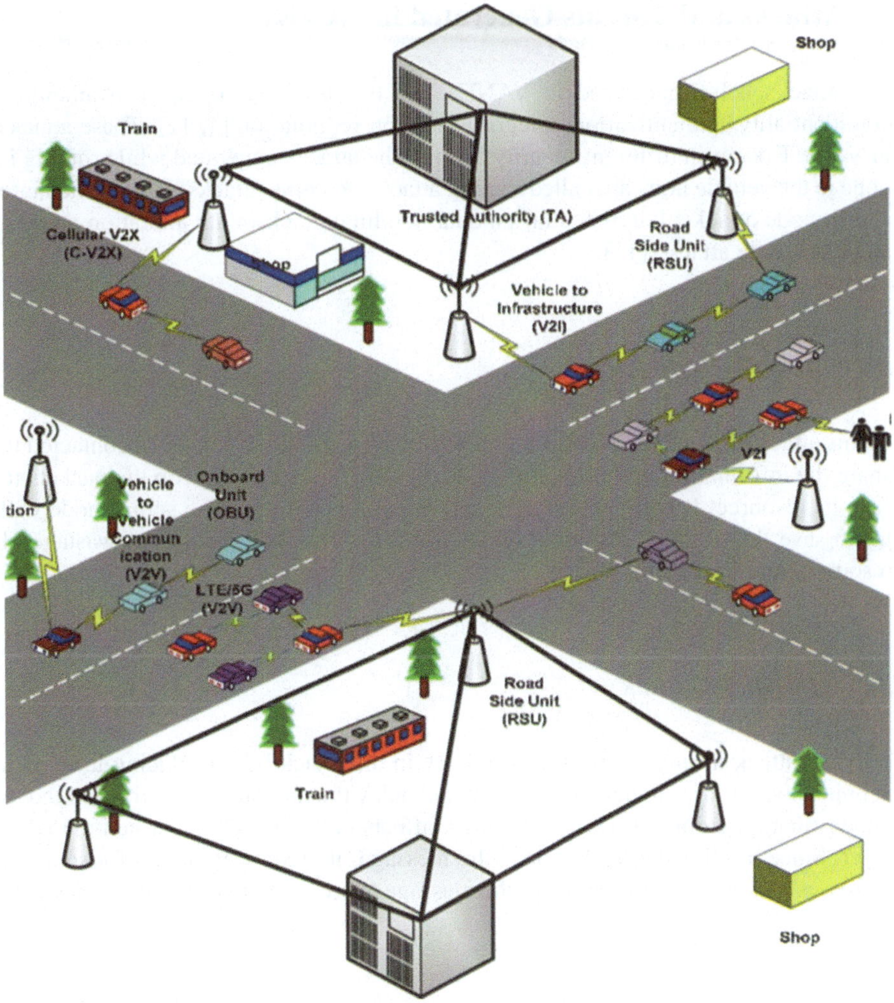

Fig. 1 VANET architecture

VANET. As was already mentioned, its main components are smart automobiles, roadside units, and trusted authorities. It is feasible for vehicles to communicate with each other, with pedestrians, with roadside equipment, with infrastructure, and even with other vehicles [11].

Vehicle-to-vehicle communication is possible because of the DSRC protocol (Dedicated Short Range Communication).

3 Attacks and Threats Generated in VANET

The attacks being generated in VANET can be classified based on availability, confidentiality, authentication, integrity, and non-repudiation [4, 12]. These attacks in VANET occur in different security layers. The attacks generated while routing is done in the vehicle node are called routing attacks. Routing attacks are classified into selfish node attacks, jellyfish attacks, data flooding attacks, data alteration attacks, and black hole attacks [13].

3.1 Selfish Node Attack

In this attack, the node will behave as a self-obsessed node. If it is been contacted to share the information with the other node, it will either send or drop the packets to save its resource. It will not share the information genuinely. The selfish node will try to save the available resource for its use [14]. This will result in a wastage of resources and bandwidth [15].

3.2 Jellyfish Attack

Jellyfish attack normally introduces a delay in the attack by introducing itself as a genuine node. It tries to deny the service and delay the message being transmitted to the destination around 0–10 s. This attack affects both the TCP layer and the UDP layer. Since a jellyfish attack delays the message, it creates wastage of resources, increases bandwidth, and replicates the message to create more traffic in the network [13].

3.3 Data Flooding Attack

Data flooding attack creates high congestion in network traffic. This malicious hacker selects an IP address that is not available in the network communication and tries to send unwanted messages to all the nodes using that IP address which creates high congestion and network traffic that delays the important or emergency messages that are being transmitted [16].

3.4 *Black Hole Attack*

A black hole attack in a VANET addresses itself as having the shortest path from the source to the destination when a route request is sent to the nodes that are available in a specific zone. The source node sends packets through the malicious node after realizing the message was a hoax. After receiving the message, the attacker will discard the packet without sending it to its intended location. A packet drop attack is another name for this black hole assault [17].

4 Prominent Issues Caused by a Black Hole Node

In VANET, numerous assaults are produced. These attacks can be divided into categories depending on assaults on confidentiality, integrity, availability, authentication, and non-repudiation. The black hole or malicious node is to blame for all of these attacks.

Black hole nodes are the main target of attack in a VANET scenario because they generate greater message transmission delays, jam communication by losing packets, or even alter data. Black hole nodes can disrupt VANET connectivity and have a negative impact on it. The proposed approach FATS- Fuzzy authentication to provide trust-based security is used to mitigate black hole attack and this approach with slight modification could be used to mitigate other attack too.

The black hole attack on VANET is depicted in Fig. 2. In order to communicate with Node D, Node A sends a Route request message to every node in the vicinity. The malevolent node, Node M, is regarded as a black hole node in this context. The time stamp and the number of hops required to get to the destination are included in the route reply message that each node sends. The malicious node M, a black node that cannot communicate with the destination node D, identifies itself by claiming

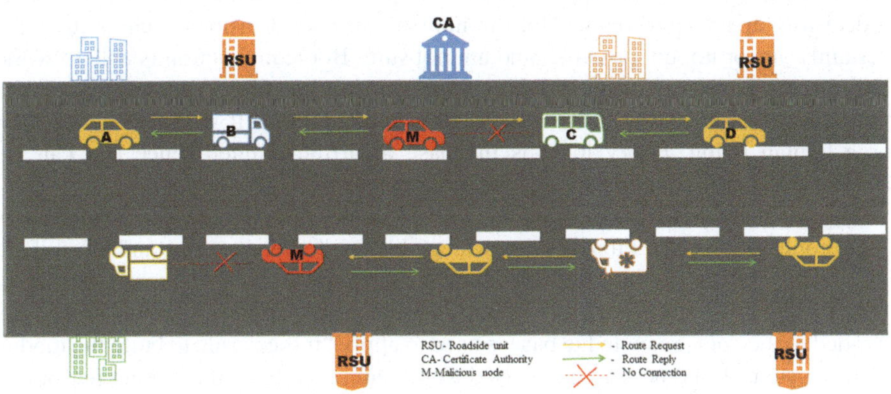

Fig. 2 Black hole attack in VANET

to have the shortest path there with the fewest hops and the earliest time stamp [18]. Node A accepts that node and transfers the data to black hole Node M. Node M drops the data packet [19]. Black hole attack can be detected based on the packet loss analysis, Traffic analysis and based on inspecting the packets being altered. This drop in the data packet increases the traffic delay of the data transfer, causes packet loss, and attacker may even inject or modify the message. If the weight of the trust factor is analyzed properly at each and every part of the node, then malicious node can be avoided from entering into the communication zone. FATS algorithm analyses the trust factor and traffic delay while transmitting the messages and detects black hole attack in VANET [20].

Ad hoc on-demand distance vector (AODV) routing protocol is used in the suggested technique. As a reactive system, AODV routing protocol offers route discovery based on demand. The AODV routing protocol locates the route and allots a path to the required source when a route request is received. The three steps of the AODV routing protocol are route discovery, data transfer, and route management. RREQ and RREP routing techniques are used by the AODV routing protocol to offer the optimum path. The source IP address, source sequence number, destination IP address, destination sequence number, and broadcast ID are all sent in RREQ query packets. The destination IP address, destination sequence number, source IP address, hop count, and lifetime of a node are included in RREP responses [21].

5 Fuzzy Logic and Its Role in the Proposed Approach

5.1 Introduction About Fuzzy Logic

A strategy that depends on the degree of truthfulness is fuzzy logic. Boolean functions can only classify anything as true or false, whereas fuzzy logic can approximate the inputs. Fuzzy logic was proposed by Lofti Zadeh [22] and is based on the concept of human approximation [23]. When a human is allowed to take a decision, apart from Yes or No, the human can take decisions such as may be, certainly yes or no, not possible, and am not sure. Boolean function is able to work only on the concept of true or false and yes or no, but fuzzy approximation will concentrate on all the possible aspects of a human being. It will act intelligently as a human being. Fuzzy logic is the base of artificial intelligence and can be implemented in both hardware and software problem [24].

Figure 3 shows the block diagram of the fuzzy logic system. The basic fuzzy logic controller has the Fuzzifier, knowledge base, inference engine, and a fuzzifier. The crisp input is fed into the fuzzifier. The fuzzifier is used to fuzzify the input based on the system model. After fuzzification of the inputs, "if then" rules are applied to the data. Knowledge base is used to apply "if then" rule to build the model of the system. Inference engine is used to get inference from the human reasoning

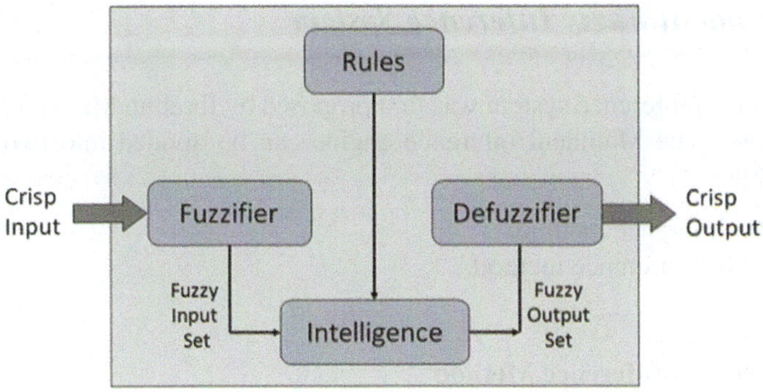

Fig. 3 Block diagram of fuzzy logic

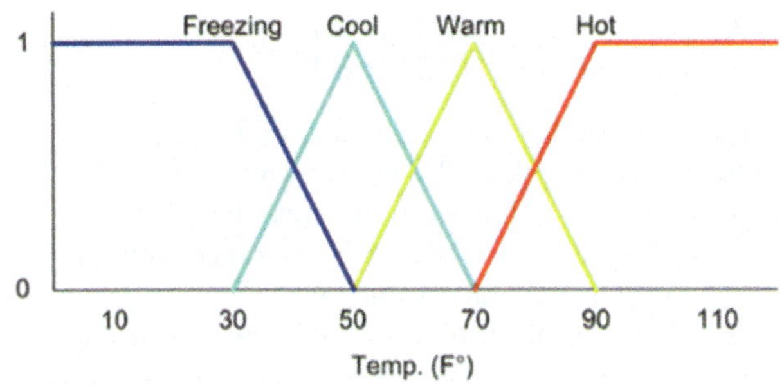

Fig. 4 Membership function of a temperature

approximation. After approximation of the model, the input is fed to the fuzzifier that are used to defuzzify the input to crisp output.

Membership function plays a significant role in quantifying the linguistic variable and in showing the fuzzy relation graphically. Figure 4 represents the membership function of the temperature. Temperature data are divided into freezing, cool, warm, and hot. In cases when there is a lot of ambiguity, fuzzy logic is applied [25].

Fuzzy logic can be divided into three types of fuzzy inference system:

- Mamdani fuzzy inference system.
- Takagi Sugeno fuzzy inference system.
- Tsukamoto fuzzy inference system.

5.2 *Mamdani Fuzzy Inference System*

Mamdani fuzzy inference system was first proposed by Ibrahim Mamdani to control steam engine. The Mamdani inference engine can be divided into two inference system [26]:

- Max-Min inference method.
- Max-product inference method.

5.2.1 Max-Min Inference Method

Consider the following rules:

Rule 1: IF x_1 is $A_1{}^1$ *and* x_2 is $A_2{}^1$ THEN y^1 is B^1.
Rule 2: IF x_1 is $A_1{}^2$ *and* x_2 is $A_2{}^2$ THEN y^2 is B^2.

Let us compute the output for $x_1 = 2.5$ and $x_2 = 3$.

Membership functions for given rules are shown below:

The highest membership value from the two input sets is assigned to the appropriate output set because it is a Max-Min inference method.

For first rule, the fuzzy membership value for x_1 would be 0.8 and for x_2 it would be 0.4. The connectives in first IF-THEN rule are connected using "and" connective. Therefore, we must find the intersection of the fuzzy values that returns the lowest value. As a result, the output y^1 will belong to fuzzy output set B1 by 0.4.

The fuzzy membership value for the second rule would be 0.3 for x_1 and 0.7 for x_2. The "or" connective is used to connect the connectives in the first IF-THEN rule. Therefore, we must use the union of fuzzy values that returns the most of them. In the fuzzy output set B2, the output y^2 will therefore have membership of 0.7. The input and output fuzzy sets and scaling of output fuzzy function are depicted in Figs. 5, 6, 7, and 8 for max-min inference method and aggregated output is shown in Fig. 9.

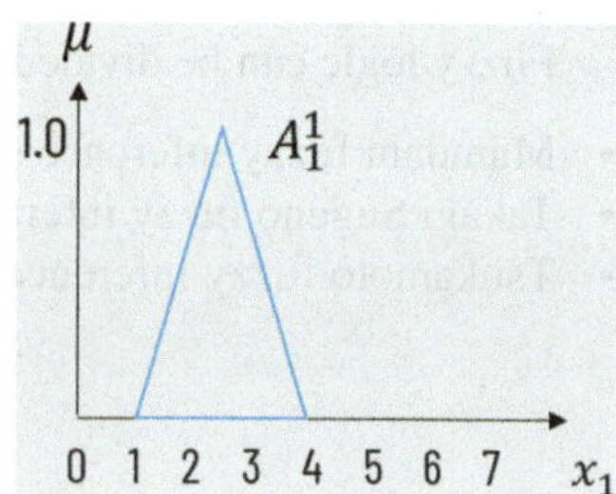

Fig. 5 Input fuzzy set $A_1{}^1$

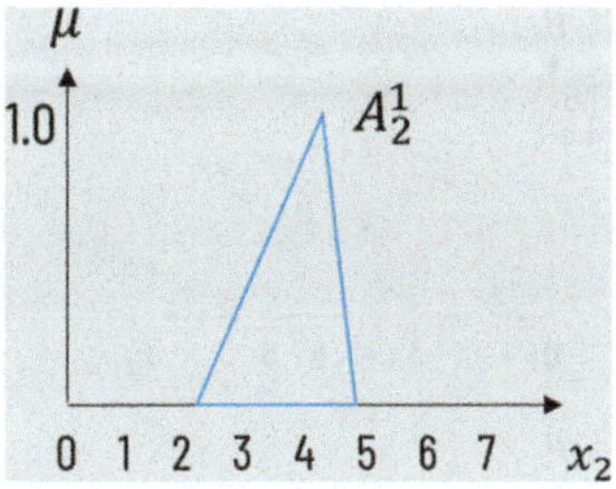

Fig. 6 Input fuzzy set A_2^1

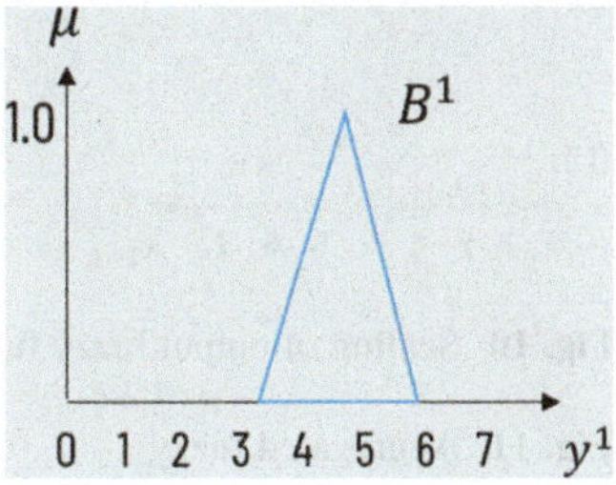

Fig. 7 Output fuzzy set B^1

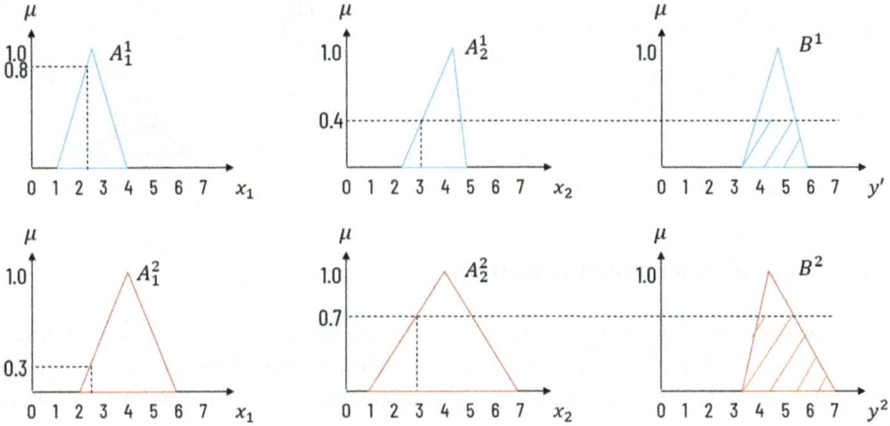

Fig. 8 Scaling of output fuzzy function

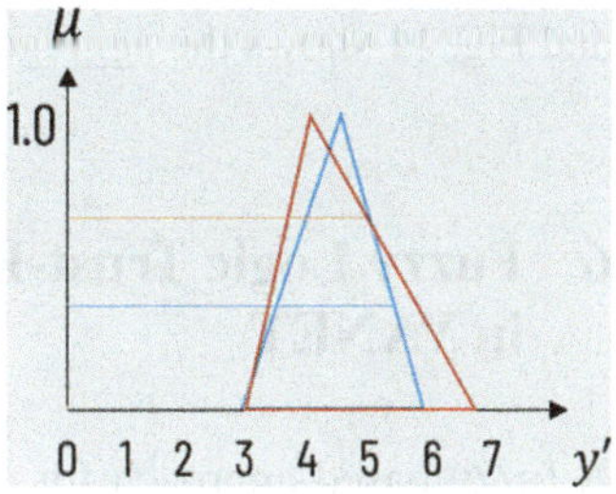

Fig. 9 Aggregated fuzzy output

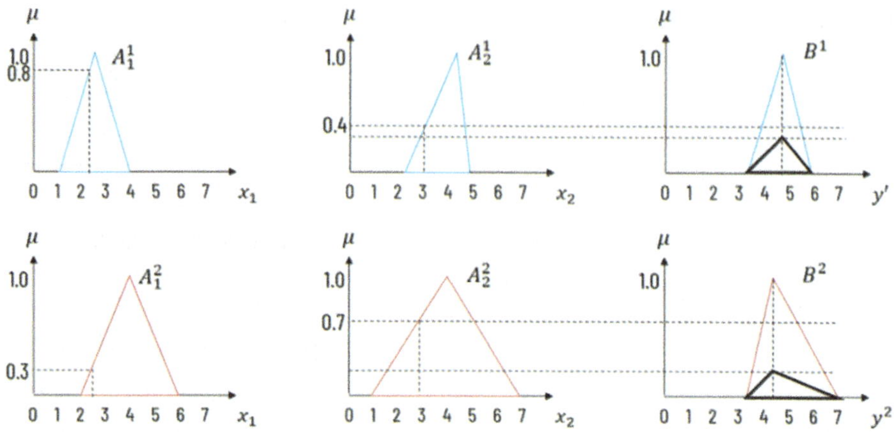

Fig. 10 Scaling of output fuzzy function

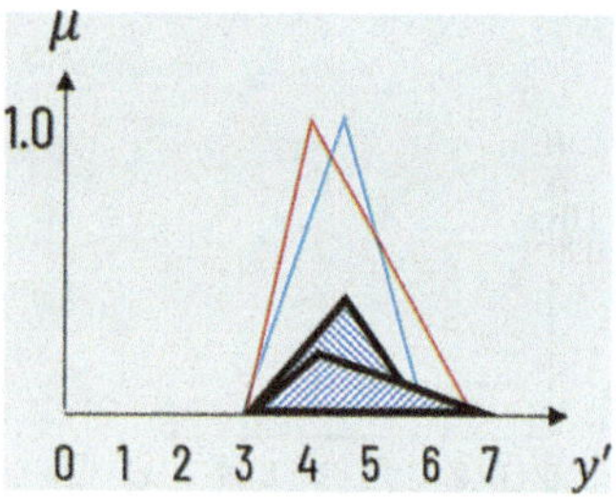

Fig. 11 Aggregated fuzzy output

5.2.2 Max-product inference method

An approach for combining fuzzy sets in fuzzy logic is the maximum product inference method. It is based on the principle that the sum of the degrees to which the components of a fuzzy set are true determines the degree to which the whole is true.

The above figure shows the representation of max-prod inference method. Figure 10 and Fig. 11 shows the scaling of output function with respect to the given input and Fig. 11 shows the aggregated fuzzy output.

6 Fuzzy Logic Trust-Based Authentication Schemes in VANET

A fuzzy-based approach for evaluating trustworthiness and managing authority in VANETs is suggested. In this chapter, the fuzzy theory was employed. To analyze the nodes' trust, it employed the behavioral attribute acquisition, vehicle device position, and comprehensive attribute weight as a parameter [27]. In the world of uncertainty, fuzzy logic is crucial. Because of VANET's high degree of mobility, it is challenging to forecast which nodes will enter and leave the network. Fuzzy logic is used to assign a trust parameter to track the node's sincerity, making it simple to foresee the node's sincerity. A trust factor based on authentication of vehicle and roadside unit was provided by Fatemehsadat et al. Direct trust, indirect trust, in-segment trust, and historical trust were taken into consideration to calculate the trust factor of the vehicle node [28]. A trust model using experience and plausibility was proposed in fuzzy logic [8].

7 Proposed Algorithm FATS (Fuzzy Authentication to Provide Trust-Based Security) for Black Hole Attack Detection

The trust factor plays a significant role in avoiding the malicious node joining the communication link. The trust factors provided by certificate authority by monitoring vehicle identity, trust factor provided by the roadside unit, trust factor calculated by the reputation of each node which is credited based on nodes behavior, the data trust and number of hops required to reach the destination is taken as input to calculate the trust degree of providing the communication link to the requestee node. The proposed algorithm FATS (Fuzzy Authentication to provide Trust-based Security) helps to detect black hole node and block the node to do any communication further. The trust degree or the weight of the node identity, weight of the trust degree provided by RSU, weight of the direct and indirect trust degree, and weight of the data trust are calculated.

Step 1: If a new node wants to join the communicating network in VANET, the node sends the RREQ message to all the nodes. After receiving the RREQ message, the trust factors are first analyzed before joining the node to the network.

Step 2: The trust factors are separately analyzed by VANET's Certificate authority. At the initial stage, vehicles' IDs are verified – whether it is a registered node or not.

Step 3: If it is not a registered node, the genuine node can block that node from joining and leaving a message as a suspicious node.

Step 4: If the node is registered, the trust factor provided by the RSU and the trust factor provided by the reputation of each node are verified.

Step 5: If the node is found to be a genuine node, after verifying the trust degree, then the communication link is provided to the new node by replying with an RREP message.

Step 6: If the malicious node still acts as a genuine node initially, a test message is forwarded to the destination node to analyze the trusted behavior of VANET. If there is a packet drop or delay in receiving the acknowledgment message, then the node can be blocked. If the acknowledgment is received, then the original messages which are in need to be transmitted are encrypted to ciphertext and forwarded to the destination node.

Step 7: When the new node is ready for a data transfer, trustworthiness of data is ensured by observing the previous data sent by the new node. This form of trust comes under historical trust.

7.1 Pseudocode for Providing a Communication Link to the New Node

Every vehicle i sends the RREQ route request message to join the communication link.

```
{
Verifiers verify the vehicle id
   If {
Node id is registered
}
Considered as a genuine node
 Else {
 Block the node from the communication link
}
}
If {
#genuine node is found by verifying vehicle's id
#Verify the node's reputation trust and RSU Trust
If the trust degree is reached above the threshold level
Accept the node
Else {
Reject node
}}
# Check the genuineness of the user by sending a test message If {
The test message is sent at the correct time and acknowledgment
is received
Provide a communication link by RREP message
Else {
 Report malicious
}
}
```

7.2 Formation of Fuzzy Rules Using Mamdani Inference System in MATLAB

Node reputation trust, roadside unit trust, data trust factor, and the number of hops required to reach the destination from the source are chosen as input parameters. All

Table 1 Calculation of trust values to provide a communication link

S. no	Node reputation trust	Roadside unit trust	Data trust	Number of hops required to reach the destination	Provide communication link
1	0	0	0	0	0
2	0	0.6	0.6	0.6	0.6
3	0	1	1	0	0.6
4	0.6	0	0	0	0
5	0.6	0.6	0.6	0.6	0.6
6	0.6	1	1	1	0.6
7	0.6	1	1	0	1
8	1	0	0	0	0
9	1	0.6	0.6	0.6	0.6
10	1	1	1	0.6	1
11	1	0.6	1	0.6	1
12	1	1	0.6	0	1

these parameters determine black hole attack in VANET and reject the malicious node if the trust factors are very low.

Mamdani FIS is used with IF-THEN rules. These rules are framed to select the genuine node and reject the black hole node in VANET. The centroid defuzzification method is used to calculate the trust parameter of the vehicle node. Table 1 discusses the various criteria for providing a communication link to the new node and rejecting the black hole node. Equation (1) gives the centroid defuzzification formula for calculating the trust factor. Figure 12 shows the surface view of the calculation of trust degree required to provide a communication link to the trusted node.

$$\text{Trustfactor} = \frac{\int \text{xi}.\mu\ (\text{xi})}{\int \mu\ (\text{xi})} \tag{1}$$

Figures 12 and 13 shows the trust metric calculation for the medium and low trust factor respectively. If there is a low trust factor, then communication link will be discarded to that node, and if there is a medium trust factor, then communication may be provided to that node. Figure 14 depicts the surface view for providing communication link.

8 Implementation of FATS

The network simulation software 2.28 is significantly used for the research work in VANET [29]. It is an open-source software in which simulation of wireless

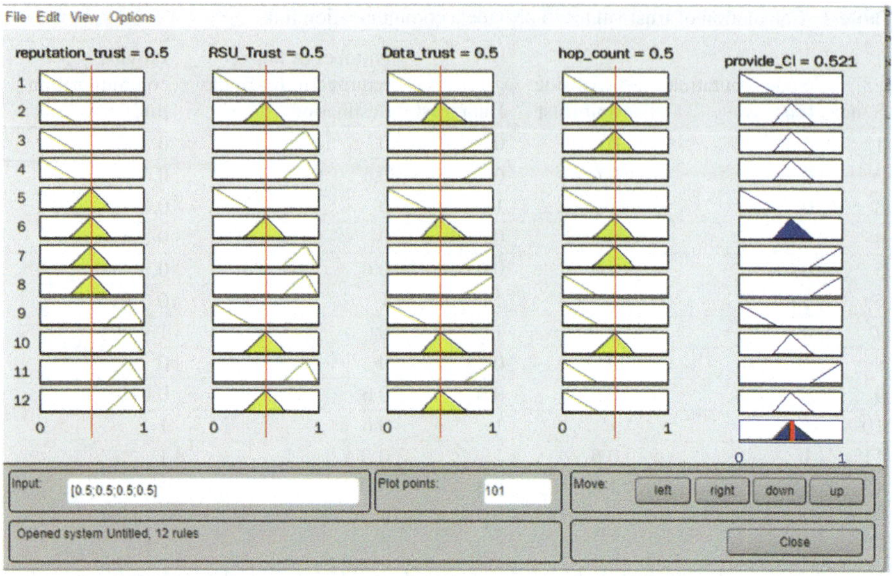

Fig. 12 Trust factor calculation for medium trust factor

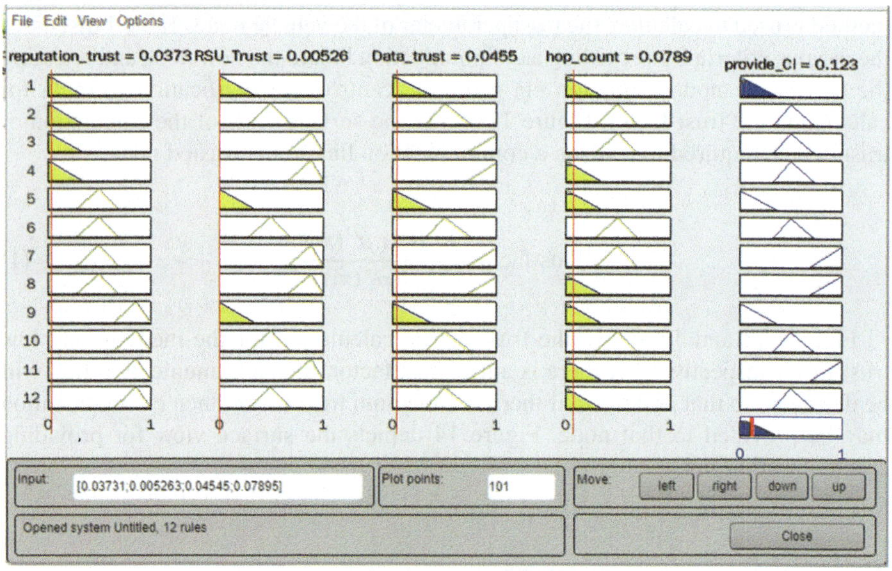

Fig. 13 Trust factor calculation for low trust factor

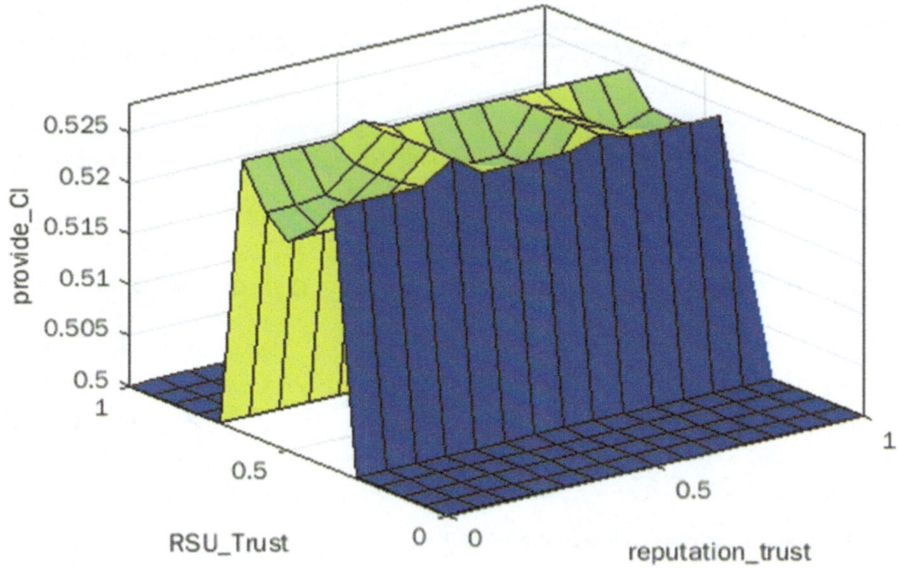

Fig. 14 Surface view for calculation of trust value to provide a communication link

applications can be done more easily. Around 100 nodes have been designed with the roadside unit and vehicle nodes.

The vehicle node moves around, and if any information is required, the user can get the information from the other node user. In this scenario, the trust factor plays a significant role; this trust factor gives the degree of whether the node which is contacted is a trusted node or not. According to Table 1, values between 0 and 0.65 are low trust values, values from 0.5 to 0.85 medium trust values, and values from 0.8 to 1 high trust values.

If the trust degree calculated is below the threshold value, the user can block that node from communication and send the request to the other node user. When the communication is initiated, messages which are transmitted are encrypted to cipher text form by using public and private key for the authentication process and key is generated to pass the information. The trust degree based on vehicle ID, the reputation of the node, and RSU trust detects black hole attack in VANET and helps the node user to discard the black node which increases the throughput and packet delivery ratio and decreases the delay ratio. Figure 3 shows the image of the ns2.28 simulation tool, which consists of vehicle nodes, roadside unit, and black hole attacks along the direction of movement of the vehicle. The fuzzy-based trusted communication detects the black hole attack and rejects the attacked node and chooses the genuine node for communication based on fuzzified trust degree. This proposed method increases the throughput and packet delivery ratio and decreases the delay time. Figure 15 shows the detection of attack in animator window of ns2.28 (Table 2).

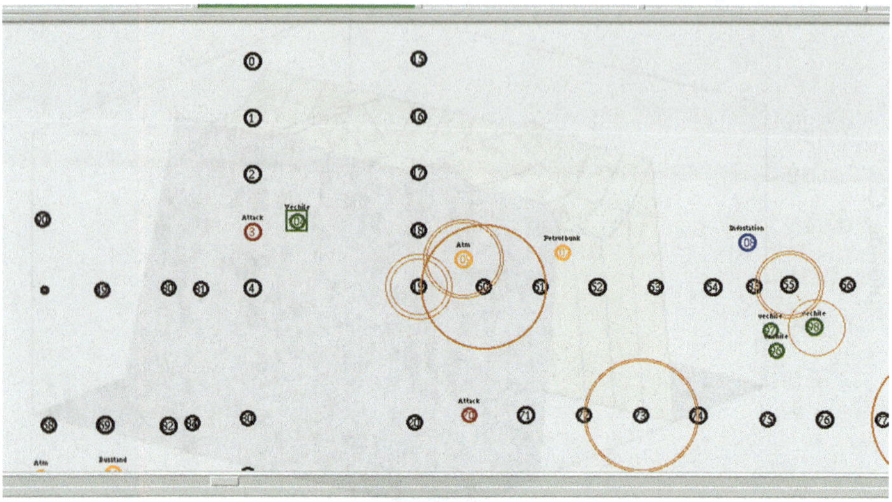

Fig. 15 Detection of black hole attack in VANET

Table 2 Depicts the simulation parameter used in this approach

Simulation tool	NS 2.28
Simulation area	1200 m × 1200 m
Routing protocol	AODV
Channel type	Wireless
Radio propagation model	Two ray ground
Packet size	64 bytes
Performance evaluation	Packet delivery ratio, throughput, and end-to-end delay

Throughput (Kbps)
Throughput is defined as the total number of packets reached at the destination to the amount of time taken.

$$\text{Throughput} = \frac{\text{Total number of packets received at the destination in bytes} \times 8}{\text{End time} - \text{Start time}}$$

Packet delivery ratio (in %) [30]
Packet delivery ratio is defined as the ratio of the total packets received at the destination to the total packets generated by the node.

$$\text{Packet delivery ratio} = \frac{\text{Packet received}}{\text{Packet generated}} \times 100$$

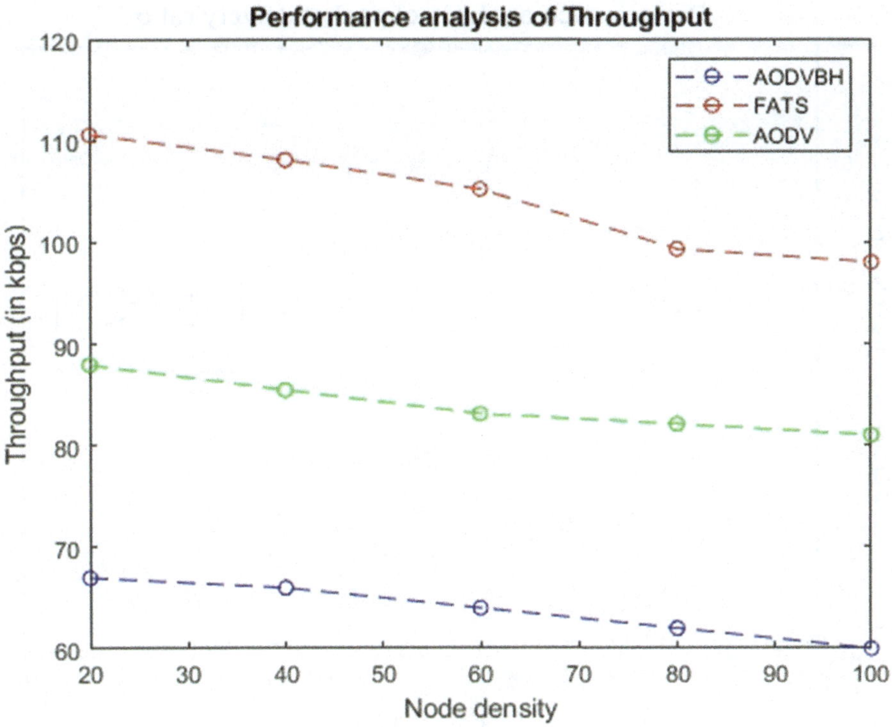

Fig. 16 Analysis of throughput performance

End-to-end delay (s)

End-to-end delay is defined as the difference between the packet sent time and packet arrival time across a network.

$$\text{End} - \text{to} - \text{end delay} = \text{Packet sent time} - \text{Packet arrival time across a network.}$$

Figures 16, 17, and 18 show the performance analysis of throughput, packet delivery ratio, and end-to-end delay performance, respectively. The performance analysis shows that Fuzzy-trusted approach increases the throughput and packet delivery ratio and decreases the delay when compared to normal AODV routing protocol and AODVBH-AODV black hole.

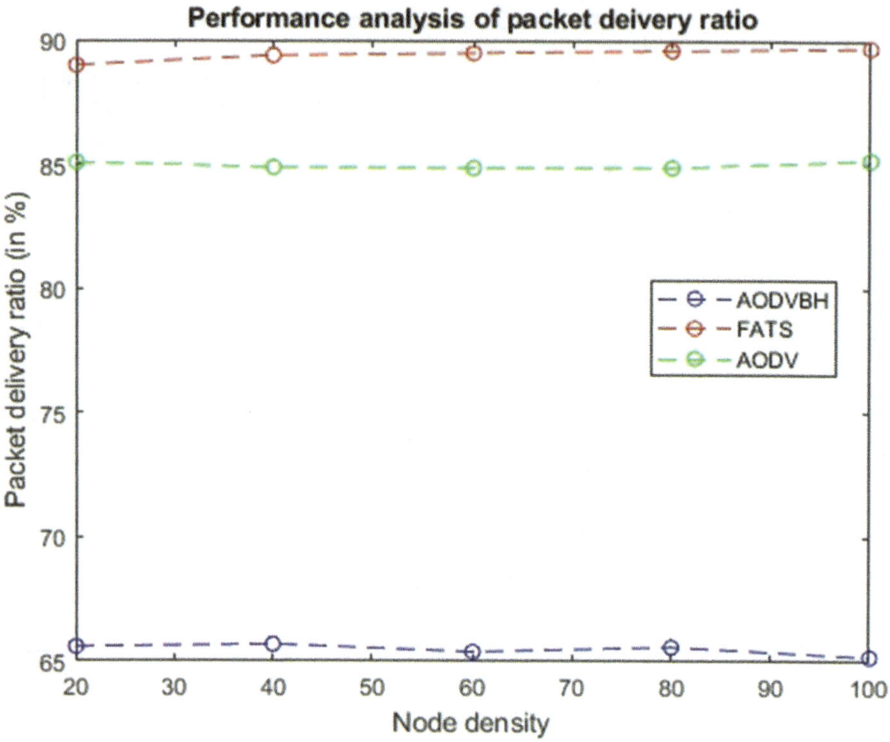

Fig. 17 Analysis of packet delivery ratio

9 Conclusion

The proposed method used FATS (Fuzzy Authentication to provide Trust-based Security) for the detection of black hole attacks in VANET. When the black hole attack is detected in VANET, the trusted authority blocks the node to join from the communication link. The trust degree plays a significant role in authenticating the user. The trust degree is calculated based on vehicle identity, roadside unit trust, node reputation trust, and data trust. This degree of trust makes the VANET user decide whether to trust the node or block the black hole node. Fuzzy logic is normally used in the scenario where there is more uncertainty. This fuzzy-based trusted authentication scheme, FATS, increases the packet delivery ratio and throughput and decreases the delay time. This proposed method increases the network lifetime and bandwidth and reduces the packet drops caused by the black hole node. Secure infrastructure is provided by this method to transfer the data packets in a secure communication network.

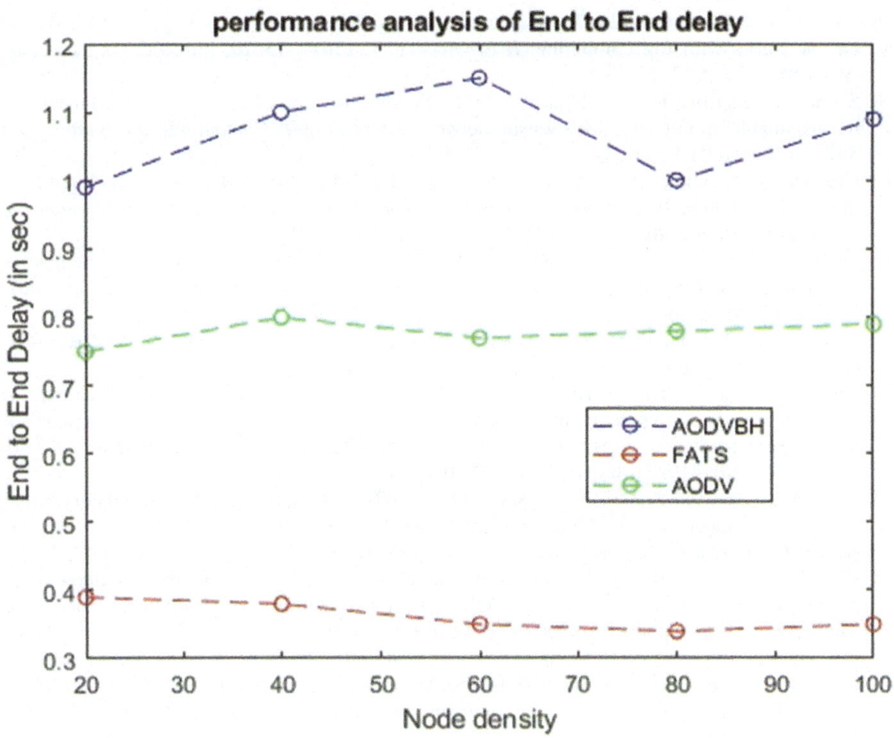

Fig. 18 Analysis of end-to-end delay performance

References

1. Cooper, C., Franklin, D., Ros, M., Safaei, F., & Abolhasan, M. (2017). A comparative survey of VANET clustering techniques. *IEEE Communications Surveys & Tutorials, 19*(1), 657–681. https://doi.org/10.1109/COMST.2016.2611524
2. Rasheed, A., Gillani, S., Ajmal, S., & Qayyum, A. (2017). Vehicular Ad Hoc Network (VANET): A survey, challenges, and applications. In *Vehicular ad-hoc networks for smart cities* (pp. 39–51). https://doi.org/10.1007/978-981-10-3503-6_4
3. Onishi, H. (2018). A survey. Engineering challenges to implement VANET security. In *2018 IEEE international conference on vehicular electronics and safety (ICVES)* (pp. 1–6). https://doi.org/10.1109/ICVES.2018.8519503
4. Sheikh, M. S., & Liang, J. (2019). A comprehensive survey on VANET security services in traffic management system. In *Wireless communications and mobile computing, 2019.*
5. Al-Sultan, S., Al-Doori, M. M., Al-Bayatti, A. H., & Zedan, H. (2014). A comprehensive survey on vehicular Ad Hoc network. *Journal of Network and Computer Applications, 37*, 380–392.
6. Azees, M., Jegatha Deborah, L., & Vijayakumar, P. (2016). Comprehensive survey on security services in vehicular adhoc networks. *IET Intelligent Transport Systems, 10*(6), 379–388.
7. Vo, M. T., Vo, A. H., Nguyen, T., Sharma, R., & Le, T. (2021). Dealing with the class imbalance problem in the detection of fake job descriptions. *Computers, Materials & Continua, 68*(1), 521–535.

8. Soleymani, S. A., et al. (2017). A secure trust model based on fuzzy logic in vehicular ad hoc networks with fog computing. *IEEE Access, 5*, 15619–15629. https://doi.org/10.1109/ACCESS.2017.2733225
9. Sachan, S., Sharma, R., & Sehgal, A. (2021). Energy efficient scheme for better connectivity in sustainable mobile wireless sensor networks. *Sustainable Computing: Informatics and Systems, 30*, 100504.
10. Ghanem, S., Kanungo, P., Panda, G., et al. (2021). Lane detection under artificial colored light in tunnels and on highways: An IoT-based framework for smart city infrastructure. *Complex & Intelligent Systems*. https://doi.org/10.1007/s40747-021-00381-2
11. Sachan, S., Sharma, R., & Sehgal, A. (2021). SINR based energy optimization schemes for 5G vehicular sensor networks. *Wireless Personal Communications*. https://doi.org/10.1007/s11277-021-08561-6
12. Gayathri, M., & Gomathy, C. (2021). A deep survey on types of cyber attacks in VANET. *JCR, 8*(1), 1029–1039. https://doi.org/10.31838/jcr.08.01.11
13. Kumar, A., Varadarajan, V., Kumar, A., Dadheech, P., Choudhary, S. S., Ambeth Kumar, V. D., Panigrahi, B. K., & Veluvolu, K. C. (2021). Black hole attack detection in vehicular ad-hoc network using secure AODV routing algorithm. *Microprocessors and Microsystems, 80*, 103352., ISSN01419331. https://doi.org/10.1016/j.micpro.2020.103352
14. Ding, Q., Zeng, X., Zhang, X., & Sung, D. K. (2019). A public goods game theory-based approach to cooperation in VANETs under a high vehicle density condition. *IEEE Transactions on Intelligent Transportation Systems, 20*(11), 3995–4005.
15. Priyadarshini, I., Mohanty, P., Kumar, R., et al. (2021). A study on the sentiments and psychology of twitter users during COVID-19 lockdown period. *Multimedia Tools and Applications*. https://doi.org/10.1007/s11042-021-11004-w
16. Liu, H., Chen, Y., Tian, H., Wang, T., & Cai, Y. (2016, October). A novel secure message delivery and authentication method for vehicular ad hoc networks. In *2016 first IEEE international conference on computer communication and the Internet (ICCCI)* (pp. 135–139). IEEE.
17. Azad, C., Bhushan, B., Sharma, R., et al. (2021). Prediction model using SMOTE, genetic algorithm and decision tree (PMSGD) for classification of diabetes mellitus. *Multimedia Systems*. https://doi.org/10.1007/s00530-021-00817-2
18. Priyadarshini, I., Kumar, R., Tuan, L. M., et al. (2021). A new enhanced cyber security framework for medical cyber physical systems. *SICS Software-Intensive Cyber-Physical Systems*. https://doi.org/10.1007/s00450-021-00427-3
19. Miao, T., Shen, J., Lai, C.-F., Ji, S., & Wang, H. (2021). Fuzzy-based trustworthiness evaluation scheme for privilege management in vehicular ad hoc networks. *IEEE Transactions on Fuzzy Systems, 29*(1), 137–147. https://doi.org/10.1109/TFUZZ.2020.3030490
20. Priyadarshini, I., Kumar, R., Sharma, R., Singh, P. K., & Satapathy, S. C. (2021). Identifying cyber insecurities in trustworthy space and energy sector for smart grids. *Computers & Electrical Engineering, 93*, 107204.
21. Upadhyaya, A. N., & Shah, J. S. (2019). Effect on AODV routing protocol under blackhole attack in VANET. *International Journal of Computer Engineering and Technology, 10*(3), 166–174.
22. Zadeh, L. (1978). Fuzzy sets as a basis of possibility. *Fuzzy Sets and Systems, 1*, 3–28.
23. Singh, R., Sharma, R., Akram, S. V., Gehlot, A., Buddhi, D., Malik, P. K., & Arya, R. (2021). Highway 4.0: Digitalization of highways for vulnerable road safety development with intelligent IoT sensors and machine learning. *Safety Science, 143*, 105407. ISSN 0925-7535.
24. Luo, Q., Cai, X., Luan, T., et al. (2018). Fuzzy logic-based integrity-oriented file transfer for highway vehicular communications. *EURASIP Journal on Wireless Communications and Networking, 2018*, 3. https://doi.org/10.1186/s13638-017-1009-x
25. Sahu, L., Sharma, R., Sahu, I., Das, M., Sahu, B., & Kumar, R. (2021). Efficient detection of Parkinson's disease using deep learning techniques over medical data. *Expert Systems*, e12787. https://doi.org/10.1111/exsy.12787

26. Codecrucks. (2021, August 22). Mamdani fuzzy inference System-Concept-CodeCrucks. *CodeCrucks*. https://codecrucks.com/mamdani-fuzzy-inference-concept/
27. Gautham, P. S., & Shanmughasundaram, R. (2017). Detection and isolation of black hole in VANET. In *2017 international conference on intelligent computing, instrumentation and control technologies (ICICICT)* (pp. 34–1539). https://doi.org/10.1109/ICICICT1.2017.8342799
28. Mirsadeghi, F., Kuchaki, M., & Gupta, R. B. B. (2020). *A trust infrastructure-based authentication method for clustered vehicular ad hoc networks*. Springer Science+Business Media, LLC, part of Springer Nature.
29. Li, J., Zhang, Y., Zhao, J., Wang, Y., Ma, X., & Wu, W. (2017). NS-2 simulation of VANET for safety applications. In *Proceedings of the 8th international conference on computer modeling and simulation - ICCMS '17*. https://doi.org/10.1145/3036331.3036349
30. Singh, R., Singh, J., & Singh, R. (2017). Fuzzy based advanced hybrid intrusion detection system to detect malicious nodes in wireless sensor networks. In *Wireless communications and mobile computing, 2017*.

AI-Based Chatbot Agents as Drivers of Purchase Intentions: An Interdisciplinary Study

Priyanka Tyagi and Ajay Jain

1 Introduction

Artificial intelligence (AI) is continuing to change how companies connect with customers [26] and is emerging as a developing and interdisciplinary area of research and innovation Liu et al., [25]. The usage of applications that interact with users in a conversational fashion and closely resemble human interaction is known as "chatbots," and it is currently one of the growing trends in artificial intelligence (AI). Chatbots have recently gained popularity as a solution to improve business's external communication with customers [19].

Online interactions have become a prominent means of communication between businesses and customers [36]. Through real-time interactions in an e-commerce environment, chatbots help businesses to improve customer experiences and meet expectations [40]. Task-oriented dialogue systems are getting more and more attention from businesses due to their importance and significance in human-computer interaction and natural language processing [42]. The usage of chatbots is appropriate to lessen the burden [5]. The adoption of chatbots by businesses for customer service has increased [15]. Researchers [38] specifically suggested that Chatbot Agents might be used to understand and engage consumers, while also providing personalization. The consumer environment in which shops offer their services has changed as a result of the adoption of immersive technology [34].

Hence, AI technology boosts consumer engagement and has a positive impact on behavioral responses, especially intentions to make purchases and intentions to share experiences with social groups. The interactive AI and MR technology create new opportunities to increase consumer involvement [37]. Moreover, artificial

P. Tyagi · A. Jain (✉)
Faculty of Management, SRM Institute of Science and Technology, Ghaziabad, Uttar Pradesh, India

R. Sharma et al. (eds.), *Data Analytics for Internet of Things Infrastructure*, Internet of Things, https://doi.org/10.1007/978-3-031-33808-3_5

intelligence (AI) also transforms the way retailers operate. The prospect of physical retail will be AI-Powered Automated Stores. Consumers in these retail stores will deal with fully automated technologies. Customer's intentions to purchase in AI-powered automated stores were influenced by perceived usability, enjoyment, personalization, and interactivity [31]. Here, the latest advancement that benefited from data-driven innovation is the digital assistant. The most important factor in developing a good mindset and purchasing intentions is AI-powered digital assistants [2]. Specifically, a crucial marketing component is to have conversational responses on the official websites of businesses that sell goods and services [39]. Managers can increase the return on marketing investments by using artificial intelligence to make conversational interactivity more intuitive for present and potential customers visiting the websites [21]. There is a lot of literature on artificial intelligence, but there is very little or no understanding of the trends, patterns, and research advancements in this field. However, there is little study focusing on Chabot marketing activities and their influence on the purchase intentions of customers.

The researcher borrowed theoretical insights from the multidisciplinary literature on chatbots to cover the research gaps and make contributions to the field of e-marketing. The goal of this study is to determine whether chatbot agents have any impact on the purchase intentions of Delhi NCR consumers. The impact of the chatbots on purchase intentions of customers may be favorable, bad, or neutral. The organization will gain from this study by being able to assess the impact of applying AI-powered chatbots and will know whether it will be advantageous to utilize them in e-marketing [11].

2 Conceptual Background and Development of Hypothesis

The chatbots are growing rapidly in a multitude of sectors, including customer service, health, education, marketing, and job assistance, and are becoming crucial entry points for digital services and information [30]. Specifically, online travel businesses have used chatbots in various customer-facing applications (L. Li et al., n.d.). Similarly, chatbots are being used rapidly in education sector. Engaging learners, personalizing learning activities, assisting teachers, and gaining an in-depth understanding of students' behavior, chatbots have the potential to revolutionize the way that education is delivered [20]. Especially among students and young adults, chatbot has a significant potential to be employed as a conversational agent to promote health literacy [28]. Chatbots are being used in transportation sector as well. One of the most significant uses of cutting-edge artificial intelligence technology, which is widely adopted in the transportation sector worldwide, is automated driving technology. Fully automated vehicles (FAV) are anticipated by policymakers to considerably reduce the frequency of accidents caused by human error and traffic congestion [35]. It is important to know the impact of chatbots when customers make online purchases. The following section discusses the literature related to Chatbot agent's informational support, trust, emotional credibility, and purchase intentions.

3 Informational Support of Chatbots and Predicting Purchase Intentions

Now chatbot agents are a popular medium to provide real-time customer service in many e-commerce platforms. In the present times, conversational bots frequently replace human chat service agents. The software agents, generally known as chatbots, are automated systems that employ natural language to communicate with human users frequently applying artificial intelligence [1]. Online commercial services and e-commerce have both benefited tremendously from the application of artificial intelligence. Online retail platforms have heavily incorporated AI technology to offer customers more precise and customized services and recommendations [41]. According to Nirala et al. [29], a chatbot has become an effective tool for responding to user enquiries in an automated, sensible, and precise manner. Researchers have used a range of soft-computing strategies to make the chatbot user-friendly, depending on the complexity of the subject domain [14]. It has been noted that chatbots have provided a wide range of services to users around the world, including ordering food, recommending products, offering insurance policy advice, offering customer care, offering financial assistance, scheduling meetings, etc.

Similarly, modern technologies are increasingly used in the well-liked virtual world. Now organizations are increasingly exploring numerous innovative applications in such virtual environments [24]. Conversational recommender systems intend to assist online users in making informed decisions by providing them with interactive help [18]. In order to communicate with their customers, businesses might use chatbots, which are technology tools with artificial intelligence. Recommender systems become essential in the modern digital age for managing information on online platforms. From social media to e-commerce, from travel to cultural consumption, automated recommendations have an impact on users' daily decisions. Recommender Systems are trained using user data to provide personalized informational support to users based on their expected preferences [12]. Now consumers can make purchases, access after-sales services, and also do informative searches using chatbots via their desktops or mobile devices [10]. Based on these prior studies, this study proposes the following hypothesis:

H_1: Informational support positively affects purchase intentions of customers.

4 Trust, Emotional Credibility, and Predicting Purchase Intentions

Businesses are also using service robots including both physical and virtual chatbots to deliver services to clients. Some businesses adopt robots that resemble humans in appearance and behavior to enhance customer's intentions [4]. Researchers [27] showed that the effectiveness of product promotion is positively correlated with emotional credibility, integrity, and trust of chatbots. Behera et al. [3] found that customers place a high value on accessibility and dependability, especially when looking for real-time information. Customers experience is flawless because of automated responses to repetitive questions on frequent problems. Trust influences the intention toward adopting cognitive chatbots. Further, Lappeman et al. [22] found that, despite the ability of a brand to influence consumers' choices, this power is not sufficient to significantly enhance consumers' self-disclosure. The three components of trust—brand trust, cognitive trust, and emotional trust—need to be combined into one bridge through education, communication, and product development. Based on the findings of the above researchers, this study proposes the following hypothesis:

H_2: Emotional credibility positively affects purchase intentions.
H_3: Trust positively affects purchase intentions.

5 Research Gap

However, there are not many studies that have specifically focused on how chatbots influence customers' intentions when they purchase online. There is plenty of literature available on the use of AI-powered chatbots in customer service and other domains. To advance the state of AI technology's current application, it is crucial to research the AI-powered chatbots in marketing.

6 Objectives

The objectives of the study are as follows:

1. To know the impact of informational support on purchase intentions.
2. To assess the impact of emotional credibility on purchase intentions.
3. To determine the impact of trust on purchase intentions.

The researcher conducted a thoughtful study in order to attain these objectives.

7 Methods

This study aims to investigate the effect of chatbot agents on online purchase intentions and inclinations to use its recommendations when making online purchases. A thoughtful study is designed to accomplish these goals.

7.1 Sampling Framework and Questionnaire Design

Indian customers served as the sample population of this study, specifically customers residing at Delhi NCR area who use chatbot agents for making online purchases. To add variety to the data, the respondents were specifically picked to have various demographic profiles. The participants were between the ages of 18 and above 51, had a good mix of genders, and represented a wide range of backgrounds, and qualifications in terms of education and employment. In this study, 215 respondents participated in an electronic survey, through which data were collected for this investigation. The survey consisted of two sections. The first section of the questionnaire included questions about the respondent's demographic characteristics, while the second section included questions about the determination of how independent variables affected dependent variables. Here, Purchase is the dependent variable, whereas informational support, emotional credibility, and trust are the independent variables (Fig. 1).

7.2 Measures

The measurement of constructs by measurement scale is an important aspect in research (Ringle & Sarstedt, [33]). The goodness of a measurement scale can be evaluated on the basis of reliability and validity. The details of measurement scale are given in Table 1.

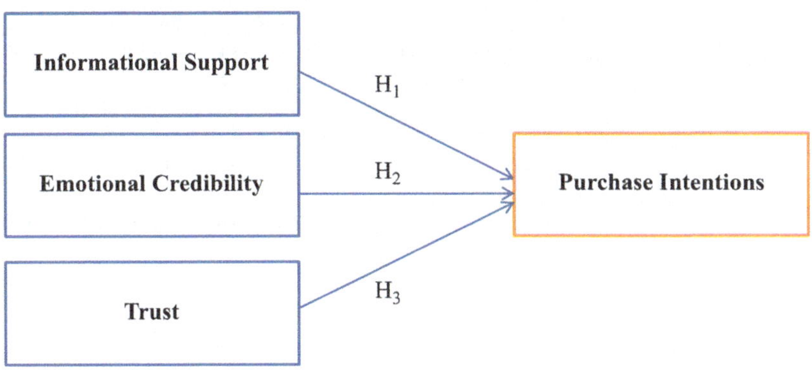

Fig. 1 Research model developed by the researchers

Table 1 Details of measurement scale

Constructs	Number of items	Items	Source
Informational support	6	I feel chatbot service agent gives me suggestions and advice about how to cope up with problems	[6]
		The chatbot service agent suggests me where I can go to get help	[6]
		I feel chatbot service agent has adequate information to answer customer's questions.	[6]
		The chatbot service agent provides information about the events happening in the organization.	[9]
		The chatbot service agent provides suggestions on the company products/services.	[9]
		I feel chatbot service agent provides useful information that helps me in purchasing.	[9]
Emotional credibility	6	I feel chatbot agent recognizes my emotions and feelings.	[32]
		I feel chatbot agent expresses feelings and emotions appropriately for the situation.	[32]
		I feel chatbot service agent uses feelings and emotions to create and organize thinking.	[32]
		I believe chatbot service agent uses feelings and emotions to facilitate problem-solving and judgment.	[32]
		The chatbot service agent knows how to respond appropriately to positive as well as negative emotions.	[32]
		I feel the chatbot agent knows how to control her own feelings and handle my feelings sensitively and effectively.	[32]
Trust	6	I believe that chatbot agents are reliable.	[8]
		The recommendations given by chatbot agents are trustworthy.	[8]
		I feel confident about the brand suggestions given by chatbot agents.	[23]
		I believe chatbot agent treats me fairly and justly.	[23]
		I believe chatbot agents are trying to make long-term commitment.	[23]
		I believe whenever chatbot agents make an important decision, I know it will be concerned about me.	[23]
Purchase intentions	4	I am intended to purchase the brands suggested by the chatbot service agent.	[40]
		It is likely that I will purchase the brands recommended by the chatbot agents in near future.	[40]
		I regularly purchase the brands recommended by chatbot agents.	[40]
		I anticipate purchasing on recommendations of chatbot agents.	[40]

8 Research Tools and Techniques

The researcher performed descriptive data analysis by using SPSS version 22. Further analysis was done by two-step methods (measurement model and structural model) proposed by Henseler et al. [17], to report and evaluate generated results. A measurement model has been used to evaluate the appropriateness of the scale items. The evaluation of collinearity among exogenous constructs, the significance and applicability of path coefficients, as well as indirect effects and total effects (if appropriate) are all included in the structural model assessment process. The coefficient of determination, or R^2, is used to measure the predictive accuracy of the model [16]. In the study, the researcher retained only the items with good factor loadings and removed the items with low factor loading, as per Chin [7]. Additionally, CR (construct reliability) was measured by using Fornell and Larcker [13] criterion. Furthermore a CR value of 0.7 or more shows acceptable level of reliability. Additionally, average variance extracted (AVE) was used in the study to assess the convergent validity, and AVE (average variance extracted) value should be above 0.50. The cross-loading values of the latent variables were used to obtain discriminant validity. The cross loading values in corresponding variables should be higher in comparison to the loadings of other constructs.

9 Measurement Model

The researchers evaluated the measurement model to examine the convergent and discriminant validity of the observed variables. The loading value of specific indicators must be above 0.6, AVE value should be above 0.5, and CR value should be above 0.7, as per Hair Jr. et al. [16]. The fitness of a model is assured by fulfilment of these criteria. The details of measurement properties of the constructs are given in Table 2.

10 Analysis of Structural Model

Estimation of structural model was performed after analyzing the validity and reliability of constructs by using measurement model. The researcher used refined items in hypothesis testing. When evaluating the reflective measures, the internal consistency was examined using the composite reliability (CR) and Cronbach's alpha, the convergent validity was examined using the indicator loading and average variance extracted (AVE), and the discriminant validity was examined using the heterotrait-monotrait ratio of correlations (HTMT). In Fig. 2, a full structural regression model is presented. The researcher found a positive impact of chatbot agent's informational support and purchase intentions ($\beta = 0.231$, $p < 0.01$).

Table 2 Parameter values of factor loadings, average variance extracted, composite reliability

Constructs	Items	Loadings	AVE	CR
Informational support	IN1	0.735		
	IN2	0.778		
	IN3	0.873	0.710	0.936
	IN4	0.949		
	IN5	0.862		
	IN6	0.843		
Emotional credibility	EMI	0.895		
	EM2	0.907		
	EM3	0.862	0.733	0.932
	EM4	0.788		
	EM6	0.822		
Trust	T2	0.786		
	T3	0.892	0.745	0.897
	T4	0.906		
Purchase intentions	P2	0.940		
	P3	0.870	0.798	0.922
	P4	0.867		

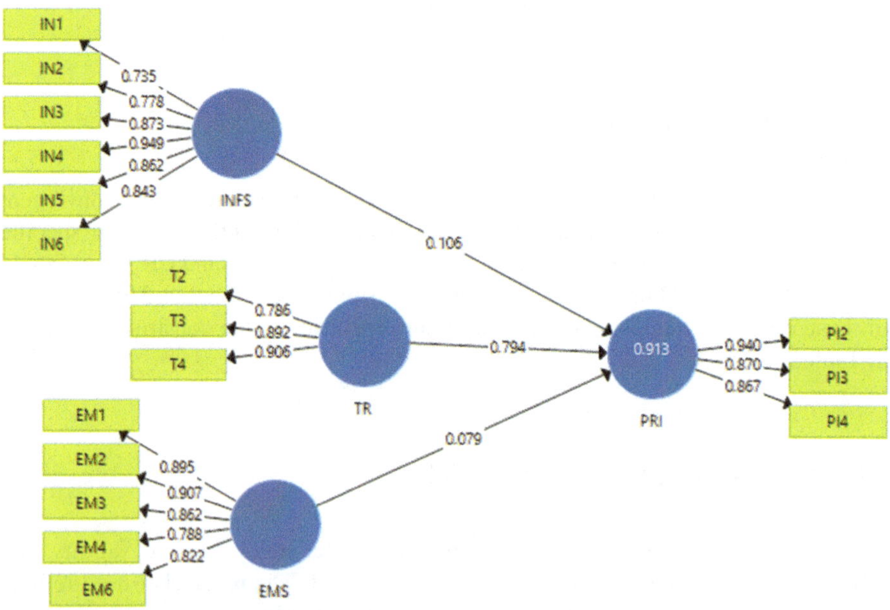

Fig. 2 Results of the measurement model

The researcher also found a positive impact of chatbot agent's trust and purchase intentions ($\beta = 0.949$, $p < 0.01$). Furthermore, results indicated that the chatbot agent's emotional credibility has a significant impact on the purchase intentions

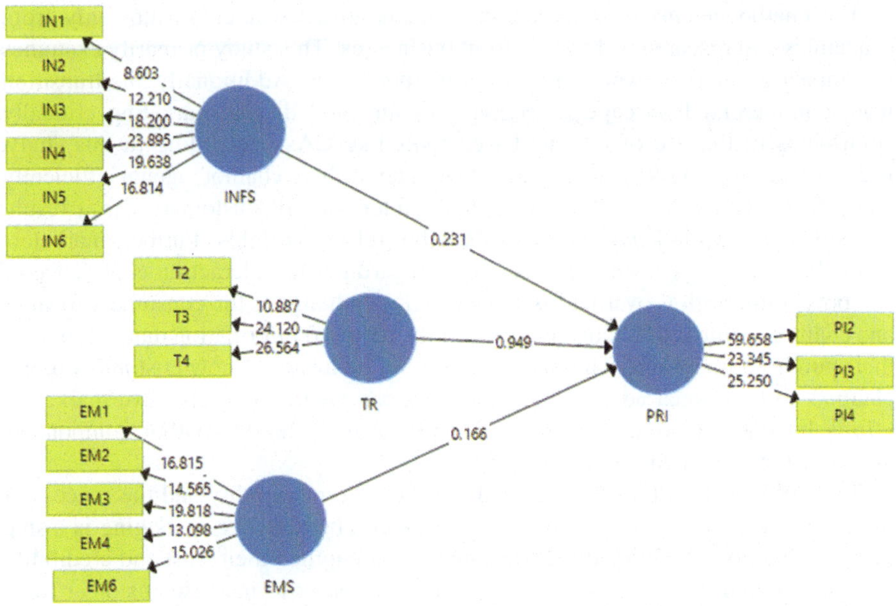

Fig. 3 Results of structural model

($\beta = 0.166$, p is less than 0.01). Where SRMR = 0.04, NFI = 0.88, $R^2 = 91.3\%$ .Therefore, H_1 and H_2 are supported (Fig. 3).

11 Conclusion and Future Research Works

Based on the results of structural model, the researcher found that chatbots agent's informational support, emotional credibility, and trust have significant impact on the purchase intentions of the Indian (NCR) customers in online markets. This study adds some pretty interesting findings to the work. AI-based chatbot agents are gaining popularity in a variety of contexts and may present a number of options for time and money savings. However, many customers continue to feel low emotional credibility with chatbots, which could lead to possible barriers to the use of chatbots. This research serves as an initial step in better understanding how chatbot impacts, and the purchase intentions of customers in the context of e-marketing can be used to increase purchase intentions with AI-based chatbot agents. The findings of this study can offer some insights into the area of integrating new technologies in e-marketing field, a field that is very competitive and where marketers are continually seeking for novel ways to increase sales in online markets. The researchers also anticipate that this study will stimulate more investigation into chatbots and the advancement of AI-based capabilities in and outside of e-marketing field.

The chatbot agents in e-marketing contexts attracted attention not only from academics and researchers but also from businesses. This study primarily examines the impact of chatbots when making online purchases. Additionally, the transition from commercial landscape to digital economy and digital marketing critically contributes to the use of internet-based tools like CAs. According to this study, it is crucial for marketers to understand and utilize chatbot agents efficiently and effectively as they influence purchase intentions of customers. As a result, this study offers marketers numerous chatbot-related insights. Further, marketers can collect feedback from their customers regarding the interaction experience of chatbots. Information overload is increasingly challenging for consumers to filter and chatbots could enable customers to get the required information that will impact their purchase intentions. The marketers should be aware of how customers prefer chatbots to traditional advertising. Marketers can create more effective marketing efforts by understanding how user perceptions of different chatbot components affect customer's intentions of using it.

Typically, businesses are concerned with long-term viability and the success of their websites. Hence, increased use of AI-based chatbots in e-marketing is a step in right direction. Additionally, businesses can strengthen their trust and credibility by assuring their CAs are at an excellent level in order to increase customer trust. The results of this study are significant and should be viewed as helpful guidance for businesses and operators of CAs as they develop their marketing strategy. The study's findings are anticipated to act as a guide for marketers and other industry participants by helping them to comprehend the function of chatbots in terms of customer intentions when they make online purchases.

Future research may go into greater detail about the connection between usage of chatbots and purchase intentions. This study provides a number of directions for future work on the chatbot agents, which could improve the way AI-based CAs interact with customers. Additionally, there is still much research to be done on the use of chatbots in context of online purchases. Further, future research can focus on a variety of contexts such as investments and financial products. If marketers are successful in influencing purchase intentions of customers in various contexts, they may be able to grow businesses in a much faster pace. Great research opportunities exist in CAs as a result of digital revolution, which could benefit not only the major market players but also customers.

12 Limitations

This study has various limitations. This study focused on the use of CAs in the context of online purchase; hence, it must be understood in light of certain limitations. This area of study is expected to expand as technology improves and as people become more exposed to it. Moreover, a larger data size is frequently advised in the field of marketing, but the sample size of this study was small, which is another drawback. Future researchers are suggested to increase the sample size and a cross-country survey can also be conducted to determine how other nation's customers feel about chatbots. Future researchers are suggested to increase the sample size and a cross-country survey can also be conducted to determine how other nation's customers feel about chatbots.

References

1. Adam, M., Wessel, M., & Benlian, A. (2021). AI-based chatbots in customer service and their effects on user compliance. *Electronic Markets, 31*(2), 427–445. https://doi.org/10.1007/s12525-020-00414-7
2. Balakrishnan, J., & Dwivedi, Y. K. (2021). Conversational commerce: Entering the next stage of AI-powered digital assistants. *Annals of Operations Research*. https://doi.org/10.1007/s10479-021-04049-5
3. Behera, R. K., Bala, P. K., & Ray, A. (2021). Cognitive Chatbot for personalised contextual customer service: Behind the scene and beyond the hype. *Information Systems Frontiers*. https://doi.org/10.1007/s10796-021-10168-y
4. Blut, M., Wang, C., Wünderlich, N. V., & Brock, C. (n.d.). *Understanding anthropomorphism in service provision: A meta-analysis of physical robots, chatbots, and other AI*. https://doi.org/10.1007/s11747-020-00762-y/Published
5. Brachten, F., Brünker, F., Frick, N. R. J., Ross, B., & Stieglitz, S. (2020). On the ability of virtual agents to decrease cognitive load: An experimental study. *Information Systems and e-Business Management, 18*(2), 187–207. https://doi.org/10.1007/s10257-020-00471-7
6. Cheng, Y., & Jiang, H. (2022). Customer–brand relationship in the era of artificial intelligence: Understanding the role of chatbot marketing efforts. *Journal of Product and Brand Management, 31*(2), 252–264. https://doi.org/10.1108/JPBM-05-2020-2907
7. Chin, W., & Marcoulides, G. (1998). The partial least squares approach to structural equation modeling. *Modern Methods for Business Research, 8*.
8. Chong, L. L., Ong, H. B., & Tan, S. H. (2021). Acceptability of mobile stock trading application: A study of young investors in Malaysia. *Technology in Society, 64*, 101497. https://doi.org/10.1016/j.techsoc.2020.101497
9. Chung, M., Ko, E., Joung, H., & Kim, S. J. (2020). Chatbot e-service and customer satisfaction regarding luxury brands. *Journal of Business Research, 117*, 587–595. https://doi.org/10.1016/j.jbusres.2018.10.004
10. de Cosmo, L. M., Piper, L., & Di Vittorio, A. (2021). The role of attitude toward chatbots and privacy concern on the relationship between attitude toward mobile advertising and behavioral intent to use chatbots. *Italian Journal of Marketing, 2021*(1–2), 83–102. https://doi.org/10.1007/s43039-021-00020-1
11. Erhan, L., Ndubuaku, M.U., Mauro, M.D., Song, W., Chen, M., Fortino, G., Bagdasar, O., & Liotta, A. (2020). *Smart anomaly detection in sensor systems: A multi-perspective review*. arXiv: Learning.
12. Fabbri, M. (2022). Social influence for societal interest: A pro-ethical framework for improving human decision making through multi-stakeholder recommender systems. *AI & SOCIETY, 38*, 995. https://doi.org/10.1007/s00146-022-01467-2
13. Fornell, C., & Larcker, D. F. (1981). Evaluating structural equation models with unobservable variables and measurement error. *Journal of Marketing Research, 18*(1), 39. https://doi.org/10.2307/3151312
14. Fortino, G., Messina, F., Rosaci, D., & Sarné, G. M. L. (2020). Using blockchain in a reputation-based model for grouping agents in the internet of things. *IEEE Transactions on Engineering Management, 67*(4), 1231–1243. https://doi.org/10.1109/TEM.2019.2918162

15. Fotheringham, D., & Wiles, M. A. (2022). The effect of implementing chatbot customer service on stock returns: An event study analysis. *Journal of the Academy of Marketing Science*. https://doi.org/10.1007/s11747-022-00841-2
16. Hair, J. F., Risher, J. J., Sarstedt, M., & Ringle, C. M. (2019). When to use and how to report the results of PLS-SEM. *European Business Review, 31*(1), 2–24. https://doi.org/10.1108/EBR-11-2018-0203
17. Henseler, J., Ringle, C. M., & Sinkovics, R. R. (2009). The use of partial least squares path modeling in international marketing. In R. R. Sinkovics & P. N. Ghauri (Eds.), *Advances in international marketing* (pp. 277–319). Emerald Group Publishing Limited. https://doi.org/10.1108/S1474-7979(2009)0000020014
18. Jannach, D. (2022). Evaluating conversational recommender systems. *Artificial Intelligence Review, 56*, 2365. https://doi.org/10.1007/s10462-022-10229-x
19. Johannsen, F., Schaller, D., & Klus, M. F. (2021). Value propositions of chatbots to support innovation management processes. *Information Systems and e-Business Management, 19*(1), 205–246. https://doi.org/10.1007/s10257-020-00487-z
20. Kuhail, M. A., Alturki, N., Alramlawi, S., & Alhejori, K. (2022). Interacting with educational chatbots: A systematic review. *Education and Information Technologies, 28*, 973. https://doi.org/10.1007/s10639-022-11177-3
21. Kushwaha, A. K., & Kar, A. K. (2021). MarkBot – A language model-driven Chatbot for interactive marketing in post-modern world. *Information Systems Frontiers*. https://doi.org/10.1007/s10796-021-10184-y
22. Lappeman, J., Marlie, S., Johnson, T., & Poggenpoel, S. (2022). Trust and digital privacy: Willingness to disclose personal information to banking chatbot services. *Journal of Financial Services Marketing*. https://doi.org/10.1057/s41264-022-00154-z
23. Lee, C. T., Pan, L. Y., & Hsieh, S. H. (2021). Artificial intelligent chatbots as brand promoters: A two-stage structural equation modeling-artificial neural network approach. *Internet Research*. https://doi.org/10.1108/INTR-01-2021-0030
24. Li, M., & Buchthal, S. (2012). Advisory services in the virtual world: An empowerment perspective. *Electronic Commerce Research, 12*(1), 53–96. https://doi.org/10.1007/s10660-012-9088-6
25. Liu, H., Peng, H., Song, X., Xu, C., & Zhang, M. (2022). Using AI chatbots to provide self-help depression interventions for university students: A randomized trial of effectiveness. *Internet Interventions, 27*, 100495. https://doi.org/10.1016/j.invent.2022.100495
26. Liu-Thompkins, Y., Okazaki, S., & Li, H. (2022). Artificial empathy in marketing interactions: Bridging the human-AI gap in affective and social customer experience. *Journal of the Academy of Marketing Science, 50*, 1198. https://doi.org/10.1007/s11747-022-00892-5
27. Low, M. P., Cham, T. H., Chang, Y. S., & Lim, X. J. (2021). Advancing on weighted PLS-SEM in examining the trust-based recommendation system in pioneering product promotion effectiveness. *Quality and Quantity*. https://doi.org/10.1007/s11135-021-01147-1
28. Mokmin, N. A. M., & Ibrahim, N. A. (2021). The evaluation of chatbot as a tool for health literacy education among undergraduate students. *Education and Information Technologies, 26*(5), 6033–6049. https://doi.org/10.1007/s10639-021-10542-y
29. Nirala, K. K., Singh, N. K., & Purani, V. S. (2022). A survey on providing customer and public administration based services using AI: Chatbot. *Multimedia Tools and Applications, 81*(16), 22215–22246. https://doi.org/10.1007/s11042-021-11458-y
30. Oh, Y. J., Zhang, J., Fang, M. L., & Fukuoka, Y. (2021). A systematic review of artificial intelligence chatbots for promoting physical activity, healthy diet, and weight loss. *International Journal of Behavioral Nutrition and Physical Activity, 18*(1). https://doi.org/10.1186/s12966-021-01224-6
31. Pillai, R., Sivathanu, B., & Dwivedi, Y. K. (2020). Shopping intention at AI-powered automated retail stores (AIPARS). *Journal of Retailing and Consumer Services, 57*, 102207. https://doi.org/10.1016/j.jretconser.2020.102207
32. Potts, C., Ennis, E., Bond, R. B., Mulvenna, M. D., McTear, M. F., Boyd, K., Broderick, T., Malcolm, M., Kuosmanen, L., Nieminen, H., Vartiainen, A. K., Kostenius, C., Cahill, B.,

Vakaloudis, A., McConvey, G., & O'Neill, S. (2021a). Chatbots to support mental wellbeing of people living in rural areas: Can user groups contribute to co-design? *Journal of Technology in Behavioral Science, 6*, 652. https://doi.org/10.1007/s41347-021-00222-6
33. Ringle, C. M., & Sarstedt, M. (2016). Gain more insight from your PLS-SEM results: The importance-performance map analysis. *Industrial Management & Data Systems, 116*(9), 1865–1886. https://doi.org/10.1108/IMDS-10-2015-0449
34. Savaglio, C., & Fortino, G. (2021). A simulation-driven methodology for IoT data mining based on edge computing. *ACM Transactions on Internet Technology, 21*(2), 30, 22 pages. https://doi.org/10.1145/3402444
35. Shin, K. J., Tada, N., & Managi, S. (2019). Consumer demand for fully automated driving technology. *Economic Analysis and Policy, 61*, 16–28. https://doi.org/10.1016/j.eap.2018.10.002
36. Steinhoff, L., Arli, D., Weaven, S., & Kozlenkova, I. V. (2019). Online relationship marketing. *Journal of the Academy of Marketing Science, 47*(3), 369–393. https://doi.org/10.1007/s11747-018-0621-6
37. Sung, E. (Christine), Bae, S., Han, D. I. D., & Kwon, O. (2021). Consumer engagement via interactive artificial intelligence and mixed reality. *International Journal of Information Management, 60*, 102382. https://doi.org/10.1016/j.ijinfomgt.2021.102382
38. Thomaz, F., Salge, C., Karahanna, E., & Hulland, J. (2020). Learning from the Dark Web: Leveraging conversational agents in the era of hyper-privacy to enhance marketing. *Journal of the Academy of Marketing Science, 48*(1), 43–63. https://doi.org/10.1007/s11747-019-00704-3
39. Xiao, W., Miao, Y., Fortino, G., Wu, D., Chen, M., & Hwang, K. (2022). Collaborative cloud-edge service cognition framework for DNN configuration toward smart IIoT. *IEEE Transactions on Industrial Informatics, 18*(10), 7038–7047. https://doi.org/10.1109/TII.2021.3105399
40. Yen, C., & Chiang, M. C. (2021). Trust me, if you can: A study on the factors that influence consumers' purchase intention triggered by chatbots based on brain image evidence and self-reported assessments. *Behaviour and Information Technology, 40*(11), 1177–1194. https://doi.org/10.1080/0144929X.2020.1743362
41. Yin, J., & Qiu, X. (2021). Ai technology and online purchase intention: Structural equation model based on perceived value. *Sustainability (Switzerland), 13*(10). https://doi.org/10.3390/su13105671
42. Zhang, Z., Takanobu, R., Zhu, Q., Huang, M., & Zhu, X. (2020). Recent advances and challenges in task-oriented dialog systems. *SCIENCE CHINA Technological Sciences, 63*(10), 2011–2027. https://doi.org/10.1007/s11431-020-1692-3

An Intelligent Model for Identifying Fluctuations in the Stock Market and Predicting Investment Policies with Guaranteed Returns

Manash Sarkar, M. N. Pratima, R. Darshan, Debkanta Chakraborty, and Maroi Agrebi

1 Introduction

Stocks are critical parts of cutting-edge financial markets. The stock market is a tough forum for funding, and it calls for ample deliberation before investing money into stocks. The stock market is one of a number of important sectors that traders are dedicated and committed to. Researchers from both the economic and technical fields continue to be interested in the topic of stock market price trend indicators and predictions. Our goal is to develop a rate trend prediction system that specialises in short-term price trend prediction [1]. Modern financial systems are based primarily on fiat or decree money, which has many advantages, benefits that are due to its divisibility, sturdiness, transferability, and abundance. In addition, fiat money is not back up with a physical commodity, and governments have the power to manipulate the value of money. This may cause many problems, such as hyperinflation and income inequality [2]. Second, although people invest

M. Sarkar (✉)
Department of Computer Science & Engineering, Atria Institute of Technology, Bengaluru, Karnataka, India

Expert Data Scientist, Limendo, Bolzano, Italy
e-mail: manash.sarkar@atria.edu; manash.sarkar@limendo.com

M. N. Pratima · R. Darshan
Department of Computer Science & Engineering, Atria Institute of Technology, Bengaluru, Karnataka, India

D. Chakraborty
Department of Computer Science and Engineering, Dr. B. C. Roy Engineering College, Durgapur, India

M. Agrebi
LAMIH UMR CNRS 8201, Department of Computer Science, Université Polytechnique Hauts-de-France, Valenciennes, France

R. Sharma et al. (eds.), *Data Analytics for Internet of Things Infrastructure*, Internet of Things, https://doi.org/10.1007/978-3-031-33808-3_6

in stocks, they are unpredictable and unreliable. The stock rate is often affected by different factors, including investor behaviour, market sentiments, dividends, market regulations, economic developments, economic guidelines, credit situations, hobby prices, trade fee levels, and global capital flows [3]. Inventory market buying and selling are parts of an exceedingly complicated and ever-changing system where people can either accrue a fortune or lose their life savings [4]. The third problem is security: protecting the amount that people are buying and selling, current process gains, and income growth from being exposed to malicious users and fraudulent schemes. A primarily sequence-based stock price [5] is the standard nonstationary stochastic system that does not require a regular maintenance for it to stay in equilibrium. A mathematical model to describe the dynamics of this sort of procedure, after which forecasts or predictions on the price of future expenses from present values and past values, has been set up thanks to the results of numerous studies on nonlinear time collection evaluation. It also forecasts stock prices along four metrics: open fee, ultimate price, best rate, and the bottom or base amount [6]. Thus, it uses all extracted or captured guidelines at any moment for its predictions, estimating future inventory or stock market price with high accuracy.

Over the past decade and continuing today, big data technology has had a significant impact on a number of industries, and it could become ubiquitous. The financial sector, like the majority of other sectors, concentrated all of its efforts on structured information research. However, statistical records saved in various sources could be obtained with the aid of technology. Big data generate a lot of power, not only through the ability to process large volumes and types of data at fast rates but also through the capacity to charge organisations. The relationship between news articles and stock prices has already been established, but it can still be strengthened by additional data. Businesses have historically made greater investments in selection support structures. Many businesses can now obtain analytical reviews that are based on OLAP (Online Analytical Processing) structures thanks to the development of enterprise intelligence tools. However, given the frequently shifting trends in consumer behaviour and consumer markets, it may become increasingly important for businesses to move beyond OLAP machine-based analysis in their analytical reviews. In their paper, they discussed about OLAP and its significant innovations in hardware and software enable businesses to take advantage of all available data formats and support businesses in gaining business insights from those statistics. Some of the most important things to concentrate on in obtaining real-time analytics on all available factual codecs are facts.

Not just for data warehousing, data technology also includes a variety of other technologies that are used to passively track recent developments in threat analysis by using large numbers of data across many domains. This chapter aims to examine the conceptual methods that draw on literature assessments of those models and improve them. In order to address the current roadblocks to this aim, similar researchers can benefit from the information in the most recent papers. In this chapter, the authors rely on fuzzy common sense and neural networks. They also use RSI (Relative Strength Index), OS (Operating System), MACD (Moving Average Convergence/Divergence), OBV (On-Balance Volume), MSE (Mean Squared

Error), MAPE (Mean Absolute Percentage Error), and RMSE (Root Mean Squared Error) to evaluate time-series data. The chapter emphasises locating implementation challenges and how these challenges can help the trader make decisions in the stock market. The study's implication is that there is too much data to examine. Additionally, this chapter is conceptual in nature. The study's findings demonstrate that the ANFIS (Adaptive Neuro-Fuzzy Inference System) is the best model for stock market prediction, outperforming all others. The predictions and expert decisions are useful in entertainment industries [7]. The expert decision for the selection of a restaurant is also implemented in the recommendation system, where a restaurant recommendation [8] is also a useful application for expert decisions in tourism industries. Additionally, the fuzzy neural model's results show that they are more effective than other models, even though it appears that when MSE is used as a proxy for a medium, MAD (Mean absolute deviation) uses significantly less data than the other models do. As a result, it will become more adept at stock prediction. One of the methods, a reinforcement-oriented forecasting framework, which converts the solution from a typical mistake-based learning approach to an intention-directed, shape-based total learning approach, includes a trial-and-error-based total approach that was used for prediction. To make predictions on the stock market the way it ought to be, many advanced prediction techniques have been developed. At one time, some approaches were generally referred to as conventional approaches, but there were no computational techniques for hazard analysis. Pass validation, which is one of the most frequently used fact-resampling techniques to estimate actual prediction blunders and to fine-tune model parameters, is one of the techniques used to validate the model—a method for analysing the stock market, assessing how risky it is, and anticipating how it will behave so that investors can profit from their investments.

In this chapter, an attempt is made to construct a time-series prediction model to predict inventory prices. This construct is more realistic because it examines the model by using not only current algorithms but also records from the benchmark, which have made it difficult to determine whether it will outperform other algorithms. Modern-day stock values accrue from sets of statistics. The statistics gathered are modelled into diverse subcomponents or data sets, which are used to train and test the algorithm. Further, a customised intelligent version is modelled and improved to provide prediction scores for all the assessment metrics. An intelligent model is proposed to anticipate the stock market and offer an expert decision on how to invest. The proposed model additionally guarantees that an investor's amount will not be much less than their capital investment after a specific amount of time. The base amount could be fixed to the funding time. To reduce uncertainty and risk, a fuzzy logic is used, and fuzzy multiobjective optimisation is likewise used to make an optimistic decision. The proposed model predicts stock and determines the term required for a good return. A real-time stock market data set is used as a test-bed data set to validate the proposed model. Cross validation is an information resampling technique that assesses the generalisation ability of predictive models and prevents overfitting. As a result, fourfold cross validation [9] is used to gauge how well the proposed or developed model performs. Similar or comparable to earlier works, the proposed model outperformed the models based on machine learning and deep learning.

The rest of the chapter is organised as follows: A literature survey is featured in Sect. 2. The impact of big data on the stock market is described in Sect. 3. The proposed model is described in Sect. 4. Sect. 5 shows how it is implemented. The results and discussion appear in Sect. 6. Finally, the conclusion and a brief mention of future works appear in Sect. 7.

2 Literature Survey

The objective of any investment is to earn return. Stock returns constitute one area for this chapter to look at, because many scholars have shown serious interest in this topic over the past several years. A quick evaluation of the literature assists in understanding the relevance of the content evaluation for inventory returns. Stock and other financial markets are complicated. They feature dramatic and dynamic behaviour and experience surprising booms and busts. The major challenge of this analysis is to build a device that can anticipate future fees in stock markets by taking samples of future costs. The experimental results of the study observed by Wei Yang [10] over a three-month screen showed that this version can effectively anticipate the route of the market with an average hit ratio of 87%. Similarly, on day-by-day prediction, the model was capable of predicting the open, excessive, low, and near fees of the preferred inventory every week and every month. An optimisation technique, namely principal component analysis (PCA), was implemented in a short time period for predicting stock market prices, and it includes preprocessing the stock market data set, engineering strategies, and a customised machine-learning, deep-learning machine for inventory market price trends and value predictions.

Several investigations have given rise to various selection guides that present buyers with superior predictions and analyses. Along with the wide availability of stock or price statistics, which was made possible through the Internet, the task of the investor has become more complicated, in that they now need to store, accumulate, examine, and filter out data to make accurate decisions from a diversity of information. This information consists of financial facts, real-time data, and opinions on how to effectively change the economic markets. It is far important to develop models that uncover the unique states of the market that will require making adjustments.

Quite a few interests have shifted to using various artificial intelligence techniques in the stock market, in addition to the statistical data that have been used to comprehend and anticipate fluctuations within the stock market. Technical analysis has also drawn the attention of many researchers as a means of reducing investment costs in the stock market. Therefore, knowing the market and being able to predict what will happen soon are essential skills for any investor. Ijegwa et al. [11] proposed solving this problem by working on an easy-to-use inference sign model that uses many variables to simplify the complicated marketplace. It gives reliable hints to traders and, as a consequence, acts as a valuable selection tool. The work employs fuzzy good judgement to perform the decision-making task, based entirely on inputs from technical analysis indicators. No longer will the best techniques, methods, and algorithms be enough to best predict the stock market, but this solution comes with shortcomings, disadvantages, and risks.

A distinction between stock cost or share cost and their respective prices must be made. Price shows the buying price and the selling price, and the stock cost is an underlying price—no matter the buying and selling prices, which do not rely on a country to determine the stock price. The cognitive styles of traders and rumours frequently make it difficult to charge a fee. The role of the investor is to evaluate the investment's values, costs, and margin of safety in order to raise money, market the products, or sell them. Understanding the underlying technical method is genuinely difficult regardless of the excellence of these practices. The rate of the stock is determined by the past patterns, using a variety of techniques and tools, as suggested by Mangale et al. [12]. The approaches are typically expressed in terms of a technical method and an essential method. The essential method is used for long-term valuations. Each inventory has a price that is independent of the inventory's price; this is known as its intrinsic value. The proposed model operates through the following stages: information gathering, feature processing, fuzzy logic mapping, and stock value calculation. The mapping of the exceptional number of evaluation factors uses fuzzy logic. If-then rules are applied to the linguistic variable. The inventory price, which is used to calculate inventory worth, is affected by the fuzzy model. The dividend discount version is used to calculate the inventory fee. The purpose of the paper proposed by Alalaya et al. [13] was to take into account the projection capabilities of fuzzy logic and neural networks, as well as to combine some of the models to address various models' implementation issues in the prediction of indexes and fees for the Amman Inventory Alternate, where other researchers have demonstrated some of the differences between these measures.

Given that the stock market depends on nonstationary economic data and that the majority of the models used in this study have nonlinear structures, the Amman Stock Exchange index prices were used as a data set to examine the unique utility models. The facts are first gathered and preprocessed and then transformed from high-frequency facts to a ratio matrix. Next, the outlier set of rules identifies any anomalies in the ratio matrix, which determines the results. Although behaviour evaluation is no longer the most effective, because of its unstable nature, it is essential for all technical and social analyses. The relationship between sentiment and stock values is then examined. The developed model can afterwards be applied to forecast future stock prices. The development of this hybrid approach for forecasting inventory prices is a step toward better forecasting. In addition to quantitative data, financial news and news articles were analysed for the neural network forecasting of the proportion charge. A novel method for time-series forecasting with a simple, linear computational algorithm was proposed by Pandey et al. [14]. The suggested method uses a quantile-based fuzzy forecasting method after first predicting the trend of the future value. The suggested method is less complex than other techniques. The method's suitability for Sensex forecasting is also investigated. We identified specific qualitative terms from our research that had favourable effects and unfavourable effects on stock price. Some of these words were frequently used in different contexts, indicating that the articles that contained them were vulnerable to price changes. After being retrieved from online databases and virtual libraries like the ACM (Association for Computing Machinery) digital library, Scopus, Kaggle, and Nifty 50, the statistical units underwent a critical analysis. Moreover, a thorough comparative analysis was completed to determine the route of structured and unstructured records for the segmentation of Nifty 50 shares. The Nifty 50 stock list was obtained from the website of the National Stock Exchange of India. The sentiment rating was derived for each tweet and each stock in the Nifty 50 by using the unstructured statistics that were downloaded from Twitter for sentiment evaluation. Twitter is a relatively new way of disseminating information, not only through short sentences but also by allowing users to highlight one specific piece of information by retweeting it.

Similar studies have characterised market performance on information: There should be no gain opportunities in stock markets, because they are defined by random walk patterns, while all new information is evaluated at any time given that prices immediately reflect that information. A crucial observation is that big data analysis should be used instead of stock market or financial fluctuations. Groups and businesses now have access to exceptional volumes and types of data thanks to cloud computing, the Internet of Things, Wi-Fi sensors, social media, and many other technologies. Innovative threat evaluation methods and applications are undergoing related developments that are tested and analysed. For an actual analysis, the data set must be preprocessed and improved. We can train the vector device on the data set and the results it produces after preprocessing the data, which was observed using the arbitrary woods technique.

To obtain precise recommendations that are based on customers' preferences, a hybrid recommendation model built on crowd search optimisation is suggested. Collaborative and content-based filtering are combined to create the hybrid recommendation described by Srakar et al. [15]. The actual time is in [16], and the data set is from the Nifty 50 stock list on Kaggle. Once investors understand the stock market and the fact that it involves a financial risk, then analysis, knowledge, advice, and prediction are all possible. Knowledge of the stock market is important because it currently makes up the majority of the stock market itself and requires a firm understanding of how stock prices are trending both now and into the future. Such knowledge could help investors to understand the upcoming inventory charge and to assess the risk. One of the resources used to train system-learning models to predict stock prices is the abundance of available stock data. Other factors that make it difficult to predict the market include a high noise-to-signal ratio and the sheer number of factors that affect inventory costs. Efforts are being made to improve and use smart devices, which include neural networks, fuzzy structures, and genetic algorithms within the discipline of financial decision-making. Thanks to the use of fuzzy rules and logic, an intelligent model has been proposed to predict stock expenses in order to guarantee returns on investment over some particular period of time, which is mathematically applied, thoroughly examined, and finally confirmed through real-time data sets.

3 Impact of Big Data in Stock Market

This section of the chapter in detail explains big data analytics, its impact on the stock market, its importance in the big data architecture, and the preparation of data sets required for further implementation, testing, and validation.

3.1 *Big Data*

Big data has emerged as crucial to the tech panorama. Big data, or massive information, analytics enables agencies to harness their facts and use them to discover new possibilities. Big data analytics, according to Modi et al. [17], is utilised in many fields for the precise prediction and analysis of numerous records. It facilitates the discovery of facts from huge amounts of information that provide guidance on how to use predictive analytics and personal behaviour analytics. Big data compile a collection of complex data sets for a method utilising conventional database management or storage equipment and traditional tools or conventional statistics-processing applications. The key tasks for this model include capturing, storing, analysing, and visualising those data. A massive investigation into facts is applied in certain disciplines, according to Attigeri et al. [18], for specific prediction analysis.

3.1.1 Big Data Architecture

Big data architectures have been developed to carry out the gathering, processing, and analysis of statistics that are too large or complex for conventional database systems. Depending on the skills of the clients and their equipment, there are different thresholds at which businesses step into the world of facts. It may mean hundreds of gigabytes for some people, while it may mean many terabytes for others. The means of handling large data sets improves along with the tools for doing so. As opposed to focusing on just the size of the facts, though they tend to be large, this era increasingly focuses on the value that can be extracted from data sets through advanced analytics (Fig. 1).

3.2 *Structure of Big Data*

According to Rajakumar et al. [19], both structured data and unstructured data have played significant roles in the decision-making processes for daily buying and selling activities. Gathering information reports is a preferred starting point for the decision-making process of a buying-and-selling activity, which leads to the creation

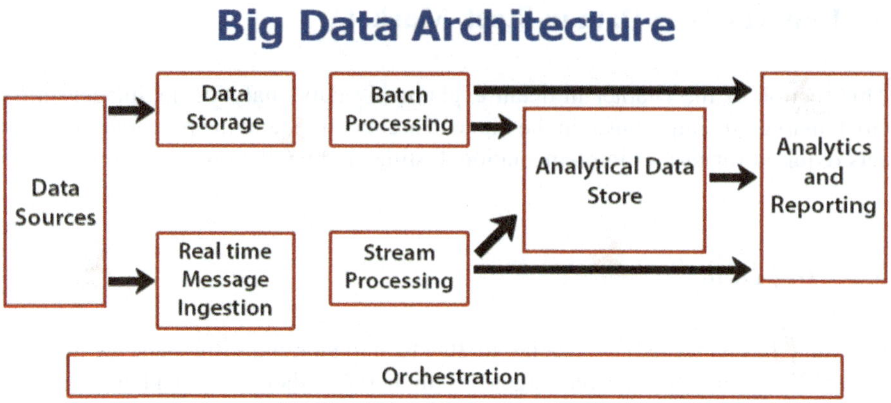

Fig. 1 Big data architecture

of the buy, keep, and sell funding method decision. A specific array of resources, including the agency database, publications, the press, radio, television, and the Internet, are used to collect information bulletins and business moves. Information from businesses include their planning, financial statistics (EPS, PE), income potentials and expectations, shares, returns, financial statements, and economic feedback. Marketers or other individuals spread exciting and informative messages. Every message is thoroughly investigated, taking into consideration variables such as economic expertise, experiences, backgrounds, and cognitive biases. On the basis of this analysis, a decision that takes into account risk tendencies and trading objectives is made. Finally, the effects or outcomes are assessed, and a recommendation to buy, hold, or sell is made. Rouf et al. [20] have used machine learning in inventory price prediction to find patterns in data. Typically, stock markets generate a sizeable amount of structured and unstructured heterogeneous statistics. It is now possible to quickly examine more-complex heterogeneous records and produce more-accurate results by using machine-learning algorithms.

3.2.1 Structured Data

Mathematics follows a prescribed format and structure, and for that reason, it is simple to use to examine and analyse the statistics. It collectively conforms to or is represented in a tabular format, with relationships in between special rows (features) and columns (records). Structured query language (SQL) databases are commonplace examples of structured records that rely on how facts may be saved, processed, and accessed.

Nifty 50 is a collection or a basket of the 50 most-energetic stocks on the national inventory alternate of India, which acts as a benchmark for the general movement of the inventory market, according to Sujata Suvarnapathaki [21]. The Nifty 50 is in the form of dependent records, and traders are regularly interested in knowing the standard behaviour of stocks. Having a prior idea of the worth of an inventory or the movement of an inventory would benefit buyers. A stock market analysis allows buyers to make informed decisions on whether to invest, not invest, or maintain a current investment. For this purpose, traders can also rely on historical data on the essential analysis parameters and the technical analysis parameters influencing the behaviour of stocks.

3.2.2 Unstructured Data

Unstructured data are statistics that come in unknown shapes, cannot be stored by using conventional methods, and cannot be analysed until they have been converted into established formats. Multimedia content material like audio files, motion pictures, videos, and pictures are examples of unstructured data. Recently, unstructured data have been growing faster than other sorts of big data. With the advent of social media, the democratisation of value, and more-efficient information systems, data has never been so easy to gather and so accessible. Each day, around 500 million tweets are publicly sent around the world, and more than one billion customers are on Facebook.

One of the most promising prospects for finance is thought to be big data, specifically for managing risk. The financial industry developed dashboards to manipulate risks by using a strong framework of social media scrutiny. The definition of *big data* is a combination of structured data and unstructured data (such as inventory costs, heartbeats, and images) produced in real time, according to Sanger et al. [22, 23]. These authors brought attention to the fact that qualitative records can be duplicated or can be reflected in statistics that have not been properly taken into account when calculating inventory fees. In my experience, tweets are similar to or fit the definition of massively produced unstructured data, and social media has become the subject of many studies in the social sciences and economics.

3.2.3 Semistructured Data

Semistructured data are a form of big data that do not conform to the formal shape needed by data models. However, these data come with a few types of organisational tags or distinctive markers that assist in cutting up or splitting semantic factors. JSON documents or XML documents store these data. This class of data includes facts that are significantly easier to examine than those in unstructured records. A range of tools can analyse and process XML documents or JSON documents, decreasing the complexity of the analysing system. Semistructured statistics can be converted by using the packages of big data from a given economic quarter, which include social media analyses, net analytics, risk management, fraud detection, and security intelligence.

According to Bach et al. [24], text mining or textual content analytics is one method for obtaining information from a vast array of big data. The goal of text content mining, which is also known as text data mining or text analytics, is to examine textual data, such as emails, reviews, texts, webpages, reviews, and legitimate documents in order to extract data, transform it into records, and make it useful for various types of decision-making. In its final form, text mining may be used for evaluation, visualisation (through maps, charts, and thought maps), integrating information from databases or warehouses, machine learning, and many other purposes. Text mining includes linguistic, statistical, and system-learning techniques.

3.3 Big Data in the Stock Market

At present, the market actions and behaviours are converting faster than ever, making the market even more unpredictable. Businesses must verify the alternative enterprise techniques and enforce them with top-of-the-line enterprise solutions, according to Lima et al. [25, 26]. In the era of top-of-the-line technological innovation, the various types of records can be improved by using the latest information technology, and data statistics is one of the most precious inputs for automated systems, according to Hasan et al. [27, 28]. Financial markets and technological evolution have affected every human interest over the past few years. Big data generation has become a fundamental part of the financial services enterprise and could continue to pressure future innovation, according to Razin E [29]. Analysing big data can enhance predictive modelling to better estimate the costs and outcomes of investments. Gaining access to big data and advanced algorithmic expertise improves the precision of predictions and the ability to mitigate the inherent risks of buying and selling efficiently. Data may be reviewed and applications advanced to regularly update information in order to make accurate predictions. Financial businesses use large numbers of data to reduce operational risk, combat fraud, significantly reduce information asymmetry issues, and meet regulatory and compliance requirements.

Contributing to and reaping the benefits from the market have not been easy, because of the market's obvious vulnerability and unpredictable nature. Stocks can quickly rise and fall in value, according to Jaweed et al. [30]. Stability is only a percentage of the earnings that are dispersed for a particular safety or market report. Usually, the more unpredictable something is, the less safe it is.

Big data are fundamentally changing both how investors make investment decisions and how the world's stock markets operate. When given records, computers are now able to make precise predictions and decisions that are similar to those of humans. This allows them to execute trades at high speeds and with high frequency. A technical analysis and a fundamental analysis of a prediction are available options. According to Sean Mallon [31], a technical evaluation is carried out by applying device learning to historical data on inventory costs, and a fundamental evaluation is carried out by conducting a sentiment analysis on social media data.

More so than ever before, social media data have significant effects and can help forecast stock market trends. The strategy involves gathering data from social media and extracting the expressed sentiments. The relationship between sentiment and stock value inventory is then examined.

3.4 *Nature of Dynamic Data in the Stock Market*

After being recorded, dynamic data continuously change or vary in order to preserve their integrity. The stock market is a true, dynamic, extremely complex system that is constantly changing. The stock price is volatile and dynamic in nature. The pipeline, the zoom info [32], and dynamic data enhance business-wide communication and alignment. Access to the same customer data is available throughout the entire organisation. Consistency in sales, marketing, and branding is easier to achieve if a data set is dynamic.

The quantitative extraction method for the financial market system indicates that the dynamic characteristics, the dynamic process [33], and the mechanism of the financial market system's dynamic characteristics are part of an evolution of monitoring abnormal behaviour in the financial market's bubble, crisis, and collapse process but are not enough. But before investing, certain factors that influence stock prices must be considered, such as market trends, price actions, etc., according to John et al. [34]. Data scientists are exploring how to predict market trends. Stock price behaviour needs to be analysed as a sequence of discrete time data. The linear model is typically used to empirically test the effectiveness of the stock market, but the price time series is chaotic.

4 Proposed Model (Fig. 2)

4.1 *Objectives*

- To implement an expert system to predict the uncertainty of market risk.
- To propose a model that guarantees that a base amount will remain unchanged while marketing in the long run, thus increasing the efficiency of trade.
- To protect against fraud by ensuring that users abide by the norms and regulations set by governments for safe trading.

4.2 *Mathematical Implementation*

The mathematical implementation is carried out by using mathematics, operations, and logic. At the beginning, a statistical analysis is performed to analyse the data.

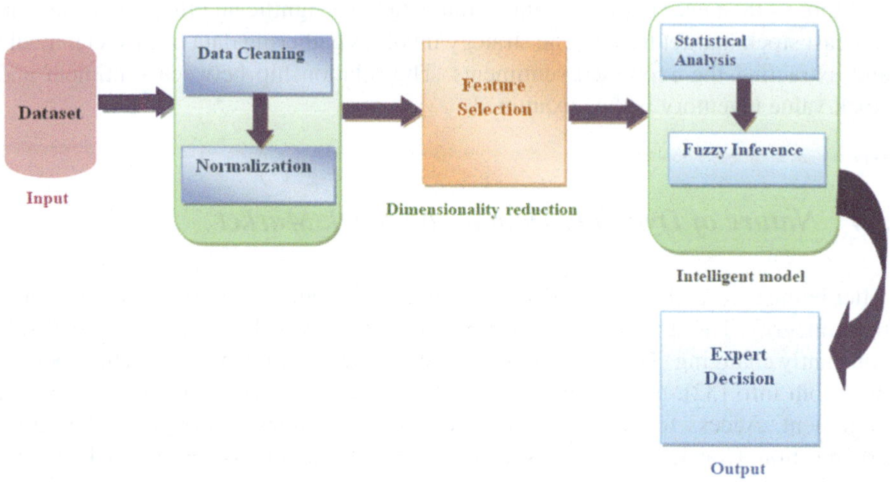

Fig. 2 Proposed Model

After analysing the data, a fuzzy inference is used to abate the uncertainty of market fluctuations.

4.2.1 Statistical Analysis

In this section, a sequential statistical analysis is performed. In this study, five companies are considered for validation. Through estimation, the expected outcome can be found in an interval that is predicted to include the unknown features of the companies' shares. The unknown features have some specific probabilities.

Let us assume that $C_1, C_2, C_3, \ldots, C_n$ constitute a random sample of companies from a set of companies. The unknown feature of a company is θ.

Two special limits, which serve as the lower and upper limits to the value of confidence, are denoted as L_1 and L_2. The confidence limit defined as

$$P(L_1 \leq \theta \leq L_2) = C \quad (1)$$

where C is the coefficient of the confidence interval.

The confidence limit for unknown features μ is considered as 95% for this research. Therefore, the normal distribution can be determined by using Eq. (2):

$$Z = \frac{\overline{x} - \mu}{\sigma/\sqrt{n}} \quad (2)$$

where $\overline{x}$is the sample and μ and σ^2/n are the mean and the variance, respectively.

In a bell curve, the area under the standard normal curve occupies 95%, and the critical region occupies 5%:

$P\left[-1.96 \leq \frac{\overline{x}-\mu}{\sigma/\sqrt{n}} \leq 96\right] = 0.95$. Here, 95% of the cases belong with the inequalities: $\left[-1.96 \leq \frac{\overline{x}-\mu}{\sigma/\sqrt{n}} \leq 96\right]$.

$\overline{x}-1.96\frac{\sigma}{\sqrt{n}} \leq \mu \leq \overline{x}+1.96\frac{\sigma}{\sqrt{n}}$. Thus, the interval $(\overline{x}-1.96\frac{\sigma}{\sqrt{n}}, \quad \overline{x}+1.96\frac{\sigma}{\sqrt{n}})$ is the 95% confidence interval for μ. The limit of μ is $\overline{x} \pm 1.96\frac{\sigma}{\sqrt{n}}$. This limit is modelled after hypothesis testing. During the hypothesis testing, H_0 and H_1 are the null hypothesis and an alternative hypothesis, respectively.

$$\begin{aligned} H_0 &: \{\theta \leq 0.95\} \\ H_1 &: \{\theta > 0.95\} \end{aligned} \tag{3}$$

4.2.2 Fuzzy Inferences

Fuzzy logic is implemented in this research to control the uncertainty in the market value of the share units. Because the share price may fluctuate within a fraction of a second, it is very difficult to use for market prediction.

Assume that S is a set of blue-chip stocks. A new company C may or may not belong in S. The objective of this research is to determine whether the stocks of the new company C are blue-chip stocks.

The membership grade of C is $\mu_{\underset{\sim}{s}}(C)$when C is mapped in S:

$$\mu_{\underset{\sim}{s}} : (C) \rightarrow \{0, 1\} \tag{4}$$

where $c \in C$.

If the c does not belong in S, then it belongs in a complementary set of S.

$$S \cap S' = \varphi \tag{5}$$

Therefore, it is determined that S and S' are disjoined sets. Considered two data sets, A and B, are in a universe of two discourses, U_1 and U_2, respectively. The multiple-input-single-output (MISO) rule base is applied with n rules.

$$\text{If } A_{i1}, \text{ and } A_{i2}, \text{ and} \ldots.\text{and } A_{in}, \text{ then } B_i \tag{6}$$

where n is the number of input variables. The jth rules can be calculated by using Eq. (7):

$$I_j = A_i \rightarrow B_i A_i = A_{i1} \cap A_{i2} \cap \ldots\ldots A_{in} = \cap_{j=1}^{n} A_{ij} \tag{7}$$

The summation of the implementation I_{Total} of n can be determined by using Eq. (8):

$$I_{\text{Total}} = R_1 \cup R_2.\cdots \cup R_N = \cup_{i=1}^{N} R_i = \cup_{i=1}^{N} A_i \to B_i \tag{8}$$

where R denotes the fuzzy relation of $U_1 \times U_2$.

5 Implementation

Implementation science is the study of techniques to improve the uptake, implementation, and translation of a study's findings into ordinary practices. The implementation of research seeks to recognise and work within the actual geographical areas of world situations, instead of looking at and manipulating conditions or removing effects.

5.1 Data Preparation

Stock market–related records are diverse, so first, similar works from a survey of economic studies are compared against stock price trends to conduct a statistical analysis in order to determine which guidelines to follow. After accumulating the records, the statistical structure of the data set is described, by following Shah et al. [35]. At this stage, the data collected from the general public and civic records are organised. A description of the data set includes the data shape, records, and tables in every class of statistics. Second, we compare similar works from the survey of economic studies to determine which instructions to follow for the collection of statistics. The data set is a group of inventory-request statistics from certain agencies (Akhtar et al. [36]). After gathering the information, we organised the records in the data set. Contributing to and benefiting from the market have by no means been essential or fundamental, because of the unpredictable nature of the market: Stocks, shares, and values are likely to quickly rise or fall in price. Recorded instability is likewise regarded as unpredictability from the instability in the actual charges for fundamental shares. As an alternative, it is represented by using a totally nonlinear dynamical framework.

In this chapter, the Nifty 50 data of columns and rows, which are data sets, are used to validate the proposed model. The information series approach is one of many critical approaches. It is carried out by filtering the statistics and preprocessing data; it is supported by training the model; and it requires checking the set of rules by using extraordinary sets of statistics. The data sets have been taken from Kaggle as real-time data sets: fifteen features from the Nifty 50 stock market records (2000–2021) have been reduced to six variables: date, low fee, excessive fee, commencing fee, final price, and extent. The data sets have been normalised, and all the capabilities are represented in their diverse units. There is no statistical inconsistency but rather data integrity.

5.2 *Data Cleaning and Data Preprocessing*

We nowadays check whether statistics have a null cost or a price and whether unknown types or styles of records have been wiped clean, crammed, or stuffed into statistical formulations (Karim et al. [37]). For example, if it changed into a null cost from a corresponding characteristic, then it is checked for a variety of values, such as discrete and classifier costs. If it is a class price, then calculate the median cost; otherwise, calculate the mean cost, and place it in the null location.

5.2.1 Data Normalisation

Normalisation changes the values of various numeric columns within the data set to one unit of measure, which further enhances the overall performance of our model (Roshan Adusumilli [38]). Normalisation shows how to effectively organise structured statistics within a database. It also includes tables presenting, depicting, and establishing relationships between them and defining guidelines for relationships. Inconsistency, uncertainty, and redundancy can be assessed by tests based entirely on these rules, thus adding flexibility to the database.

5.3 *Fuzzy Inference*

Fuzzy rules are used to ensure that certain steps are taken in the event of ambiguity or uncertainty (Abbasi et al. [39]). It can rework many concepts, variables, and ambiguous and obscure systems into mathematical models and pave the way for argumentation, inference, manipulation, and decision-making in the event of uncertainty. A dynamic gadget that uses fuzzy sets, fuzzy logics, and/or analogous mathematical frameworks is referred to as a fuzzy machine. Some approaches can be used to construct a fuzzy inference device, wherein one or more inputs or antecedents can be used to generate one or more outputs or consequents. Technical indicators are used to reveal fee movements (adjustments in rates); they form the input parameters in the fuzzy machine. Inventory statistics are taken from a special inventory exchange and used for the work, as can be seen in Table 1. The past values of open, high, low, and ultimate charges and the volumes of a particular stock are recorded for a series of days and stored in a database to train the machine.

Table 1 Actual values for various features

Date	Opening rate	Low rate	High rate	Closing rate	Volume
(a) Actual values for Axis Bank					
02-04-2018	506.75	492.2	508.5	499.05	7,120,498
03-04-2018	494	493.7	506	501.55	6,053,789
04-04-2018	504.5	490.25	504.5	492.25	6,047,813
05-04-2018	498.4	495.2	504.65	503.1	6,682,587
06-04-2018	502.95	495.2	502.95	500.2	3,812,133
Standard deviation		2.1149468	2.08044		
(a) displays the actual values for the features of Axis Bank from 02-04-2018 to 06-04-2018, for which the standard deviation of the low rate and that of the high rate have been calculated					
(b) Actual values for Axis Bank					
09-07-2018	515.1	515.1	529.3	525.25	6,295,104
10-07-2018	527.95	522	534.9	532.55	5,281,760
11-07-2018	532.4	530.8	542.35	536.55	6,432,494
12-07-2018	539.7	535.3	545.7	538.35	4,128,221
13-07-2018	539.7	519.45	541	523	4,169,212
Standard deviation		8.3132	6.52687		
(b) displays the actual values for the features of Axis Bank from 09-07-2018 to 13-07-2018, for which the standard deviation of the low rate and that of the high rate have been calculated					
(c) Actual value for Axis Bank					
03-12-2018	632	621.6	633	624.8	5,914,250
04-12-2018	624.95	618.6	627.55	623.25	8,382,005
05-12-2018	620.9	613.4	623.95	614.95	9,728,083
06-12-2018	610.25	598.25	613.85	600.05	7,519,849
07-12-2018	602.4	592.35	609.5	603.95	12,262,327
Standard deviation		12.8738	9.71279		
(c) displays the actual values for the features of Axis Bank from 03-12-2018 to 07-12-2018, for which the standard deviation of the low rate and that of the high rate have been calculated					

In this chapter, the above data set tables are taken from Axis Bank and are related to the stock market features that are used for building a model. They are divided into three sets of random samples taken over a period of eight months, and afterwards, the respective standard derivations of all fifteen random samples are calculated

5.3.1 Axis Bank

The Table 2 depicts the mean of the standard deviation calculated in Table 1 at three intervals of random samples taken over a period of eight months in 2018. The mean of the standard deviation of the low and high rates has been calculated to be 4.61682, which is used to establish a model.

Table 2 Mean and standard deviation for three intervals (Axis Bank)

Intervals	Low rate	High rate
1st interval	2.1149468	2.08044
2nd interval	8.3132	6.52687
3rd interval	12.8738	9.71279
Standard deviation	5.40016	3.83348
Mean		4.61682

Table 3 Mean and standard deviation for three intervals (Tata Steel)

Intervals	Low	High
1st interval	10.9415	12.6731
2nd interval	6.81887	9.12113
3rd interval	7.12811	5.16207
Standard deviation	2.29614	3.72795
Mean		3.012045

Table 4 Mean and standard deviation for three intervals (Titan)

Intervals	Low	High
1st interval	25.8135	17.1588
2nd interval	13.7267	8.46692
3rd interval	15.8859	17.133
Standard deviation	6.44066	5.01083
Mean		5.725745

5.3.2 Tata Steel

Similarly, the standard deviations of the low and high rates have been calculated for Tata Steel's stock market features at three intervals, which appear in Table 3, depicting the mean of the standard deviation calculated at three intervals of random samples taken over a period of seven months in 2018. The mean of the standard deviation of the low and high rates has been calculated to be 6.61882 and 3.012045, which is used to establish a model.

5.3.3 Titan

Similarly, the standard deviations of the low and high rates have been calculated for Titan's stock market features at three intervals, which appear in Table 4, depicting the mean of the standard deviation calculated at three intervals of random samples taken over a period of twelve months in 2018–2019. The mean of the standard deviation of the low and high rates has been calculated to be 5.725745, which is used to establish a model.

Table 5 Threshold values of three companies' data sets

Company	Low rate	High rate
Axis Bank	5.40016	3.83348
Tata Steel	2.29614	3.012045
Titan	6.44066	5.725745
Mean	4.71232	4.190423

5.3.4 Threshold Value

The threshold value calculations for the dataset features of low rate and high rate for the AXIS BANK, TATASTEEL, and TITAN firms are shown in Table 5. The results are 4.71232 for the mean low rate across the three companies and 4.190423 for the high rate, which are then utilised to create fuzzy rules.

5.3.5 Parameters

Standard deviation: Σ
Lesser than: $<$
Greater than: $>$
Greater than or equal to: $\geq$
Lesser than or equal to: $\leq$
Equal to: $=$

5.3.6 Fuzzy Rules

Fuzzy rules are developed on the basis of the threshold values in Table 5.

Rule 1: *IF* s.d Σ of low rate is <4.71232 *THEN* returned amount will be $<$ assured amount (loss).

Rule 2: *IF* s.d Σ of low rate is $=$ to 4.71232 *THEN* returned amount will be $=$ assured amount.

Rule 3: *IF* s.d Σ of low rate is >4.71232 and <4.190423 *THEN* returned amount will be $\geq$ assured return.

Rule 4: *IF* s.d $\Sigma$of high rate is >4.190423 *THEN* the return amount will be surely be $>$ assured return (profit).

6 Results and Discussion

In Fig. 3, the graphs depict the opening rate, low rate, high rate, and closing rate against the volume for Axis Bank, taken from fifteen random samples at three intervals, which are further used for computation.

In Fig. 4, the graphs depict the opening rate, low rate, high rate, and closing rate against the volume for Tata Steel, taken from fifteen random samples at three intervals, which are further used for computation.

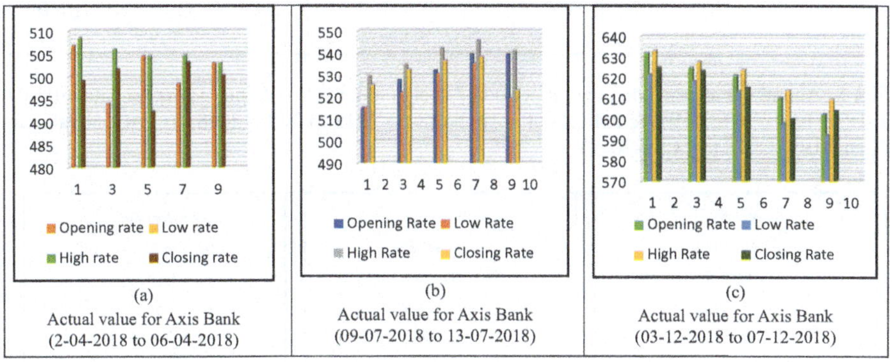

Fig. 3 Actual values for Axis Bank over a period of seven months. (**a**) Actual value for Axis Bank (2-04-2018 to 06-04-2018). (**b**) Actual value for Axis Bank (09-07-2018 to 13-07-2018). (**c**) Actual value for Axis Bank (03-12-2018 to 07-12-2018)

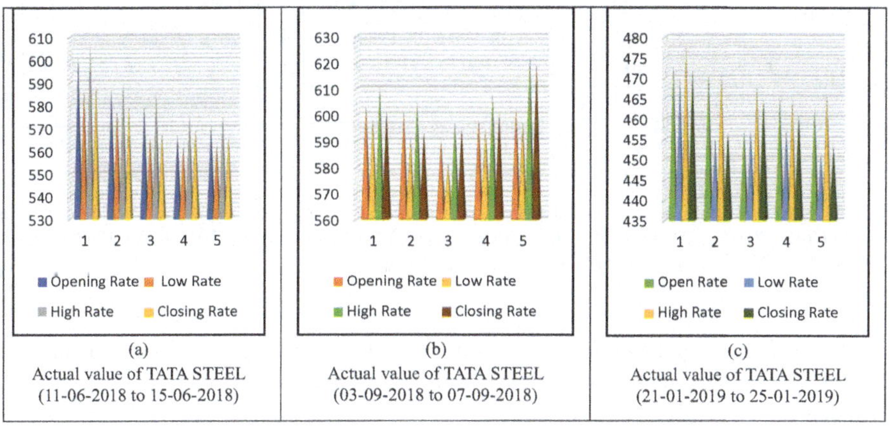

Fig. 4 Actual values of Tata Steel over a period of seven months. (**a**) Actual value for Tata Steel (11-06-2018 to 15-06-2018). (**b**) Actual value for Tata Steel (03-09-2018 to 07-09-2018). (**c**) Actual value for Tata Steel (21-01-2019 to 25-01-2019)

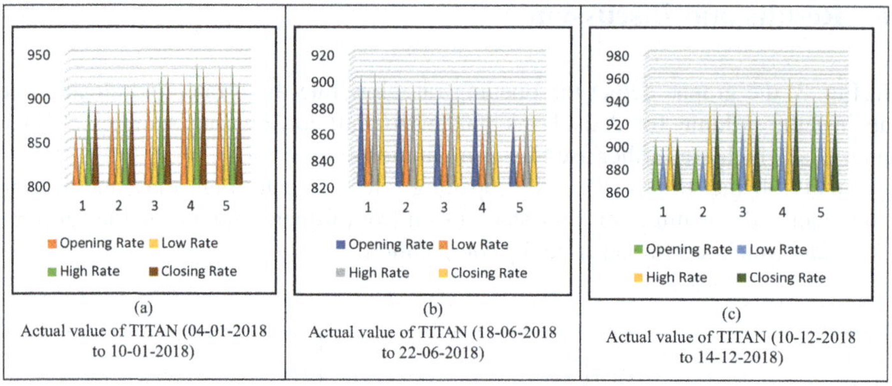

(a) Actual value of TITAN (04-01-2018 to 10-01-2018)

(b) Actual value of TITAN (18-06-2018 to 22-06-2018)

(c) Actual value of TITAN (10-12-2018 to 14-12-2018)

Fig. 5 Actual values for Titan over a period of twelve months. (**a**) Actual value for Titan (04-01-2018 to 10-01-2018). (**b**) Actual value for Titan (18-06-2018 to 22-06-2018). (**c**) Actual value for Titan (10-12-2018 to 14-12-2018)

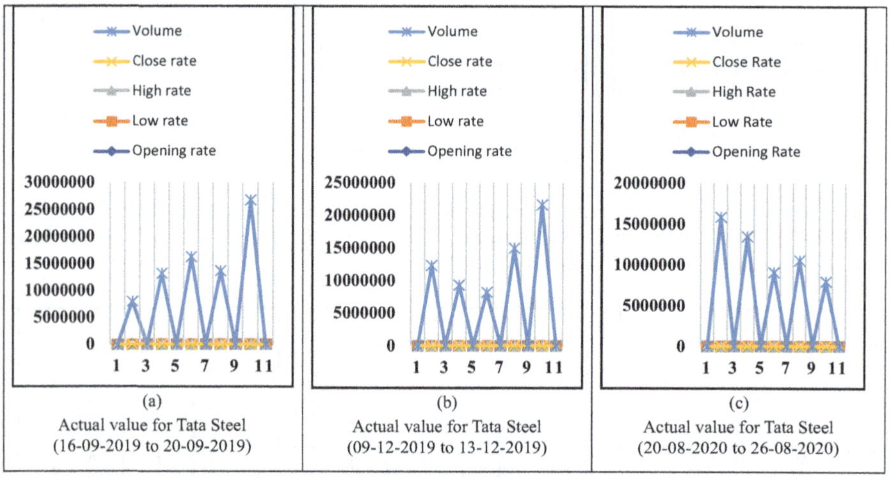

(a) Actual value for Tata Steel (16-09-2019 to 20-09-2019)

(b) Actual value for Tata Steel (09-12-2019 to 13-12-2019)

(c) Actual value for Tata Steel (20-08-2020 to 26-08-2020)

Fig. 6 Actual values the Tata Steel over a period of eleven months. (**a**) Actual value for Tata Steel (16-09-2019 to 20-09-2019). (**b**) Actual value for Tata Steel (09-12-2019 to 13-12-2019). (**c**) Actual value for Tata Steel (20-08-2020 to 26-08-2020)

In Fig. 5, the graphs depict the opening rate, low rate, high rate, and closing rate against the volume for Titan, taken from fifteen random samples at three intervals, which are further used for computation.

In Fig. 6, the graphs depict the opening rate, low rate, high rate, and closing rate against the volume for Tata Steel, taken from fifteen random samples at three intervals, which are further used for computation.

In Fig. 7, the graphs depict the opening rate, low rate, high rate, and closing rate against the volume for Titan, taken from fifteen random samples at three intervals, which are further used for computation.

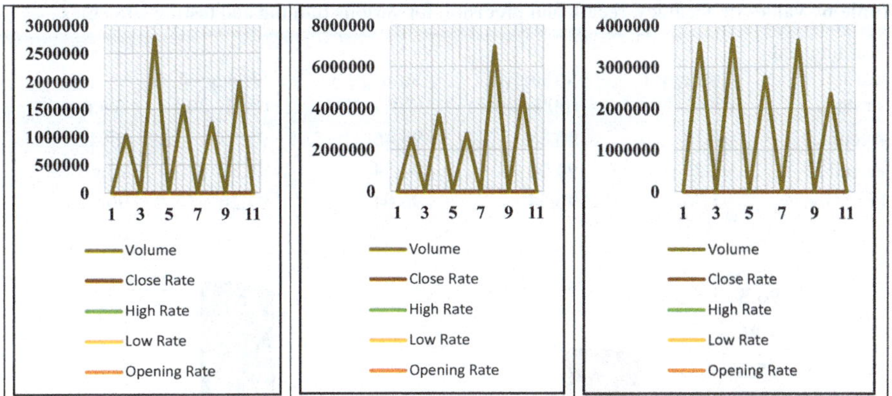

(a) Actual value for Titan (21-01-2019 to 25-01-2019)

(b) Actual value for Titan (20-04-2020 to 24-04-2020)

(c) Actual value for Titan (20-04-2020 to 24-04-2020)

Fig. 7 Actual values for Titan over a period of sixteen months. (**a**) Actual value for Titan (21-01-2019 to 25-01-2019). (**b**) Actual value for Titan (20–04-2020to 24-04-2020). (**c**) Actual value for Titan (20-04-2020 to 24-04-2020)

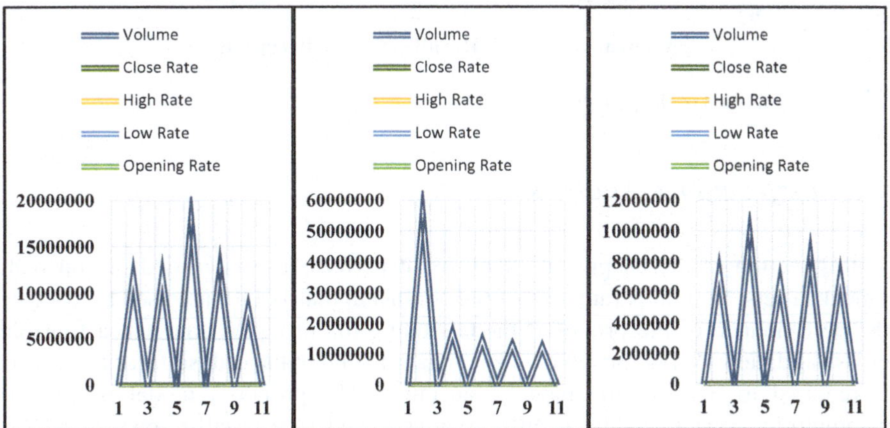

(a) Actual value for Axis Bank (07-01-2019 to 11-01-2019)

(b) Actual value for Axis Bank (31-07-2019 to 06-08-2019)

(c) Actual value for Axis Bank (16-12-2019 to 20-12-2019)

Fig. 8 Actual values for Axis Bank over a period of twelve months. (**a**) Actual value for Axis Bank (07-01-2019 to 11-01-2019). (**b**) Actual value for Axis Bank (31-07-2019 to 06-08-2019). (**c**) Actual value for Axis Bank (16-12-2019 to 20-12-2019)

In Fig. 8, the graphs depict the opening rate, low rate, high rate, and closing rate against the volume for Axis Bank, taken from fifteen random samples at three intervals, which are further used for computation.

Table 6 Values of accuracy, recall, and precision for various training and testing strategies

Training and testing strategy	Strategy 1 (50%–50%)	Strategy 2 (60%–40%)	Strategy 3 (70%–30%)	Strategy 4 (80%–20%)	Average
Accuracy	92.78	96.88	96.46	95.55	95.42
Recall	91.22	94.06	94.54	94.82	93.66
Precision	95.76	96.24	96.04	97.26	96.325

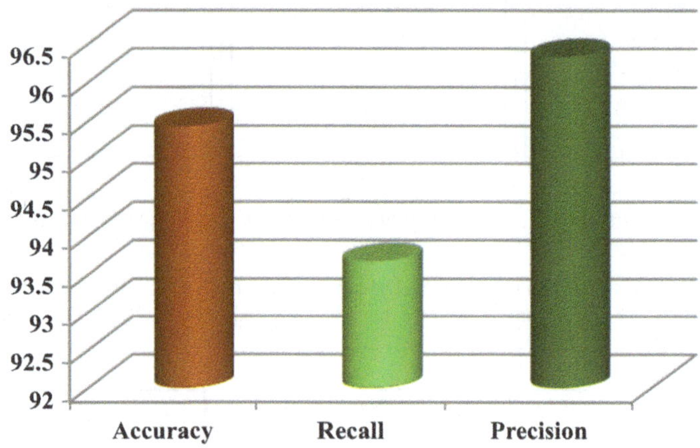

Fig. 9 Visualisation of the performance metrics

6.1 Performance Analysis

In this section of the chapter, an analysis of performance is carried out on real-time data sets. Several parameters of the confusion matrix [40] are used to evaluate the performance of the proposed model. Using the provided data set, a fourfold cross validation [41] is carried out. The data set is not balanced, so error correction is required to improve the model's accuracy. The proposed model is used as a monitoring system in the healthcare sector, and recall and precision are key performance indicators. Table 6 features the performance metrics for four strategies, the ratios from the training and testing data, and the accuracy, recall, and precision values for various training and testing strategies.

The average values of the performance metrics are visualised in Fig. 9.

7 Conclusion

One of the most significant stages in the financial market is online share trading, where people can invest their money as capital on long- or short-term bases.

The purpose of the research in this chapter is to discover an expert solution that yields a guaranteed return on investment. Numerous blue-chip businesses are in the stock market. Blue-chip companies typically trade on a significant stock exchange, such as the New York Stock Exchange, though this is not strictly required. They are frequently listed in market indexes. The stocks are very liquid because both individual and institutional investors frequently trade them on the market. Because there will always be a buyer on the other side of the transaction, investors who suddenly need money can sell their stock with confidence. Thanks to this research, an intelligent framework for choosing investment firms that offers guaranteed returns can be suggested. To create a pilot surveyor to examine the data set, a statistical data analysis is carried out. To reduce the uncertainty in market fluctuations, fuzzy logic is used. The accuracy, recall, and precision values for the model's performance are 95.42, 93.66, and 96.325, respectively. In the future, this model might incorporate blockchain technology to increase the security of online trading.

References

1. Shen, J. & Omair Shafiq, M. (2020). Short-term stock market price trend prediction using a comprehensive deep learning system. Springer Open, *Journal of Big Data, 7*, Open access, Article number: 66.
2. Tanwar, S., Patel, N. P., Patel, S. N., Patel, J. R., Sharma, G., & Davidson, I. E. (2020). Deep learning-based cryptocurrency price prediction scheme with inter-dependent relations. *IEEE Access, 9*, 1345–1356.
3. Liu, H., Qi, L., & Sun, M. (2022, June). Short-term stock price prediction based on CAE-LSTM method, ResearchGate. Hindawi access, *Wireless Communications and Mobile Computing, 2022*(S1), 1–7. https://doi.org/10.1155/2022/4809632
4. Banerjee, S., Neha Dabeeru, R., & Lavanya. (2020). Stock market prediction. *International Journal of Innovative Technology and Exploring Engineering (IJITEE), 9*(9), 506–509. ISSN: 2278–3075.
5. Li, H., Dagli, C.H., & Enke, D. (2007, May). Short-term stock market timing prediction under reinforcement learning schemes. In *IEEE Xplore Conference on Approximate Dynamic Programming and Reinforcement Learning. ADPRL 2007. IEEE International Symposium*. pp. 233–237. https://doi.org/10.1109/ADPRL.2007.368193
6. Rao, P. S., Srinivas, K., & Krishna Mohan, A. (2020, May). A survey on stock market prediction using machine learning techniques. https://doi.org/10.1007/978-981-15-1420-3_101
7. Sarkar, M., Roy, A., Badr, Y., Gaur, B., & Gupta, S. (2021). An intelligent music recommendation framework for multimedia big data: A journey of entertainment industry. *Studies of Big Data, Springer Nature Singapore, 2*, 39–67.
8. Roy, A., Banerjee, S., Sarkar, M., Darwish, A., Elhosen, M., & Hassanieen, A. E. (2018). Exploring New Vista of intelligent collaborative filtering: A restaurant recommendation paradigm. *Journal of Computational Science, Elsevier, 27*, 168–182.
9. Berra, D. (2018, January). Cross-validation, ResearchGate. Reference module in life sciences, pp. 1–3. https://doi.org/10.1016/B978-0-12-809633-8.20349-X
10. Yang, W. (2007). Stock price prediction based on fuzzy logic. In *2007 International Conference on Machine Learning and Cybernetics*, pp. 1309–1314. https://doi.org/10.1109/ICMLC.2007.4370347

11. Vo, M. T., Vo, A. H., Nguyen, T., Sharma, R., & Le, T. (2021). Dealing with the class imbalance problem in the detection of fake job descriptions. *Computers, Materials & Continua, 68*(1), 521–535.
12. Sachan, S., Sharma, R., & Sehgal, A. (2021). Energy efficient scheme for better connectivity in sustainable mobile wireless sensor networks. *Sustainable Computing: Informatics and Systems, 30*, 100504.
13. Ghanem, S., Kanungo, P., Panda, G., et al. (2021). Lane detection under artificial colored light in tunnels and on highways: An IoT-based framework for smart city infrastructure. *Complex & Intelligent Systems*. https://doi.org/10.1007/s40747-021-00381-2
14. Sachan, S., Sharma, R., & Sehgal, A. (2021). SINR based energy optimization schemes for 5G vehicular sensor networks. *Wireless Personal Communications*. https://doi.org/10.1007/s11277-021-08561-6
15. Priyadarshini, I., Mohanty, P., Kumar, R., et al. (2021). A study on the sentiments and psychology of twitter users during COVID-19 lockdown period. *Multimedia Tools and Applications*. https://doi.org/10.1007/s11042-021-11004-w
16. Azad, C., Bhushan, B., Sharma, R., et al. (2021). Prediction model using SMOTE, genetic algorithm and decision tree (PMSGD) for classification of diabetes mellitus. *Multimedia Systems*. https://doi.org/10.1007/s00530-021-00817-2
17. Priyadarshini, I., Kumar, R., Tuan, L. M., et al. (2021). A new enhanced cyber security framework for medical cyber physical systems. *SICS Software-Intensive Cyber-Physical Systems*. https://doi.org/10.1007/s00450-021-00427-3
18. Priyadarshini, I., Kumar, R., Sharma, R., Singh, P. K., & Satapathy, S. C. (2021). identifying cyber insecurities in trustworthy space and energy sector for smart grids. *Computers & Electrical Engineering, 93*, 107204.
19. Singh, R., Sharma, R., Akram, S. V., Gehlot, A., Buddhi, D., Malik, P. K., & Arya, R. (2021). Highway 4.0: Digitalization of highways for vulnerable road safety development with intelligent IoT sensors and machine learning. *Safety Science, 143*, 105407, ISSN 0925-7535.
20. Sahu, L., Sharma, R., Sahu, I., Das, M., Sahu, B., & Kumar, R. (2021). Efficient detection of Parkinson's disease using deep learning techniques over medical data. *Expert Systems*, e12787. https://doi.org/10.1111/exsy.12787
21. Suvarnapathaki, S. (2022). Using unstructured data with structured data for segmentation of nifty 50 stocks. *JETIR, 9*(6).
22. Sanger, W., & Warin, T. (2016). High frequency and unstructured data in finance: An exploratory study of Twitter. *JGRCS 2016, 7*(4).
23. Tetlock, P. C., Saar-Tsechansky, M., & Macskassy, S. (2008). More than words: Quantifying language to measure firms' fundamentals. *The Journal of Finance, 63*(3), 1437–1467.
24. Peji'c Bach, M., Krsti, Ž., Seljan, S., & Turulja, L. (2019). Text mining for big data analysis in financial sector - A literature review. *Sustainability, 11*, 1277. https://doi.org/10.3390/su11051277
25. Lima, L., Portela, F., Santos, M. F., Abelha, A., & Machado, J. (2015). *Big data for stock market by means of mining techniques*. Springer Science and Business Media. https://doi.org/10.1007/978-3-319-16486-1_67
26. Santhosh Baboo, L., & Renjith Kumar, P. (2013). Next generation data warehouse design with big data for big analytics and better insights. *Global Journal of Computer Science and Technology, 13*(7).
27. Morshadul Hasan, M., Popp, J., & Olah, J. (2020). Current landscape and influence of big data on finance. *Journal of Big Data, 7*, Article No: 21.
28. Choi, T.-M., & Lambert, J. H. (2017, August 11). *Advances in risk analysis with big data.*https://doi.org/10.1111/risa.12859
29. Razin, E. (2015, December 3). Big buzz about big data: 5 ways big data is changing finance. *Forbes*.
30. Jaweed, M. D., & Jebathangam, J. (2018). Analysis of stock market by using Big Data Processing Environment. *International Journal of Pure and Applied Mathematics, 119*(10), 81–86.

31. Mallon, S. *Big data analytics has potential to massively disrupt the stock market*. https://www.smartdatacollective.com/big-data-analytics-has-potential-to-massively-disrupt-stock-market/
32. The Pipeline, The ZoomInfo. https://pipeline.zoominfo.com/marketing/dynamic-data
33. Yang, P., & Hou, X. (2022). Research on dynamic characteristics of stock market based on big data analysis. *ResearchGate, Hindawi, Discrete Dynamics in Nature and Society, 2022*, Article ID 8758976, 1–8. https://doi.org/10.1155/2022/8758976
34. John, J., & Joseph, B. (2022). Stock price prediction using LSTM with dynamic data sets. *Proceedings of the National Conference on Emerging Computer Applications (NCECA), 4*(1), 625–628. https://doi.org/10.5281/zenodo.6938228
35. Shah, A., Patel, P., & Vora, D. (2020). Dynamic approach to stock trades using ML techniques. *International Research Journal of Engineering and Technology (IRJET), 07*(12), 608–610, e-ISSN: 2395-0056.
36. Akhtar, M. M., Zamani, A. S., Khan, S., AliShatat, A. S., Dilshan, S., & Samdan, F. (2022). Stock market prediction based on statistical data using machine learning algorithms. *Journal of King Saud University – Science, Science Direct, 34*(4), 101940.
37. Karim, R., Alam, M. K., & Hossain, M. R. (2021, August). Stock market analysis using linear regression and decision tree regression. In *IEEE Conference: 2021 1st International Conference on Emerging Smart Technologies and Applications (eSmarTA)*. https://doi.org/10.1109/eSmarTA52612.2021.9515762
38. Adusumilli, R. (2019). Machine learning to predict stock prices, Published in towards Data Science, Dec 26.
39. Abbasi, E., & Abouec, A. (2008). Stock price forecast by using neuro-fuzzy inference system. *World Academy of Science, Engineering and Technology, 46*, 320–323.
40. Visa, S., Ramsay, B., Ralescuand, A., & van der Knaap, E. (2011). Confusion matrix-based feature selection. In *Proceedings of the 22nd Midwest Artificial Intelligence and Cognitive Science Conference*, Cincinnati, pp. 16–17.
41. Jung, Y., & Hu, J. (2015). A k-fold averaging cross-validation procedure. *Journal of Nonparametric Statistics, 27*(2), 1–13.

Sandwiched Metasurface Antenna for Small Spacecrafts in IoT Infrastructure

Boutaina Benhmimou, Niamat Hussain, Nancy Gupta, Rachid Ahl Laamara, Sandeep Kumar Arora, Josep M. Guerrero, and Mohamed El Bakkali

1 Introduction

To date, thousands of very small and ultra-small spacecrafts have been launched into the low earth orbit (LEO) in order to make these low-cost engineering available and more popular for students, engineers, and researchers [1, 2]. For instance, CubeSats, as a part of miniaturized satellites, have dimensions of $10 \times 10 \times 10\,\text{cm}^3$ (1 U) and a low mass of about 1.3 kg for 1 U configuration [3, 4]. The other CubeSat configurations can be designed for low-cost space missions as multiple of 1 U structure. However, CubeSats and the smallest spacecrafts such as FemtoSats and ZeptoSats introduce major technical challenges facing the design

B. Benhmimou · R. A. Laamara · M. El Bakkali
Faculty of Sciences of Rabat, Mohammed Five University of Rabat (UM5R), Agdal, Rabat, Morocco
e-mail: boutaina.benhmimou@um5r.ac.ma; r.ahllaamara@um5r.ac.ma

N. Hussain
Department of Smart Device Engineering, Sejong University, Seoul, South Korea
e-mail: niamathussain@sejong.ac.kr

N. Gupta
ECE, LKCE, I.K.G. Punjab Technical University, Jalandhar, India
e-mail: nancygupta@lkcengg.edu.in

S. K. Arora (✉)
School of Electronics & Electrical Engineering, Lovely Professional University, Phagwara, Punjab, India
e-mail: sandeep.16930@lpu.co.in

J. M. Guerrero
Center for Research on Microgrids (CROM), Department of Energy Technology, Aalborg University, Aalborg East, Denmark
e-mail: joz@et.aau.dk

R. Sharma et al. (eds.), *Data Analytics for Internet of Things Infrastructure*, Internet of Things, https://doi.org/10.1007/978-3-031-33808-3_7

of their subsystems and hence limit their lifetimes. They usually need ingenious propulsion, attitude control, and high-performance communication systems [5–7]. This research work will focus on the part of antenna design for very small and ultra-small spacecraft communications [8]. Satellite area requires an antenna fit those dimensions, and keeping the best behavior. Antenna design for these small configurations must be of very low profile, high gains and wide bands [9]. From this point, planar antennas can be proposed as best choice for use on these emerging space missions [10–17]. They may present low-profile antenna structures, have good compactness, are easy to integrate with other component fabrication, and exhibit significant robustness [18–23]. In addition, they are cheap, have low dispersion, and present low radiation losses. However, their low gain antennas cannot cover larger distances so unable to achieve long duration space missions. To overcome these limitations of low gain, some techniques of gain enhancement can be applied in order to increase the radiated energy outside the spacecrafts and in direction of targeted earth stations.

In this research work, we used sandwiched Metasurface below a 50 Ω probe-excited, very small–crossed antenna in order to forward the back radiation and hence increase the antenna peak gain at X-band. As 2D configuration of metamaterial, metasurfaces have received great attention in recent years due to their fantastic electromagnetic characteristics. Meta-material can be used as a range of artificial metallic cells with negative permittivity and permeability in the miniaturization of antenna designs, gain and bandwidth enhancements, and so on [24]. In this study, we have discussed and analyzed the effectiveness of sandwiched metasurface for increasing the peak gain of a very small probe-excited crossed-ring antenna without increasing its physical dimensions, thereby making the proposed antenna design very suitable for very small and ultra-small spacecrafts such as FemtoSats and ZeptoSats [25]. This study also presents the advantage of minimizing interferences with other spacecraft subsystems, thereby limiting the sources of risks inside the spacecraft body.

Proposed work aims to evaluate the proposed antenna implemented with 10 × 10 small unit cells inside the antenna dielectric for improving gain and bandwidth. In addition to those, more directive radiation characteristics were also observed from the use of sandwiched metasurface antenna compared to the source antenna alone such as 3D shape of radiation pattern at the targeting operating frequencies.

This research work is organized as follows: Section 2 describes the geometry of proposed sandwiched metasurface. Section 3 summarizes and discusses the effects of sandwiched metasurface with suitability of the overall design for very small/ultra-small spacecrafts. This section, therefore, shows how the sandwiched metasurface improves the performances of proposed probe-excited crossed antenna. Section 4 concludes the study.

2 Antenna Design and Geometrical Analysis

The size of the proposed antenna is 29 × 29 mm^2 and is implemented on Rogers RO 4003 dielectric material with a relative permittivity of $\varepsilon_r = 3.55$, loss tangent of tan δ ≈ 0.002, and thickness of 2 mm. It consists of a miniaturized crossed antenna integrated with a sandwiched metasurface of 10 × 10 very small unit cells placed between the source antenna and the ground plane at a distance of 0.4 mm below the excited radiating element (see Figs. 1 and 2). The proposed sandwiched metasurface antenna is designed using ANSYS HFSS simulator and is optimized and miniaturized using Quasi Newtonian method (QNM) with a package of that electromagnetic software [26].

The source antenna is excited using 50 Ω coaxial probe whose length in 6 mm and radius 0.05 mm and is connected to the antenna geometrical center. Unit cells chosen in the design of proposed metasurface are square closed rings that can be manipulated using few parameters and therefore optimized using simple QNM program (refer to Fig. 3). The length of the optimized unit cells is 4 mm and width 2 mm, and they are arranged at an inter-cell distance of 1 mm. As shown in Fig. 3, both strips of the proposed unit cell structure are slanted by an angle of 45° in order to occupy very small area on the bodies of ultra-small and very small–sized spacecrafts such as FemtoSats. Our proposed antenna approach focuses, therefore, on the return loss and gain enhancement of a probe-excited crossed antenna at an X-band operating frequency of 8.4 GHz with the suitability for very small and ultra-small spacecrafts such as FemtoSats [27–29].

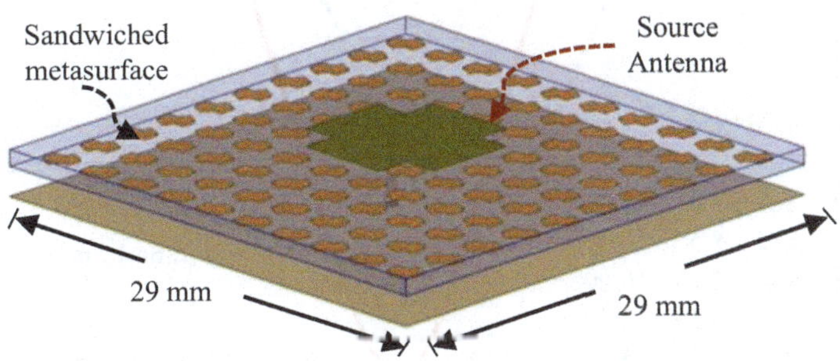

Fig. 1 3D HFSS model of the proposed antenna system

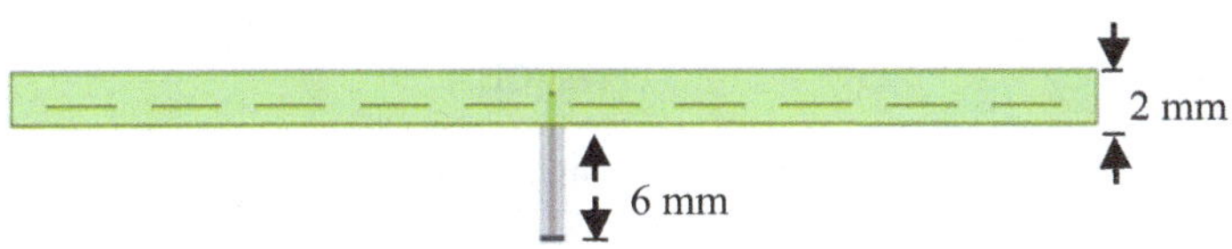

Fig. 2 Side view of the proposed crossed antenna

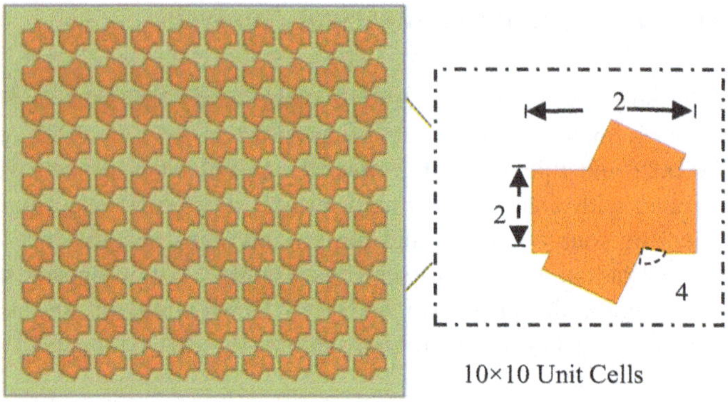

Fig. 3 Configuration of the proposed sandwiched metasurface

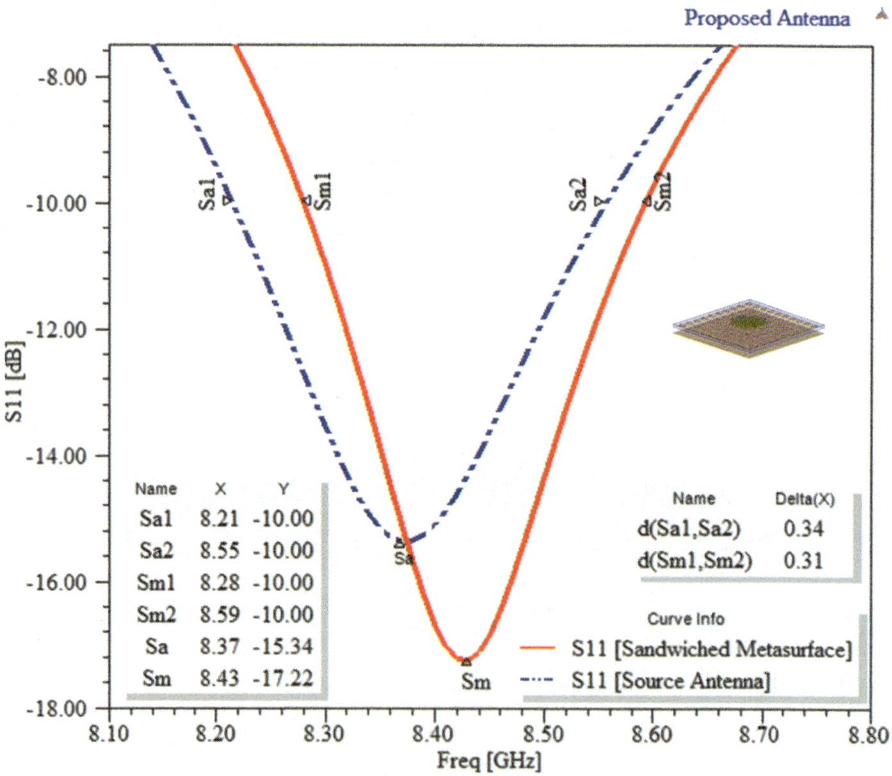

Fig. 4 Reflection coefficients of the proposed configurations versus frequency

3 Results, Data Analysis, and Discussions

As mentioned previously, this study aims to design a very small antenna system for use on very small and ultra-small spacecrafts. Therefore, the proposed antenna design faces several challenges due to the requirement of very limited geometrical size. The results of this study present a significant solution for very small and ultra-small spacecraft communications. Figures 4 and 5 depict the reflection and VSWR coefficients of both source antenna alone and being sandwiched with metasurface, respectively. It is shown that the second one minimizes the reflection coefficient (increases return loss) around an operating frequency of 8.4 GHz and gives −10 dB impedance bandwidth of about 340 MHz at X-band despite the very small area occupied by the proposed sandwiched metasurface antenna.

These results are proved by a VSWR smaller than 2.0 around the same working frequency. Therefore, the achieved bandwidths are suitable for use by very small and ultra-small spacecrafts such as FemtoSats and CubeSats where antennas are implemented outside the box. However, Fig. 6 shows that the source antenna

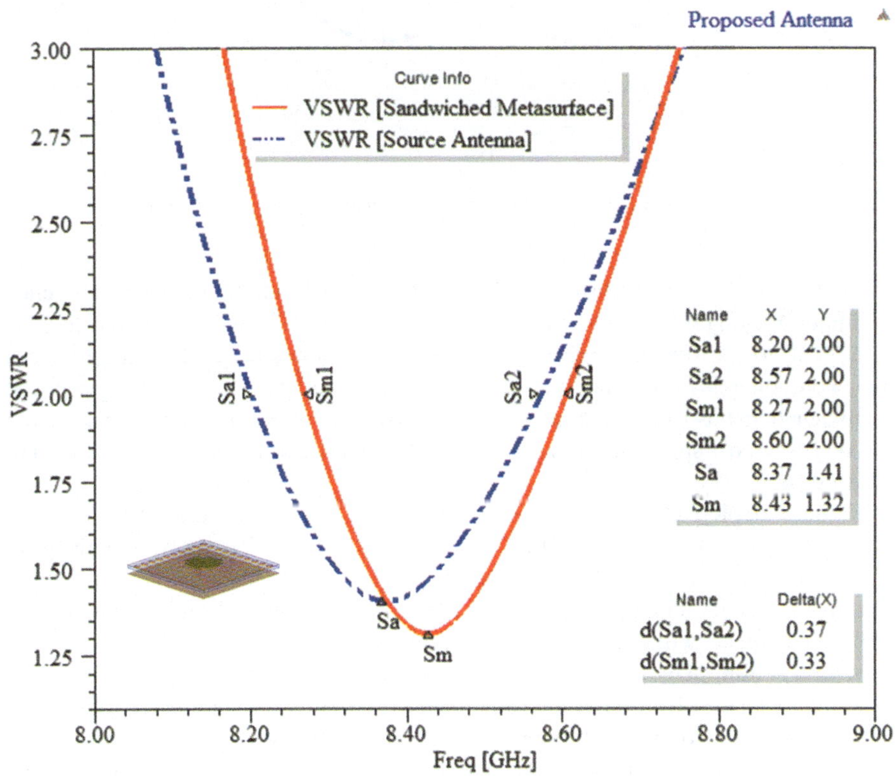

Fig. 5 VSWR coefficients of the proposed configurations versus frequency

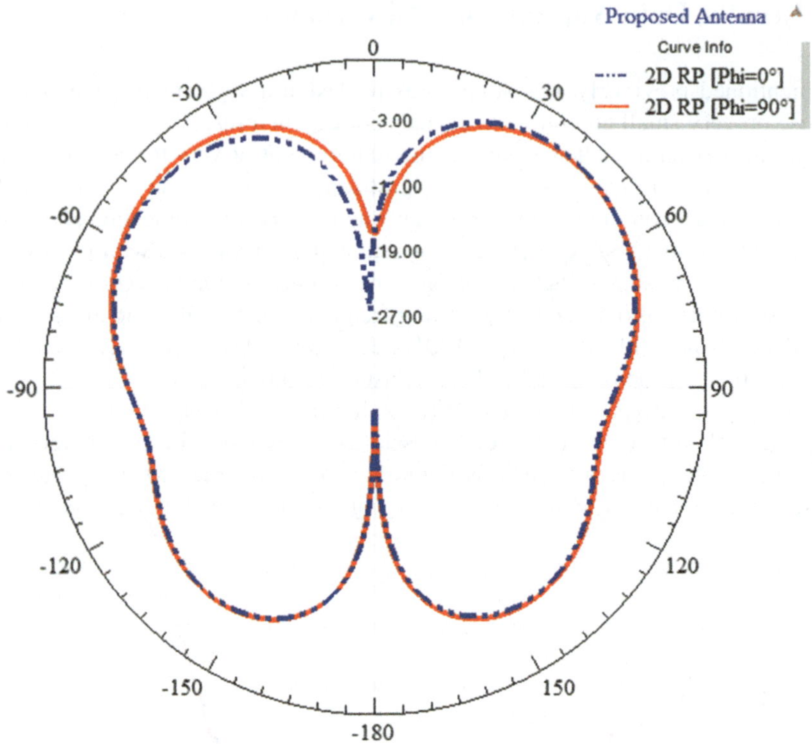

Fig. 6 2D radiation pattern of the proposed crossed patch antenna alone at 8.4 GHz

without sandwiched metasurface achieves bidirectional radiation pattern and hence an important quantity of electromagnetic energy can be used for increasing the peak gain. Moreover, the high level of back lobes generated makes interferences with components inside the spacecraft too important. Henceforth, the proposed sandwiched metasurface increases the effectiveness of proposed antenna design for very low-cost space missions such as those that use very small and ultra-small spacecrafts.

Figure 7 depicts the 2D gain of both source antenna alone and sandwiched with metasurface antenna at our operating frequency of 8.4 GHz. It is shown that the peak gain, a significant characteristic of satellite and spacecraft antenna, is significantly affected by sandwiching the proposed metasurface between source antenna and the ground plane without increasing the antenna thickness. The back-lobe level, therefore, is minimized, and so interferences are limited. The peak gain is improved from 4.67 to 5.80 dBi around the same operating frequency, and hence the chance of success of mission is much improved. This might be due to the reason that when the antenna is tuned to the resonant frequency of the sandwiched metasurface, the radiated electrical field is spread over a larger radiating aperture enhancing its gain and thereby minimizing the back-lobe radiation. This is because the meta-surface

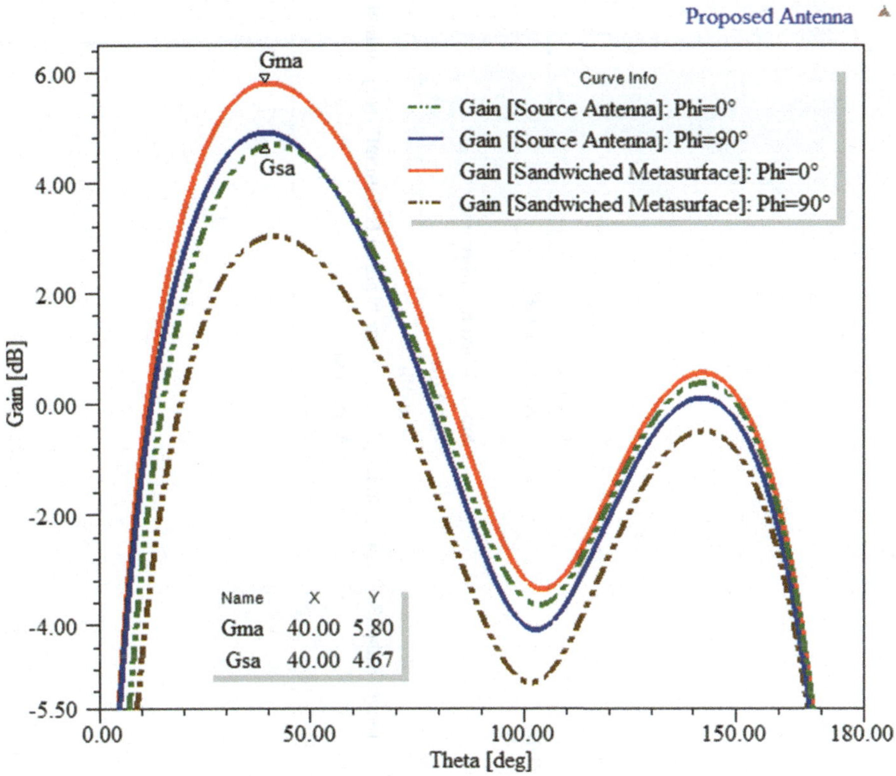

Fig. 7 Peak gains of the proposed configurations at 8.4 GHz

picks up the radiated power of the antenna and then re-radiates outside the spacecraft body.

To complement our study, in Table 1, our antenna is compared to similar CubeSat antennas designs and is found that our design can achieve the best gain at X-band.

The presented characteristics are of great importance to the very small and ultra-small spacecrafts because a high gain of almost 6.0 dBi with very large beam width angle and very high stiffness are highly suitable for these very low-cost space missions using ultra small and very small spacecrafts [32, 33].

4 Conclusions and Future Works

A sandwiched metasurface is applied in order to increase return loss and gain of a very small probe–excited crossed antenna at X-band for very small and ultra-small spacecrafts. The developed sandwiched metasurface is created through the use of 10×10 crossed unit cells which are inclined by an angle of 45° in order to order

Table 1 Comparison with FemtoSat metasurfaced antennas

Reference	[30]	[31]	**Our work**
Frequency	X-band (8.2GHz)	V-band (60 GHz)	**X-band (8.4 GHz)**
Material	FR4	FR4	**Glass**
Approach	Metasurface (10 × 10 cells)	Metamaterial (7 × 7 cells)	**Sandwiched metasurface**
Return loss	~20.0 dB	~12.0 dB	**~20.0 dB**
Gain	~5.0 dBi	×	**~6.0 dBi**
FemtoSat model	3× 3 × 3 cm^3"Indian KalamSat"-suitable-	3× 3 × 3 cm^3"Indian KalamSat"-not suitable-	**3× 3 × 3 cm^3"Indian KalamSat"-very suitable-**
Stiffness	Medium	Medium	**Very high**
Weight (mg)	High	High	**Low**

to improves each of return loss, bandwidth and peak gain without editing physical length and width of proposed source antenna. The sandwiched metasurface antenna presented in this paper achieves a wide bandwidth high gain of about 340 MHz and 6.0 dBi despite its very small physical size of 29×29 mm^2 at X-band. The proposed antenna, therefore, is ultra-lightweight, has very low profile, and is cheap and can be used for configuring very small/ultra-small spacecraft antennas.

Consequently, the use of sandwiched metasurface in the design of very small/ultra-small spacecrafts, where areas for each subsystem are too limited, has high potential and offers a cost-effective solution for these smart space missions.

If we want to communicate the proposed space craft to communicate with earth segments, one unit of proposed sandwiched metasurface antenna can be mounted per face in order to achieve coverage close to 360° around the same operating frequency.

Acknowledgments The authors would like to thank Prof. Gurjot Singh, Prof. Alexander Kogut, and Prof. Yadgar Ibrahim Abdulkarim for their valuable support in preparing this research work.

References

1. El Bakkali, M. (2020, July). *Planar antennas with parasitic elements and Metasurface superstrate structure for 3U CubeSats* [PhD. Thesis]. Sidi Mohamed Ben Abdellah University, city of Fez, Morocco.
2. Rahmat-Samii, Y., Manohar, V., & Kovitz, J. M. (2017). For satellites, think small, dream big: A review of recent antenna developments for CubeSats. *IEEE Antennas and Propagation Magazine, 59*(2), 22–30.
3. Rivera, M., & Boyle, A. (2013). Space for all: Small, cheap satellites may one day do your bidding. *NBC NEWS, Innovation, 5*, 11 am ET.
4. An international network of 50 CubeSats Project [online] Available: https://www.qb50.eu/. Accessed 10/09/2019.
5. El Bakkali, M., Bekkali, M. E., Gaba, G. S., Guerrero, J. M., Kansal, L., & Masud, M. (2021). *Fully integrated high gain S-band triangular slot antenna for CubeSat communications*. electronics.
6. Crisp, N., Smith, K., & Hollingsworth, P. (2014). Small satellite launch to LEO: A review of current and future launch systems. *Transactions of the Japan Society for Aeronautical and Space Sciences, Aerospace Technology Japan, 12*(ists29), Tf39–Tf47. https://doi.org/10.2322/tastj.12.Tf39
7. Perez, T. R., & Subbarao, K. (2016). A survey of current femtosatellite designs, technologies, and mission concepts. *JoSS, 5*(3), 467–482.
8. Tristancho, J., & Gutierrez-Cabello, J. (2011, October 16–20). A probe of concept for FEMTO-SATELLITES based on commercial-of-the-shelf. In *Digital avionics systems conference (DASC), 2011 IEEE/AIAA 30th* (pp. 8A2-1–8A2-9).
9. Sundaramoorthy, P. P., Gilland, E., & Verhoeven, C. J. M. (2010, September). Systematic identification of applications for a cluster of femto-satellites. In *61st International Astronautical Congress, Prague, Czech Republic.*
10. Mohammadi, M., Kashani, F. H., & Ghalibafan, J. (2021). Backfire-to-endfire scanning capability of a balanced metamaterial structure based on slotted ferrite-filled waveguide. *Waves in Random and Complex Media, 31*(6), 1211–1225.

11. Vo, M. T., Vo, A. H., Nguyen, T., Sharma, R., & Le, T. (2021). Dealing with the class imbalance problem in the detection of fake job descriptions. *Computers, Materials & Continua, 68*(1), 521–535.
12. Sachan, S., Sharma, R., & Sehgal, A. (2021). Energy efficient scheme for better connectivity in sustainable mobile wireless sensor networks. *Sustainable Computing: Informatics and Systems, 30*, 100504.
13. Ghanem, S., Kanungo, P., Panda, G., et al. (2021). Lane detection under artificial colored light in tunnels and on highways: An IoT-based framework for smart city infrastructure. *Complex & Intelligent Systems*. https://doi.org/10.1007/s40747-021-00381-2
14. Sachan, S., Sharma, R., & Sehgal, A. (2021). SINR based energy optimization schemes for 5G vehicular sensor networks. *Wireless Personal Communications, 127*, 1023. https://doi.org/10.1007/s11277-021-08561-6
15. Priyadarshini, I., Mohanty, P., Kumar, R., et al. (2021). A study on the sentiments and psychology of twitter users during COVID-19 lockdown period. *Multimedia Tools and Applications, 81*, 27009. https://doi.org/10.1007/s11042-021-11004-w
16. Azad, C., Bhushan, B., Sharma, R., et al. (2021). Prediction model using SMOTE, genetic algorithm and decision tree (PMSGD) for classification of diabetes mellitus. *Multimedia Systems, 28*, 1289. https://doi.org/10.1007/s00530-021-00817-2
17. Priyadarshini, I., Kumar, R., Tuan, L. M., et al. (2021). A new enhanced cyber security framework for medical cyber physical systems. *SICS Software-Intensive Cyber-Physical Systems, 35*, 159. https://doi.org/10.1007/s00450-021-00427-3
18. Priyadarshini, I., Kumar, R., Sharma, R., Singh, P. K., & Satapathy, S. C. (2021). Identifying cyber insecurities in trustworthy space and energy sector for smart grids. *Computers & Electrical Engineering, 93*, 107204.
19. Singh, R., Sharma, R., Akram, S. V., Gehlot, A., Buddhi, D., Malik, P. K., & Arya, R. (2021). Highway 4.0: Digitalization of highways for vulnerable road safety development with intelligent IoT sensors and machine learning. *Safety Science, 143*, 105407. ISSN 0925-7535.
20. Sahu, L., Sharma, R., Sahu, I., Das, M., Sahu, B., & Kumar, R. (2021). Efficient detection of Parkinson's disease using deep learning techniques over medical data. *Expert Systems*, e12787. https://doi.org/10.1111/exsy.12787
21. Ali, E. M., Soruri, M., Dalarsson, M., Naser-Moghadasi, M., Virdee, B. S., Stefanovic, C., Pietrenko-Dabrowska, A., Koziel, S., Szczepanski, S., & Limiti, E. (2022). A comprehensive survey on antennas on-chip based on metamaterial, metasurface, and substrate integrated waveguide principles for millimeter-waves and terahertz integrated circuits and systems. *IEEE Access, 10*, 3668–3692.
22. Babaeian, F., Virdee, B. S., Aïssa, S., Azpilicueta, L., See, C. H., Althuwayb, A. A., Huynen, I., Abd-Alhameed, R., Falcone, F., & Limiti, E. (2020). A comprehensive survey on "Various decoupling mechanisms with focus on metamaterial and metasurface principles applicable to SAR and MIMO antenna systems". *IEEE Access, 8*, 192965–193004.
23. Virdee, B. S., Azpilicueta, L., Naser-Moghadasi, M., Akinsolu, M. O., See, C. H., Liu, B., Abd-Alhameed, R. A., Falcone, F., Huynen, I., Denidni, T. A., & Limiti, E. (2020). A comprehensive survey of "Metamaterial transmission-line based antennas: Design, challenges, and applications". *IEEE Access, 8*, 144778–144808.
24. Holloway, C. L., Kuester, E. F., Gordon, J. A., O'Hara, J., Booth, J., & Smith, D. R. (2012). An overview of the theory and applications of metasurfaces: The two-dimensional equivalents of metamaterials. *Antennas and Propagation Magazine, IEEE, 54*, 10–35.
25. G. Minatti, M. Faenzi, , E. Martini, Caminita, F., De Vita, P., Ovejero, D. G., Sabbadini, M., and Maci, S.: Modulated metasurface antennas for space: Synthesis, analysis and realizations, IEEE Transactions on Antennas and Propagation, vol. 63, no. 4, pp. 1288–1300, 2015.
26. ANSYS HFSS simulator. [online] Available: http://www.ansys.com/products/electronics/ansys.
27. "KalamSat – World's smallest satellite built by Indian teen to be launched by NASA on June 21". Zee News. May 15, 2017. Retrieved May 16, 2017.

28. Lastovicka-Medin, G. (2016, June). Nano/pico/femto-satellites: Review of challenges in space education and science integration towards disruptive technology. In: [Online] (pp. 357–362) IEEE. Available from: https://doi.org/10.1109/MECO.2016.7525781
29. El Bakkali, M., Gaba, G. S., Tubbal, F. E., & El Idrissi, N. E. A. (2018). High gain patch antenna array with parasitic elements for CubeSat applications. In *IEEE Indian conference on antennas and propogation* (pp. 1–5).
30. Bakkali, M. E., Tubbal, F., Gaba, G. S., Kansal, L., & El Idrissi, N. A. High gain miniaturized MPA with MSS for FemtoSat communications. In A. Luhach, D. Jat, K. Hawari, X. Z. Gao, & P. Lingras (Eds.), *Advanced informatics for computing research. ICAICR 2019* (Communications in Computer and Information Science) (Vol. 1076). Springer. https://doi.org/10.1007/978-981-15-0111-1_32
31. Karthikeya, G. S., & Suraj, H. S. (2016). mmWave metamaterial inspired coaxial-fed microstrip antenna array for Femtosat. In *2016 Loughborough Antennas & Propagation Conference (LAPC)* (pp. 1–5). https://doi.org/10.1109/LAPC.2016.7807518
32. Omari, F., Hussain, N., Benhmimou, B., Gupta, N., Laamara, R. A., ..., Abdulkarim, Y. I., & El Bakkali, M. (2022, October 3rd – 5th). Only-metal ultra-small circular slot antenna for 3U CubeSats. In *13th international conference on computing, communication and technologies (13th ICCCNT)* (pp. 1–6).
33. Benhmimou, B., Hussain, N., Gupta, N., Laamara, R. A., Guerrero, J. M., ..., El Bakkali, M., Arpanaei, F., & Alibakhshikenari, M. (2022, October 3rd–5th). Miniaturized transparent slot antenna for 1U and 2U CubeSats: CRTS space missions. In *13th international conference on computing, communication and technologies (13th ICCCNT)* (pp. 1–6).

Development of Laser-Beam Cutting-Edge Technology and IOT-Based Race Car Lapse Time Computational System

B. Thiyaneswaran, E. Ganasri, A. H. Hariharasudan, S. Kumarganesh, K. Martin Sagayam, and Ahmed Alkhayyat

1 Introduction

The objective of racing contest is to determine the winner of the race. The participant who completes the race with predetermined number of lapses in a shortest time will be announced the winner of the race. Racing originated during 1920–1930s in Europe in order to entertain people. The basic set of rules for formula 1 racing was formed according to the FIA standardized racing rules in 1946. Other than the world championship series, many other non-championship races were also held after 1983. Every race has four drivers and the drivers can be switched if the race is long, and along with this, the support staff of each team will also be present who plays a vital role in the team's success.

Each formula 1 racing has its unique characteristics – has iconic tracks or street circuits, has different predetermined set of lapses, and has any specific distance. The number of lapses will be decided according to the length of the racing circuit. At present RF (Radio Frequency) transmitter and receiver are used for calculating the lapse time in a race which makes high cost for fabrication. It can be used for multi car racing but in single car racing this project can be used where it has an advantage

B. Thiyaneswaran (✉) · E. Ganasri · A. H. Hariharasudan
Department of ECE, Sona College of Technology, Salem, Tamilnadu, India

S. Kumarganesh
Department of ECE, Knowledge Institute of Technology, Salem, Tamilnadu, India

K. M. Sagayam
Department of ECE, Karunya Institute of Technology and Sciences, Coimbatore, Tamilnadu, India

A. Alkhayyat
College of technical engineering, The Islamic University, Najaf, Iraq
e-mail: ahmedalkhayyat85@iunajaf.edu.iq

R. Sharma et al. (eds.), *Data Analytics for Internet of Things Infrastructure*, Internet of Things, https://doi.org/10.1007/978-3-031-33808-3_8

over the pre-existing system. One major advantage of using embedded systems is data can be transmitted to various devices as they are connected to internet.

In this research, the lapse count, race starting time, race ending time, time taken for each lapse, and their starting and ending time can be calculated using embedded systems and can be uploaded to cloud via internet. The devices that can communicate over internet are known as Internet of Things. As mentioned earlier, time is a very important factor in racing and it should be sensed accurately. For sensing time, we used LiDAR (light detection and ranging) sensor also known as laser sensing.

Existing race car lapse calculation systems have two sensors: one in the car and the other on the track. The exact position of the vehicle is estimated using global positioning system (GPS). Unique sensors on the vehicle enable a signal back again each time when the vehicle crosses the start-finish line. There is the special type of sensors used to track the racing. The embedded system along with the sensors finds the start-finish event. The signal computes the duration of lapse time and increments the lapse count by 1. This is how the lapse and lap times are measured for a car.

Every car running on the track would have similar sensors on board, which give the relative position of the cars. On track, sensors estimate the lap times. But this is costly, and for individual race car lapse calculations, investing large amounts is unnecessary. This ideology is different from the others as it uses a distance calculating sensor to calculate the lapse and it is also precise. Also, the fabrication cost of this project is low. Moreover, LiDAR is a sensing technology and uses air as medium, helps in fast collection of data, and has high accuracy.

Furthermore, the values obtained are not influenced by the amount of light present, but exact values even in darkness and bright light can be calculated using this method. The system will be suitable for fast moving object detection and faster response rate. This system along with the instant display and also has the added advantage of updating the collected data to the cloud. The data can be updated instantly to the cloud and people from different places can view the data without any huge delay time. Thus, using LiDAR and embedded technology in racing is efficient.

2 Literature Review

A vehicle identification framework based on light detection and ranging sensor was proposed by Xianjian et al. [1], which exhibited good stability under simple working conditions and was cost-efficient. Jiaxing et al. [2] proposed a method in which vehicle speed is considered key variable for detection and validation. LiDAR technologies have significant potential in quick detection of fast-moving objects. The method suggested in [2] proposes a tracking framework from roadside LiDAR to detect and track vehicles with the aim of accurate vehicle speed estimation. Chao Deng et al. [3] proposed a model utilizing LiDAR technology for path planning and vehicle control of racing car that ensures safe driving of the car. The result of actual experiment shows that the proposed method [3] can quickly detect the target of the racing car.

Mohamed Zied Chaari et al. [4] proposed an idea of wirelessly charging Internet of Things (IoT) devices. This solution powers up sensors and devices wirelessly via radio frequency (RF) energy. Thiyaneswaran et al. [5] proposed a breathing-level monitoring process that utilizes embedded system. The way of receiving data may useful for developing proposed system. Changyoung et al. [6] proposed that wireless power transmission can be achieved through various methods, but it is important to focus on power transmission distance and efficiency. Their paper deals with multi-antenna design that focuses on the efficiency of power transmission.

Shengnan et al. [7] proposed a model to automatically observe and manipulate vehicle tracking system with time lapse, focuses on the safety of vehicles and lapse calculation. Kenneth [8] proposed a paper on the feasibility of time-lapse ground-penetrating radar for deep excavation works and to identify differences that are usually not notable by naked eye but by signal processed images. Ki-Woong Park et al. [9] have proposed a paper which deals with the problem that the fast-moving object cannot be calculated in time series. It captures the pictures in time-lapse concept which has the great advantage. It periodically captures images of specific point over an extended period and replays it quickly [10].

According to Mingcong Cao et al. [11], autonomous vehicles have widely developed with radars, camera, and LiDAR technology in recent days; however, accurate measurement of data is important for the safety of riders. Q-learning-based Gaussian mixture model is a promising solution for LiDAR fault tolerant. Emil et al. [12] stated that accurate determination of distance between objects and their shapes and sizes using current technologies is highly challenging. This can be overcome by LiDAR system, which utilizes the principle of measuring the time of flight of an optical signal to efficiently calculate the distance between the sensor and object in an effective manner [13].

Kiho Lim et al. [14] stated that for an advanced autonomous driving and vehicular ad-hoc networks, it is important to have effective vehicle-to-vehicle communication. Sharing traffic information collected by sensor with improves driving safety. In this paper LiDAR technology is used in v2v communication. Shuo Gu et al. [15] stated that most of the existing vehicle detection models are of single modal based on either LiDAR or camera. They used multi-modal-based LiDAR-camera fusion to increase the performance of vehicle detection, which works only in the presence of day light, whereas Lidar can work even in the absence of day light [16].

According to Jing Chen et al. [17], autonomous driving has become remarkable in industrial sectors and for personal uses. Using a greater number of LiDARs in order to reduce the blind spot of the LiDAR is cost-effective. So, based on the kinetic behavior of the vehicle, dynamic analysis was performed and the angle of LiDAR detection changes with the rotation of the steering wheel. Jing Huang et al. [18] proposed that since autonomous driving vehicles have been commercialized, it is important to promote more efficient and safer autonomous driving technologies.

LiDAR technologies is one of the most effective sensors utilized for the lane detection of roads and road curbs. To reduce the trade-off between time consumption and object detection, LiDAR technologies can be used [19].

Tao Yang et al. [20] proposed that LiDAR is one of the important technologies used in autonomous vehicles. Thus, as a critical sensor, LiDAR needs to work in terrible weather condition such as rain, fog, and snow. Popular near-infrared (NIR) ToF LiDAR also can be used to obtain accurate results in extreme weather conditions. Birgit Schlager et al. [21] have proposed the fault detection algorithm based on LiDAR sensor used in autonomous driving vehicle. They used an automotive LiDAR sensor to calculate the deviation between the actual distance and the ideal plane representing the target [22].

Lane information is an essential part of high-resolution traffic data, which was proposed by Ciyun Lin et al. [23]. High-density onboard LiDAR cannot be used to process low-density road-side data. Using ground recognition and lane marking point extraction, low density data can also be measured in autonomous driving. Jasmina Zubaca et al. [24] have proposed the system that state the capabilities of the algorithm can be applied to any race track, regardless of their curves, shape of the track, gate position, width etc. It also has the advantage of low computational effort, which enables fast tuning during racing events [25].

Alexander Liniger et al. [26] proposed three different designs to avoid collision accidents during racing games. In the first method, the collision avoidance constraints are followed only by the follower; in the second, the players are conscious about the collision constraints; and in the third, the game is designed to promote blocking. This research shows that the proposed games can have different racing behaviors and generate interesting racing situations. Leyao Huang [27] have proposed the LiDAR-based simultaneous localization and mapping (LiDAR-SLAM). It uses the sensors to build the map of the surrounding environment and observing the environmental features. Localization with high accuracy and practicability is a complex and hot issue in recent years [28].

3 Proposed Method

The method proposed in this chapter is applicable when one car is raced individually and its lapse count needs to be calculated. It can also be used to calculate the speed of the racing car. This system uses LiDAR technology and is advantageous compared to other systems in terms of accuracy and fast detection of objects. The data acquired from racing can also be displayed instantly and can be immediately updated to the cloud so that people from different places can access and view the data. LiDAR technology is advantageous compared to other preexisting techniques such as manual method used to calculate lapses in racing, ultrasonic sensors, fully automatic timing (FAT), radio frequency identification (RFID), and IR transmitter and receiver. The lapse cutting edge calculation of race car performed using laser beam is more efficient compared to other technologies. Unlike other forms of lights such as flashlights, the beam from laser lights stays focused and will not spread out. Thus, laser beams are very narrow, bright, and can travel very long distances. The bunch of laser light waves travel together with their peaks all lined up, or in phase [29].

3.1 Block Diagram

The block diagram of the proposed system is shown in Fig. 1. The microcontroller used for this project is Arduino UNO board, which is an open-source platform used for building embedded projects. It has an Integrated Development Environment (IDE) along with physical programmable circuit [30]. The sensor used for tracking the car is LiDAR (Light Detection and Ranging) sensor, which measures the time difference between the transmitted and reflected pulses, which is used to calculate the distance between the obstacle and the sensing sensor. LiDAR is one of the important components in this project because it emits laser beam which expands less with travel distance. The sensor transmits and receives laser pulse in nanoseconds, so a very large amount of data can be fetched in a very short period of time, which is one of the major benefits of using this sensor in racing.

$$D_{\mathrm{s}} = c \times \frac{T}{2} \tag{1}$$

D_{s} is the distance between the sensor and object.
c is the speed of light in vacuum.
T is the measured time between emitting and receiving the signal.

The LiDAR sensor is a popular sensor that enables the user to know the exact distance of the object present on the surface of the earth. The LiDAR follows a basic principle in which it emits the laser light into the environment which hits an obstacle

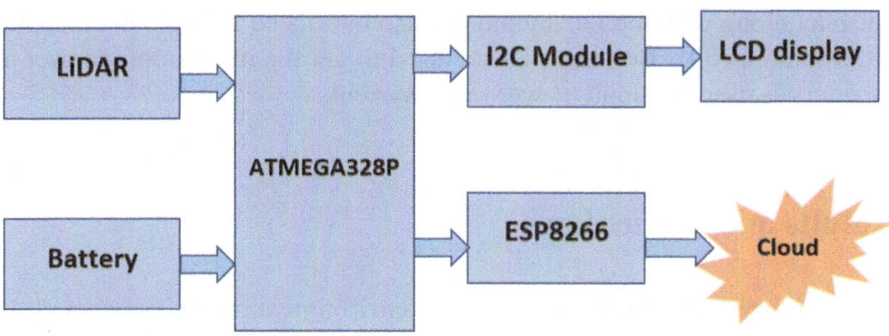

Fig. 1 Block diagram of developed system

on the surface of the earth. The time duration it takes to return to the receiver part of the LiDAR sensor is computed to find the distance. The distance can be calculated using Eq. 1. The performance of the sensor depends on the speed at which the light travels which is about 186,000 miles per second, and the process of measuring the exact distance by LiDAR seems to be unbelievably faster than any other existing technologies.

The LiDAR sensor operates in a range of 0.1–12 m, needs a voltage supply of minimum 5 volts, and has a frame rate of 10 to 1000 Hz. This sensor can be used for obstacle detection, obstacle avoidance, assisted landing, terrain following, vehicle position sensing, etc. It is compact in size and is light in weight. An LCD screen is used to display the output, if this is furnished in real entity then TFT display or any other large display can be used. The data are then uploaded to the google cloud so that people from different places can access them.

When the process is initiated, the LiDAR sensor starts working and gets inputs. When the car cuts the laser beam, the lapse count gets incremented and this will continue until the total number of lapses required to complete the race is reached. Once the race is completed, the LCD displays the output. Initially, two push buttons are used to get the total number of lapse count needed to complete the race.

The flowchart of the proposed system is shown in Fig. 2. It consists of control switch. When the control switch is turned on, the other switch starts taking input. Each time when it is pressed, the lapse count value gets incremented. After counting the total number of lapses, turn off the control switch. Now the microcontroller will have the total number of lapse counts needed to complete the car race.

3.2 Simulation of Proposed System

The simulation of the proposed system is shown in Fig. 3. Figure 3 shows the circuit diagram of the system and is a simple representation of the components of an electrical circuit. It also shows the relative position of all the elements and their connections to one another. Arduino UNO is the central unit of the circuit, which is the microcontroller used in this system. A LiDAR sensor is connected to the system and is used to measure the distance between the car and the sensor. An LCD display along with the I2C module is also interfaced with the microcontroller board (Arduino UNO). Two switches are used to get the input from the user and a 9-volt battery is used to supply power to the system.

4 Results and Discussion

The developed system is tested in a practical environment before launching it in the market to ensure that the system is free of error.

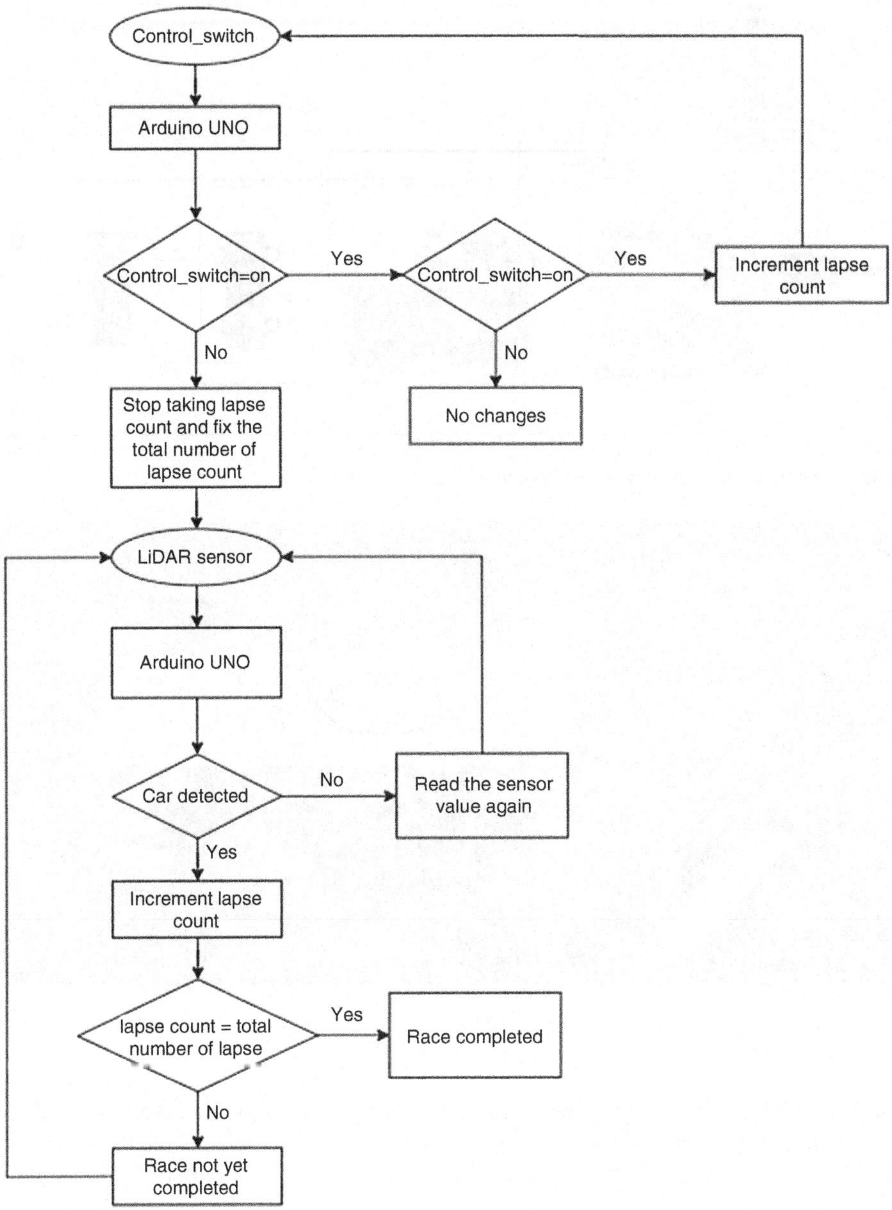

Fig. 2 Flowchart of the system

The hardware of the system shown in Fig. 4 contains two switches for getting the final lapse count needed to end the race. The switches are connected to the resistor for regulating the power supply. Arduino UNO board is used as a microcontroller

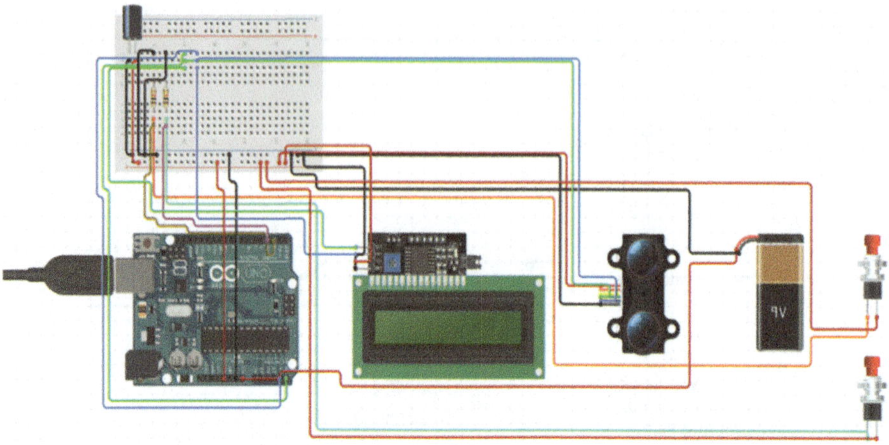

Fig. 3 Simulation of the developed system

Fig. 4 Hardware implementation of the system

and an LCD display is connected to the I2C module. I2C module is used to establish communication between two or more ICs. This display overcomes the drawback of LCD 16 × 2 parallel LCD display, which intern reduces the number of pins used in Arduino.

The output of the proposed system is shown in Fig. 5. First, the system instructs the user to give the total lapse count (Fig. 5a), on which the user starts entering the lapse count by pressing the push button. Afterward, the count gets incremented and the total lapse count is displayed for reference (see the count 7 shown in Fig. 5b). When the car starts moving, the system starts calculating the lapse count. Every time a lapse is completed, it is displayed on the LCD along with the duration of that lapse.

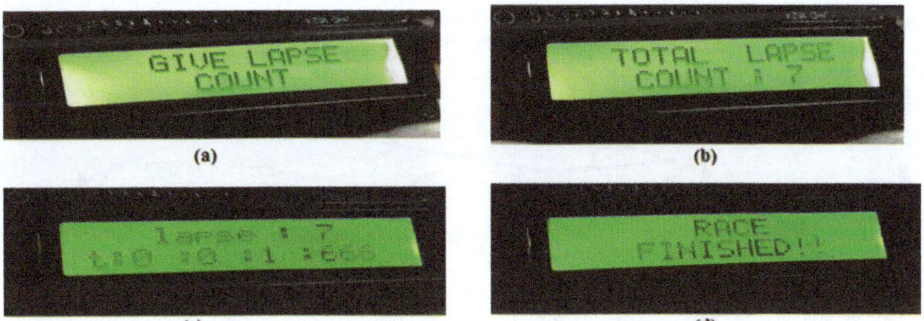

Fig. 5 Display Output view of proposed system

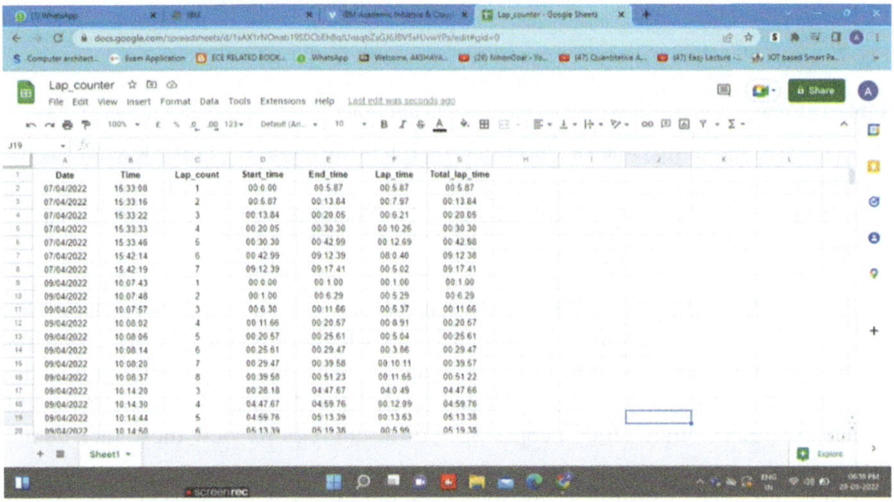

Date	Time	Lap_count	Start_time	End_time	Lap_time	Total_lap_time
07/04/2022	15:33:08	1	00:0.00	00:5.87	00:5.87	00:5.87
07/04/2022	15:33:16	2	00:5.87	00:13.84	00:7.97	00:13.84
07/04/2022	15:33:22	3	00:13.84	00:20.05	00:6.21	00:20.05
07/04/2022	15:33:33	4	00:20.05	00:30.30	00:10.26	00:30.30
07/04/2022	15:33:46	5	00:30.30	00:42.99	00:12.69	00:42.98
07/04/2022	15:42:14	6	00:42.99	09:12.39	08:0.40	09:12.38
07/04/2022	15:42:19	7	09:12.39	09:17.41	00:5.02	09:17.41
09/04/2022	10:07:43	1	00:0.00	00:1.00	00:1.00	00:1.00
09/04/2022	10:07:48	2	00:1.00	00:6.29	00:5.29	00:6.29
09/04/2022	10:07:57	3	00:6.30	00:11.66	00:5.37	00:11.66
09/04/2022	10:08:02	4	00:11.66	00:20.57	00:8.91	00:20.57
09/04/2022	10:08:06	5	00:20.57	00:25.61	00:5.04	00:25.61
09/04/2022	10:08:14	6	00:25.61	00:29.47	00:3.86	00:29.47
09/04/2022	10:08:20	7	00:29.47	00:39.58	00:10.11	00:39.57
09/04/2022	10:08:37	8	00:39.58	00:51.23	00:11.65	00:51.22
09/04/2022	10:14:20	3	00:28.18	04:47.67	04:0.49	04:47.66
09/04/2022	10:14:30	4	04:47.67	04:59.76	00:12.09	04:59.76
09/04/2022	10:14:44	5	04:59.76	05:13.39	00:13.63	05:13.38
09/04/2022	10:14:50	6	05:13.39	05:19.38	00:5.99	05:19.38

Fig. 6 Output in google spread sheet

Once the lapse count gets incremented and reaches the final count, the race gets completed. Fig. 5c shows the display of seventh race lapse. Finally, the completion status of race is displayed on the LCD display. Fig. 5d shows the finished status. The whole procedure is repeated once the system is restarted.

Figure 6 shows the data of the race acquired using the proposed model, i.e., car-lapse cutting-edge technology using laser beam. These data are instantly uploaded to cloud, preferably google sheets, for ease of access, which contains the date and time at which the race took place, the race lapse count, its starting and finishing time, and the duration of each lapse.

Each trail is set with a specific lapse count. The graph is plotted between time in minutes and lapse count. Figure 7 shows the lapse count and time taken for each lapse in three different trails.

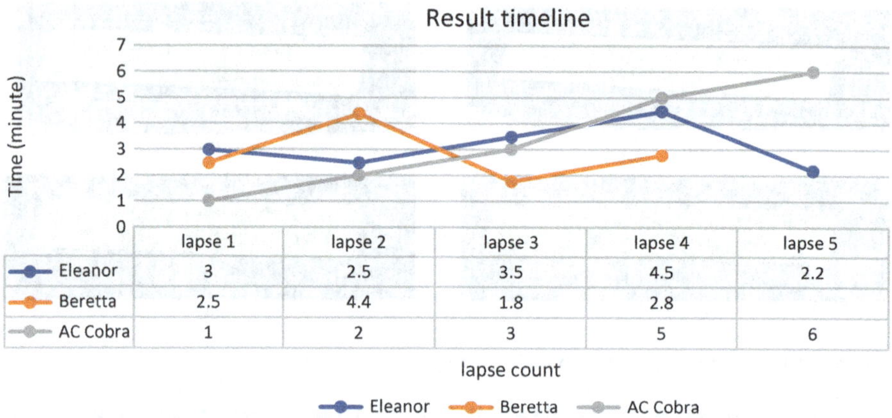

Fig. 7 Sample timeline and table of results

Table 1 Comparison between manual method and ultrasonic and LiDAR lapse technologies

	Manual calculation	Ultrasonic sensor	LiDAR technology
Max distance coverage	Not accurate	4 m	12 m
Time delay	More than a second	20–50 milliseconds	1–10 nanoseconds

Table 1 shows the comparison between manual calculation, ultrasonic sensor, and LiDAR sensor. The table shows that LiDAR technology is superior compared to the other two methods in terms of time delay, accuracy, and maximum distance coverage. This data is instantly uploaded to cloud without any time delay; a link is then created through which people from different regions view these race details precisely.

Using ultrasonic sensor for calculating racing lapses is not a good idea. As the ultrasonic sensor works on the principle of emitting sound waves at very high frequency in which cannot hear it and receives then back. The time gap between transmitting and receiving a signal is used to compute the distance between the sensor and the object. Here, since the signal is a sound wave, a time delay is noted. This can be overcome by utilizing LiDAR sensors, in which the laser beams emitted are used to calculate the distance between the sensor and the object (in nanoseconds of time delay).

Another method used in vehicle tracking is IR ID chip which acts as the key of IR sensor as it has the car identification number. Each IR sensor is fit in the car and it is embedded with one IR ID chip which has the identification number of the vehicle. It also has an RF transmitter that frequently transmits a RF signal through an antenna toward the direction in which the car is moving. The receiver in the control unit fit at a specific point in the race track accepts the signal transmitted by the transmitter and sends the data to the system present in the nearby control room for getting the full details such as lapse count, lapse time etc.

Table 2 Comparison of lapse values calculated using IR transmitter and LiDAR technology

Lapse count	Lap time using IR transmitter (minutes)	Time delay (seconds)	Lap time found using LiDAR technology (minutes)	Time delay (nanoseconds)
1	2:30.18	0.1	2:30.17	0.1
2	3:42.11	0.2	3:42.09	0.2
3	4:11.23	0.2	4:11.21	0.2
4	5:21.33	0.1	5:21.32	0.1
5	6:18.09	0.2	06:18.07	0.1
6	7:37.17	0.1	07:37.16	0.1

Table 3 Comparison of recent technologies and LiDAR technology

Features	Radio frequency identification (RFID)	IR transmitter and receiver	LiDAR technology
Reader range	0–2 m	0–5 m	0–12 m
Integration	Difficult	Difficult	Easy
Memory storage	No	Yes	Yes
IOT enabling	No	Yes	Yes
Response	Identification	Identification + positioning	Identification + positioning + speed
Cost of fabrication	Expensive	Expensive	Affordable

This system has an advantage when it comes to multiple car tracking, but this system is not accurate for single cars and cannot sense fast-moving vehicles. Table 2 compares the lapse values calculated using IR transmitter and LiDAR technology . It is to be noted that LiDAR technology calculates time delay in nanoseconds, whereas IR transmitter and receiver calculates the time delay in seconds.

Table 3 shows a comparison of the recent technologies and LiDAR technologies and shows why LiDAR technologies are advantageous than the other two technologies. An RFID tag, a microchip, is attached to the car to track the position of the vehicle. The tag picks up the signal from an RFID reader and scanner and then returns the signal, usually with some additional information such as start and finish time of the racing vehicle. The main disadvantage of this system is it cannot track fast-moving vehicle like race cars. These show that LiDAR technology more efficiently detects fast-moving objects like racing car with less time delay compared to other existing technologies.

Figure 8 shows a comparison of the efficiency of different technologies used for lapse calculation. The time delay, i.e., the minimum time the object should be present in the sensor range so that the sensor can track its presence, and accuracy of the object are represented in the chart given in Table 4. Radio Frequency Identification (RFID) is an advanced technology that uses wireless transformation of data between the tag or car and a reader or device to automatically track or identify the physical location of each object. The system's transmission range is restricted to few meters from the reader and the tag should be in clear line of sight.

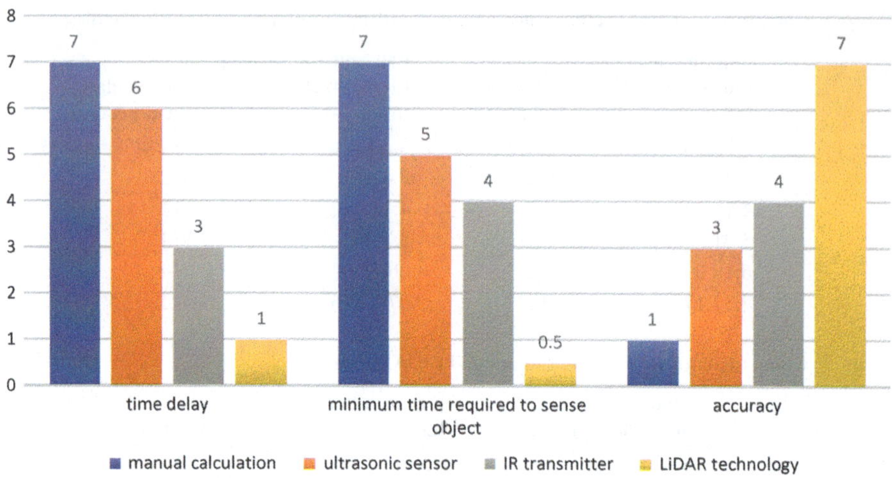

Fig. 8 Comparison of different technologies utilized for lapse calculation

Table 4 Data obtained using the proposed system

S.no	Date	Time	Lap count	Start time	End time	Lap time	Total lap time
1	17/10/2002	15:33:08	1	00:0.00	00:5.87	00:5.87	00:5.87
2	17/10/2002	15:33:16	2	00:5.87	00:13.84	00:7.97	00:1.84
3	17/10/2002	15:33:22	3	00:13.84	00:20.05	00:6.21	00:20.05
4	17/10/2002	15:33:33	4	00:20.05	00:30.30	00:10.26	00:30.30
5	17/10/2002	15:33:46	5	00:30.30	00:42.99	00:12.69	00:42.98
6	17/10/2002	15:42:14	6	00:42.99	09:12.39	08:0.40	09:12.38

Table 4 shows the date at which the event occurred, time, lapse count, start and end time of each lapse, duration of each lapse, and the total race time calculated with this system.

The accuracy and efficiency of the developed system were analyzed by comparing it with a fully automatic timing (FAT) system. Fully automatic timing (FAT) is a popular racing timing system that helps obtain accurate race results, i.e., accurate to 0.01 of a second. This system needs a start signal, running time, and capture device to be digitally synchronized to ensure accuracy. This system is designed in such a way that it is activated automatically by an initiation signal rather than manually. The start signal is generated by a start sensor integrated with a gun that is used to start the race. The finish signal is generated by a ribbon or string and is recorded electronically to remove any human error. The results obtained using the developed system is shown in Table 5, which show that LiDAR technology is accurate and does not need any external activation system for initiation like FAT.

This work addresses the common difficulties encountered during lapse calculation. The proposed system enables uninterrupted monitoring and storing of race-related information, and the data obtained are displayed on an LCD display

Table 5 Accuracy of the developed system compared to FAT

Lapse count	FAT	LiDAR technology
1	00.5.11	00:5.11
2	00:6.45	00:6.45
3	00:4.98	00:4.97
4	00:5.21	00:5.21
5	00:6.11	00:6.10
6	00:5.55	00:5.55
7	00:6.32	00:6.31

for instant viewing. This model is cheap, efficient, and accurate in terms of time measurements.

5 Conclusion

The proposed system was developed by utilizing laser beam-based LiDAR. ATMEGA328 controller was used to access LiDAR technology. When the racing element crossed the beam, LiDAR was activated by triggering. In this work, a laser beam–based cutting-edge technology for calculating race car lapses was fabricated and tested. The results of this system were accurate, and the system was able to detect time delay even in nanoseconds in comparison with fully automatic timing (FAT) system. The data obtained were also uploaded to cloud using Internet of Things (IOT). LiDAR technology finds many advantages in racing sector where even fast moving objects can be detected effectively. The developed system can track a maximum of 12 m track width with maximum 10 ns computation.

References

1. Jin, X., Yang, H., & Li, Z. (2022, January). Vehicle detection framework based on LiDAR for autonomous driving. In *5th CAA international conference on vehicular control and intelligence (CVCI)*.
2. Zhang, J., Xiao, W., Coifman, B., & Mills, J. P. (2020). Vehicle tracking and speed estimation from roadside Lidar. *IEEE Journal of Selected Topics in Applied Earth Observations and Remote Sensing, 13*, 5597.
3. Deng, C., Liu, G., Jia, A., Wen, X., Ma, K., & Ying, B. (2021, September). Study on LiDAR obstacle detection for FSAC racing car. In *4th international conference on intelligent autonomous systems (ICoIAS)*.
4. Chaari, M. Z., & Al-Rahimi, R. (2021, May). Energized IoT devices through RF wireless power transfer. In *International symposium on electrical and electronics engineering (ISEE)*.
5. Thiyaneswaran, B., Bhuvaneshwaran, V., Dharun, M., Gopu, K., & Gowsikan, T. (2020). Breathing level monitoring and alerting by using embedded IOT. *Journal of Green Engineering, 10*(6), 2986–2994.
6. An, C., & Ryu, H.-G. (2020, December). Multiple antennas design for the RF wireless power transfer system. In *IEEE wireless power transfer conference (WPTC)*.

7. Dong, S., Liu, X., Lin, Y., Arai, T., & Kojima, M. (2018, October). Automated tracking system for time lapse observation. In *IEEE international conference on mechatronics and automation (ICMA)*.
8. Kenneth, P. A. K. C. W., Ren, G., Li, J., & Lai, W. W.-L. (2018, August). Feasibility study of time lapse ground penetrating radar as monitoring measures for deep excavation works. In *17th international conference on ground penetrating radar (GPR)*.
9. Park, K.-W., Choi, D., & Jeon, W.-J. (2018, November). Applying time-lapse concepts onto storage system for long-term system trace analysis: Technical challenges and blueprints. In *IEEE first international conference on artificial intelligence and knowledge engineering (AIKE)*.
10. Thiyaneswaran, B., Anguraj, K., Kumarganesh, S., & Thangaraj, K. (2020). Early detection of melanoma images using gray level co-occurrence matrix features and machine learning techniques for effective clinical diagnosis. *International Journal of Imaging Systems and Technology, 31*(2), 682–694.
11. Cao, M., Wang, R., Chen, N., & Wang, J. (2021). A learning-based vehicle trajectory-tracking approach for autonomous vehicles with LiDAR failure under various lighting conditions. *IEEE/ASME Transactions on Mechatronics, 27*(2), 1011.
12. Tudor, E., Vasile, I., Popa, G., & Gheti, M. (2021, May). LiDAR sensors used for improving safety of electronic-controlled vehicles. In *12th international symposium on advanced topics in electrical engineering (ATEE)*.
13. Sharma, R., Gupta, D., Polkowski, Z., & Peng, S.-L. (2021). Introduction to the special section on big data analytics and deep learning approaches for 5G and 6G communication networks (VSI-5g6g). *Computers & Electrical Engineering, 95*, 107507. https://doi.org/10.1016/j.compeleceng.2021.107507. ISSN 0045-7906.
14. Lim, K., & Tuladhar, K. M. (2019, February). LIDAR: Lidar information based dynamic V2V authentication for roadside infrastructure-less vehicular networks. In *16th IEEE Annual Consumer Communications & Networking Conference (CCNC)*.
15. Gu, S., Yang, J., & Kong, H. (2021, October). A cascaded LiDAR-camera fusion network for road detection. In *IEEE international conference on robotics and automation (ICRA)*.
16. Singh, P. D., Dhiman, G., & Sharma, R. (2022). Internet of things for sustaining a smart and secure healthcare system. *Sustainable Computing: Informatics and Systems, 33*, 100622. https://doi.org/10.1016/j.suscom.2021.100622. ISSN 2210-5379.
17. Chen, J., Zhang, H., Lu, Y., & Zhang, Q. (2020, December). The research on control and dynamic property of autonomous vehicle adaptive Lidar system. In *International conferences on internet of things (iThings) and IEEE green computing and communications (GreenCom) and IEEE cyber*.
18. Huang, J., Choudhury, P. K., Yin, S., & Zhu, L. (2021). Real-time road curb and lane detection for autonomous driving using LiDAR point clouds. *IEEE Access, 9*, 144940.
19. Sharma, R., & Arya, R. (2021). A secure authentication technique for connecting different IoT devices in the smart city infrastructure. *Cluster Computing, 25*, 2333. https://doi.org/10.1007/s10586-021-03444-8
20. Yang, T., Li, Y., Ruichek, Y., & Yan, Z. (2021). Performance modeling a near-infrared ToF LiDAR under fog: A data-driven approach. *IEEE Transactions on Intelligent Transportation Systems, 23*(8), 11227.
21. Schlager, B., Goelles, T., & Watzenig, D. (2021). Effects of sensor cover damages on point clouds of automotive Lidar. *IEEE Sensors*.
22. Sharma, R., & Arya, R. (2021). Secure transmission technique for data in IoT edge computing infrastructure. *Complex & Intelligent Systems, 8*, 3817. https://doi.org/10.1007/s40747-021-00576-7
23. Lin, C., Guo, Y., Li, W., Liu, H., & Dayong, W. (2021). An automatic lane marking detection method with low-density roadside LiDAR data. *IEEE Sensors Journal, 21*(8), 10029.
24. Sharma, R., Kumar, R., Sharma, D. K., et al. (2021). Water pollution examination through quality analysis of different rivers: A case study in India. *Environment, Development and Sustainability, 24*, 7471. https://doi.org/10.1007/s10668-021-01777-3

25. Rai, M., Sharma, R., Satapathy, S. C., et al. (2022). An improved statistical approach for moving object detection in thermal video frames. *Multimedia Tools and Applications, 81*, 9289. https://doi.org/10.1007/s11042-021-11548-x
26. Ha, D. H., Nguyen, P. T., Costache, R., et al. (2021). Quadratic discriminant analysis based ensemble machine learning models for groundwater potential modeling and mapping. *Water Resources Management, 35*, 4415. https://doi.org/10.1007/s11269-021-02957-6
27. Dhiman, G., & Sharma, R. (2021). SHANN: An IoT and machine-learning-assisted edge cross-layered routing protocol using spotted hyena optimizer. *Complex & Intelligent Systems, 8*, 3779. https://doi.org/10.1007/s40747-021-00578-5
28. Verma, R., & Sharma, R. (2022). Dual notched conformal patch fed 3-D printed two-port MIMO DRA for ISM band applications. *Frequenz, 76*, 287. https://doi.org/10.1515/freq-2021-0242
29. Sharma, N., & Sharma, R. (2022). Real-time monitoring of physicochemical parameters in water using big data and smart IoT sensors. *Environment, Development and Sustainability*. https://doi.org/10.1007/s10668-022-02142-8
30. Balashanmugam, T., Sengottaiyan, K., Kulandairaj, M. S., & Dang, H. (2022). An effective model for the iris regional characteristics and classification using deep learning Alex network. *IET Image Processing, 00*, 1–12. https://doi.org/10.1049/ipr2.12630

A Study of Cloud-Based Solution for Data Analytics

Urvashi Gupta and Rohit Sharma

1 Introduction

In this era, organizations worldwide have been integrating various new cloud technologies for digitizing, analyzing data, and ultimately gain the predictions and insights that drive the businesses. New opportunities in various fields of healthcare, e-commerce, and marketing can be seen with the changing time as Internet of Things is being integrated along with big data analysis and cloud computing [1]. All firms nowadays need to be computerized, but big investments in infrastructure, IT resources (strong connections, computers, programs, and memory), and IT employees to administer them are not feasible. Cloud computing offers a Web framework for autonomously gathering, using, and managing computational resources [2]. Through its tightly connected resources and equipped data centers, cloud computing offers a wide range of services to users and businesses. Available resources may be dynamically provided to end users to meet their demands. Users or organizations can use the whole pool of compute resources without worrying about the source, scalability, or resource constraints of cloud services. Hence, consumers and organizations do not need to make any initial investments to access cloud resources; instead, they only need to pay for the services they use [3, 4].

Combining technology and medicine has been facilitated by cell phones and IoT (Internet of Things) that works at a very fast pace and keeps on developing the healthcare being given to people around the world [5]. Data is proving to be a highly efficient and effective tool in improving health. The advancements in the field of data science and data analytics using AI/ML are being used to achieve adherence and success in the field of healthcare, telecommunication, BFSI, retail trade, and

U. Gupta · R. Sharma (✉)
Department of Electronics and Communication Engineering, SRM Institute of Science and Technology, Delhi NCR Campus, Ghaziabad, India
e-mail: ug3398@srmist.edu.in

R. Sharma et al. (eds.), *Data Analytics for Internet of Things Infrastructure*,
Internet of Things, https://doi.org/10.1007/978-3-031-33808-3_9

e-commerce [6]. This real-world data coming from multiple heterogeneous sources poses significant challenges in their storage, analysis, and study. Considering the volume (where and how to store the huge data flowing from multiple sources), variety (data flowing in all forms, structured and unstructured, etc.), and velocity (the speed required for the data to be stored and processed) of this data [7], it is important to use new-age components of cloud computing and big data technologies for processing [8]. Cloud operators provide a catalogue of components to build an end-to-end encrypted data platform for ingesting, storing, and analyzing, but there is always a question of which tools and technologies are to be provisioned and how the architecture will be built using all these components.

The area of cloud computing has had remarkable expansion in recent years; according to Gartner [9], public cloud is anticipated to reach $411 million by the end of 2020. As more and more businesses migrate to the cloud, it is getting harder to choose a cloud service provider that will be suitable for respective needs over the long run from a wide pool of cloud service providers. There are numerous cloud service providers now, but there is no common standard and their growth is happening in parallel. Many of these providers are focused on computing power and offering end customers CPU, storage, database, and networking services. Some network operators prioritize cost cutting, while the others prioritize continuous service quality and adaptability. These many aspects have made it extremely difficult to select an appropriate service provider depending on the needs of the customer or company [10, 11].

Section 2 of this chapter presents the methodology required for data analytics. This section discusses architecture of AWS and GCP-based platform of the overall data analytics solution for business industry. Section 3 of this chapter offers a comparative analysis of features offered by the major two-cloud service providers – Amazon (AWS) and Google's (GCP) cloud computing platform.

Section 4 covers the major challenges faced in data analytics using different cloud platforms, to handle, understand, and reconcile the data flowing from structured and unstructured source systems for adoption of cloud-based data analytics framework for improving the business services. Section 5 of this chapter is the conclusion.

2 Methodology

The cloud platform plays vital role for data ingestion. In this section, the architecture of data informatics platform using AWS and GCP is discussed. A standard data analytics platform can be built using multiple services offered by AWS and GCP cloud platform. Figure 1 illustrates a high-level architecture of a data analytics system. This figure shows the various stages of data input from data source till the final visualization layer available to the end users of the analytics system.

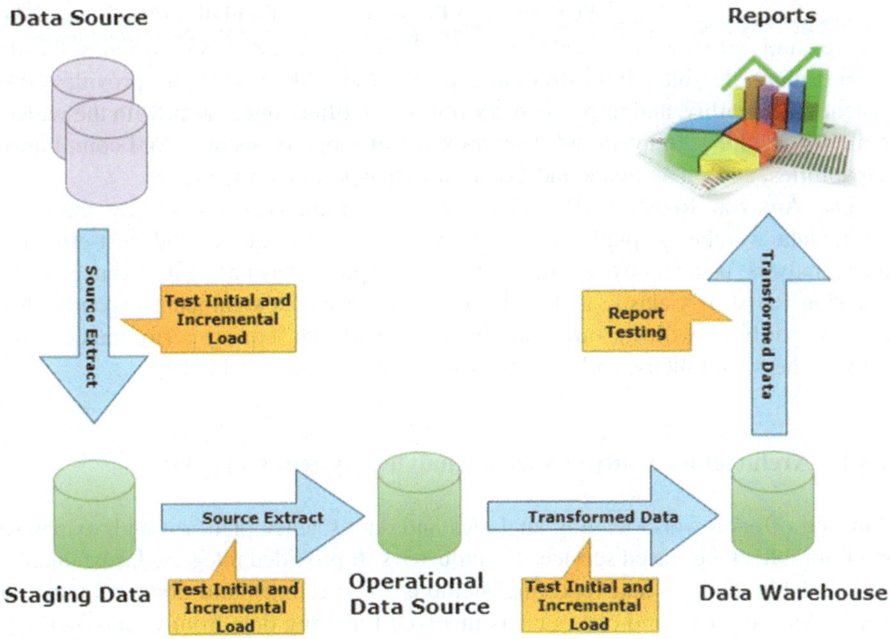

Fig. 1 Data visualization block diagram

2.1 *Amazon Web Services (AWS) Cloud Platform for Data Analytics*

With over 200 fully featured services accessible from data centers all over the globe, Amazon Web Services (AWS) is the most complete and commonly utilized cloud platform in the world. Millions of customers, including the fastest-growing start-ups, biggest enterprises, and top government agencies, rely on AWS to save costs, enhance agility, and speed up innovation. AWS has more services and features offered by any other cloud provider, including traditional infrastructure technologies such as computation, storage, and databases, as well as emerging technologies such as machine learning and artificial intelligence, Data Lake and analytics, and the Internet of Things. This enables migrating existing programs to the cloud quicker, simpler and less expensive, as well as building practically anything. AWS offers a variety of big data analytics frameworks that allow you to quickly and easily create and deploy big data analytics applications.

Big Data Analytics Framework (Amazon EMR) for running large-scale distributed data processing jobs delivers capabilities like real-time data processing, clickstream analytics and big data analytics. It has built-in connectors with other AWS services [12] that is simple to use, has a scalable cluster, highly accessible, and supports open-source frameworks like Apache Hadoop and Spark [13–16].

The Amazon S3 (Large Data Storage Framework) is a widely utilized, scalable, secure, and reliable large data storage platform. Amazon S3 provides elastic management, flexible administration, and robustness; the framework provides consistency, scalability, and in-place query options; it offers integration with the widest range of vendor solutions; and it offers a broad range of security and compliance capabilities, as well as easier and faster data transport [13–15, 17, 18].

The Amazon Redshift (Data Warehousing Framework) is a fully managed AWS data warehouse platform that allows for quick, simple, and cost-effective data analysis using conventional SQL and popular business intelligence tools. Amazon Redshift's elastic and scalable cluster provides significant expandability and compatibility with various SQL clients, as well as fast query performance, cost-effectiveness, simplicity, and a highly secure foundation [15, 17, 19].

2.1.1 Architecture Study of a Data Analytics System Using AWS

Amazon offers a world leading analytics and data science platform with extensive configurable cloud-based services for industries. It provided integrated, sustainable, and scalable analysis and reporting methods for meeting multi-domain industrial needs. Various steps and components involved for using data from various sources for analytics and ML modeling are mentioned below. Figure 2 shows the architecture of amazon web services (AWS).

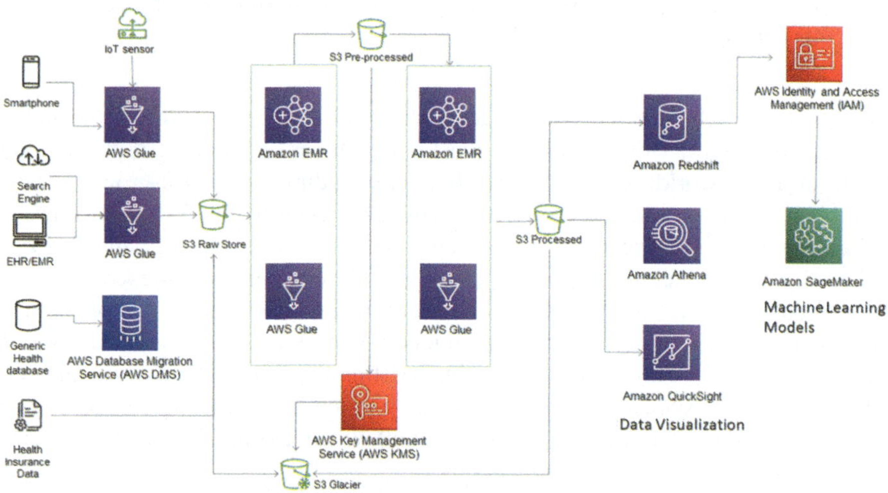

Fig. 2 Architecture of AWS (Amazon Web Services)

2.1.2 Data Ingestion and Processing

Data required for building a data analytics system will come from variety of sources. The range of data varies from social media data, images, files, documents, photographs, generic health databases, smart phones, wearable devices, electronic health, medical records, and government agencies [20]. While the data in the Data Lake would be stored in a source-native format for easy consumption for data scientist who would be working on developing machine learning models, consuming applications like AWS Glue and AWS EMR jobs would transform the heterogeneous data from various sources in different formats to a single homogenous format to make the best use. The data would need to be cleaned, enriched, correlated, and aggregated to be made available to different users for varied use cases in data analytics [21, 22].

The transformation layer is responsible for identifying the structure of data it needs, as an output and then reading the data from the Data Lake (stored in native formats) and transforming it to the relevant structures, determined based on the use cases. The data transformation layer is built using Glue and/or EMR based on the use cases. Glue is used for smaller workloads where the time to process data is smaller and is primarily batch in nature. AWS EMR is used for any real-time processing and/or batch processes that require larger volumes of data where the runtime is high, and Glue would become cost prohibitive. The data after the transformation phase will be stored in the S3 in a processed data store in well-defined structures to enable further usability by the downstream applications [23].

2.1.3 Data Preparation

Data Modelling Different departments and users would need access to different data and KPIs at multiple levels of aggregation. The KPIs and the aggregation will be performed by the transformation tier, and the data will be made available in a ready-to-use format in the tables. The dimension model is designed and deployed using AWS Redshift, which provides parallel query processing, helping to reduce the time of analysis. The model will be designed as per the requirements and use cases defined. Ad hoc query and analysis will be facilitated using Redshift Spectrum, to contain the costs at Athena and a high number of ad hoc queries fired [23, 24].

Master Data and Metadata Data Catalogue is the source containing information related to data residing in the Data Lake and its corresponding metadata contains source and structure of data. AWS provides Glue as a tool to generate the data source catalogue and update it with the changes in the sources [20]. The catalogue created using Glue is available for view and update using the console and SDKs. Apart from Glue, cataloguing would be maintained using the ETL tool used.

Data Classification, Encryption, and Access The security considerations are very critical when we are dealing with confidential data such as patient medical records, old history, kin relationships, and their medical data. Data security in data analytics includes aspects like the following:

- Data encryption in transit and at rest [25, 26]
- Access control
- Network isolation
- Audit Trail and Access logs [27, 28]

Data at rest in S3 would be encrypted and secured using S3 encryption capabilities. The encryption is done using the keys from AWS KMS. Access to the lake and other AWS services is restricted using IAM Roles and Users [29, 30]. The following measures are taken to ensure security and limit access control:

- IAM accounts with the least privileged policy of providing only the rights required for performing the required tasks.
- Password policy, MFA enabled for IAM users.
- Security groups and a network access list are created to limit access to resources.
- Access control, bucket policies, and IAM policy-based access to data on S3 buckets.
- AWS CloudTrail would be responsible to maintain an Audit Trail and Access logs for all the users [24].

Data Storage The data acquired from the data sources will be persisted and maintained in a Data Lake in AWS S3. This Data Lake architecture will be keeping in mind the data formats and various usage of the Data Lake [31]. Data Lake will be tiered based on the processing stage of the data. The different zones that will be built within the lake are as follows:

- Landing Zone: Also called the Bronze Zone, it is the place where all the raw data from different source systems land in its as-is format. This is the place where further downstream tools use data based on their implementation use cases [32].
- Refinery Zone: This is an intermediate zone called the Silver Zone, primarily with some limited processing like optimizing storage and data cleaning. It is the place where medical data can be discovered, explored, and experimented with for hypothesis validation and tests [33].
- Production Zone: Also called the Gold Zone, clean, well-structured data is saved in the best manner to inform crucial business choices and promote efficient operations, often known as the Gold Zone. An operational data store that feeds standard data warehouses and data marts is frequently included [23]. Transformation rules will be applied to the data lying in the Lake to create tables/views that will be used by various users for their concerned use cases. All the data stored in Lake will be stored in compressed formats.
- *Data Archival, Backup, and Recovery* – The platform employs the S3 Glacier Instant Retrieval storage class for archive material that requires quick access, such as photographs, audio-visual assets, or genomics data. This is an archive storage class that provides the lowest cost storage with millisecond retrieval.

- *Data Visualization* – The dashboards built in AWS Quick sight can contain important Key Performance Indicators (KPIs) for monitoring the variances of data and their significance as per the expected calculations.

2.1.4 AI/ML Workbench

ML Model Evaluation Evaluation of a machine learning model is a major part of the overall AI/ML strategy development process. It is necessary to estimate how best the chosen model represents the datasets which have been captured and how well this data can be used for machine learning model training and output generation in future [34]. The overall objective is to enhance the efficiency of the model, which required tuning of model parameters to increase accuracy and observing the confusion matrix for increasing the number of true positives and negatives.

ML Model Training As mentioned before, the training of machine learning model is a crucial step as it helps in getting the correct outcome. The data is initially split into three sections "Training data," "Validation data," and "Testing data." The training data is the data which is used by the machine learning algorithm to learn how to process the fed information. The validation dataset is used for cross validating the skill of machine learning model on unseen dataset. This data is used to fit the parameters of classification algorithms. A set of unseen data is used from the training dataset for tuning the parameters of model classifier [34].

The training data set can be studied using the AWS Athena tool which can be used to analyze data using interactive SQL queries. In the overall data set, a training set is used for building the model, while the testing (or validation) dataset is to validate the model built. The training dataset and the testing dataset are mutually exclusive of each other so that it does not create co-relation issues. The validation and testing datasets can be inclusive of each other and can form a single dataset [35].

ML Model Deployment After the model has been trained and evaluated, it is important to create value from the machine learning codes sitting in notebooks. Model deployment can be done by containerization of applications using docker. This clean containerized code can be put behind the Application Program Interface to interface with the machine learning models. This step is then followed by creation of a front-end application for end users. AWS Sage maker is an effective cloud computing service which can be used for creating machine learning models on refined data sets and can be used for generating insights and foresights for data analytics [35].

Amazon Web Services (AWS) has announced the full launch of Amazon Health Lake, a HIPAA-eligible service for healthcare and life sciences organizations for large-scale ingestion, storage, query, and analysis of their health data. Health Lake enables healthcare organizations to use machine learning to store, transform, and query health data in the cloud to fully understand the health status of patients and populations. This service is part of AWS for Health, which provides a suite of cloud-based services for healthcare, biopharmaceutical, and genomic users to personalize patient care, rapidly innovate, and bring new therapies to market.

2.2 Google Cloud Platform (GCP) for Data Analytics

Google offers a large suite of cloud computing services under the name Google Cloud Platform. GCP includes a large variety of services for computation, application development, storage, and big data analytics. GCP suite of services is ever evolving and Google keeps introducing cloud services based on user demand and to maintain competitive advantage over AWS and MS Azure. Some standard frameworks and services offered by GCP for big data analytics, big data storage, and big data warehouse framework are as follows:

Framework for Big data Analytics (Service Name – Google Cloud Dataproc) is a fully scalable and automated cloud-based Apache Hadoop and Spark service for speedy, simplified, and economical cluster management operations. It is an open-source framework and offers major features and benefits such as fast cluster scaling and cost-effectiveness [15, 16, 36].

Framework for Big Data Storage (Service Name – Google Cloud Storage) is an inline object storage framework which can complete several tasks, such as real-time data processing, data archiving (Cold and real-time), and data analytics. It offers high availability at low pricing, along with streamlined data transition with enhanced security for critical resources [15, 18, 36].

Framework for Big Data Warehousing (Service Name – Google Big Query) is a highly scalable, fast, and low-price data warehouse for big data analytics. It provides a simple way for infrastructure set-up, which can be scaled up seamlessly and can generate quick insights into the data [15, 36].

2.2.1 Architecture Study of a Data Analytics System Using GCP

Google provides several cloud-based services to help researchers realize the value of data by providing comprehensive insights for IT industry and life sciences solutions. Researchers may use tools like Big Query, Cloud Machine Learning Engine, and TensorFlow to apply analytics and artificial intelligence to data. Users may utilize these technologies to create auto-detect patterns, anticipate clinical outcomes, and evaluate enormous volumes of data efficiently [37, 38]. From machine learning to data analytics, Google Cloud supports a broad range of different datasets used in any data analytics system. Figure 3 shows the architecture of Google Cloud Platform (GCP) for data analytics. Various steps and components involved for using data from various sources for analytics and machine learning modelling are mentioned below in the architecture.

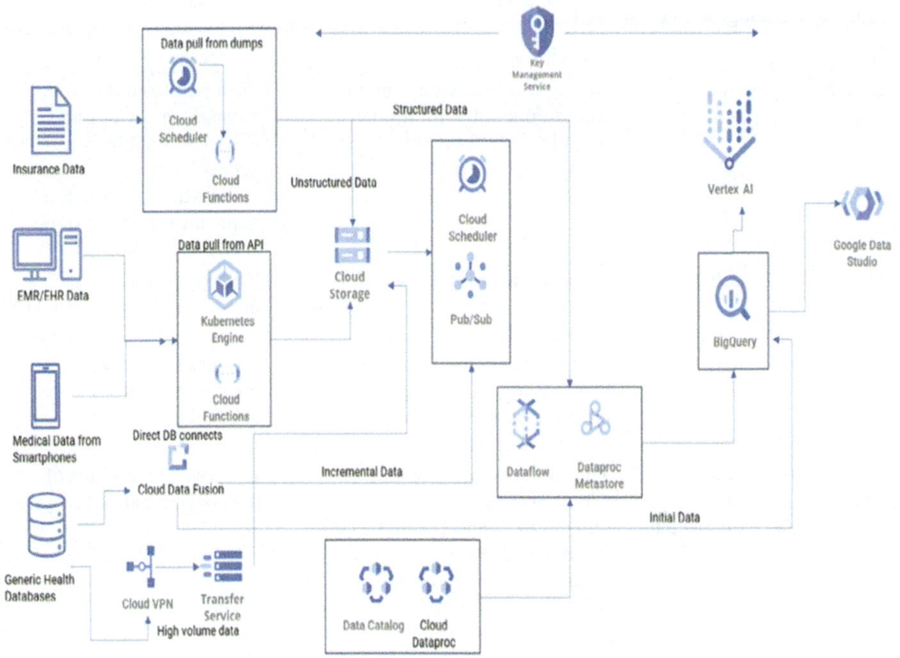

Fig. 3 Architecture of GCP (Google Cloud Platform) for data analytics

2.2.2 Data Ingestion and Processing

The data ingestion using Google Cloud Platform (GCP) with description and its use cases is listed in Table 1.

Unstructured and semi-structured data storage can be handled with the data flow into Google Cloud's Object Storage, namely, Cloud Storage (provides transition to storage classes for any workloads easily with multiple redundancy options), before sending the data for transformations [36]. Cloud Scheduler is an enterprise-grade task scheduler that is fully managed. It enables the scheduling of nearly any process, including batch, data processing, and cloud infrastructure operations, among others. To save time and effort, it may automate everything, even retries in the case of a failure. Cloud Scheduler may also be used as a single pane of glass to manage all of automation jobs in one place [39].

Cloud Pub/Sub [39] is used for messaging and ingestion for event-driven systems and streaming analytics, and it enables for scalable, in-order message delivery with pull and push modes. Auto-scaling and auto-provisioning from 0 to hundreds of GB per second are supported. It also provides individual quotas and pricing for publishers and subscribers, as well as global message routing to make multi-region systems simpler to operate.

Cloud Dataproc [39] runs Apache Spark, Apache Flink, Presto, and more than 30 open-source tools and frameworks that are all part of this fully managed and

Table 1 Data ingestion using GCP

Method	Description	Use cases
Data pull from DB	An event-driven script running on Cloud Function with timing provided by Cloud Scheduler	Data pull from different databases provides an export and not direct access Data extracted from SaaS applications onto a staging server for pickup
Data pull from API	Batch scripts running on GKE or event-driven scripts running on Cloud Function	SaaS or business applications expose an API
Direct DB	One-time data transfer using Cloud Data Fusion and subsequent updates through Debezium connector [39]	Database (backend) connections URLs available
High volume data from smartphones, wearable devices, search engine data	Transfer services available to Cloud Storage [39] Direct DB connect from Generic Databases directly Install Docker agents on on-premises and initiate Transfer jobs from Google Cloud Based on gsutil commands	Usually for terabytes of data or more at once

Table 2 Factors for choosing Dataproc or Dataflow

Factors	Dataproc	Dataflow
DevOps requirement	Provisioning clusters manually	Serverless, i.e., automatic provisioning of clusters
SDK based on	Apache Spark and Hadoop	Apache Beam
Use cases	For data science/ML ecosystem, largely batch	For batch and stream processing
Nature of data	Suitable for small volumes of large chunks of data	Suitable for large volumes of small chunks of data

scalable service. Cloud Dataflow [19] is an Apache Beam SDK-based serverless, fast, and cost-effective unified stream and batch data processing solution. Table 2 depicts the various instances in which an appropriate tool can be selected.

Cloud Composer [39] is an Apache Airflow-based workflow orchestration solution that is fully managed. It is used to create, schedule, and monitor pipelines in hybrid and multi-cloud settings. It is based on the Apache Airflow that runs on Python, allowing it to be free of lock-in and simple to use.

2.2.3 Data Preparation

Data Modelling Big Query [36] is a serverless, highly scalable, and cost-effective multi-cloud data warehouse. It is designed for corporate flexibility. It analyzes petabytes of data at breakneck rates using ANSI SQL, with no operational overhead (Pay-as use model with cost mostly determined by the amount of data queried.) It democratizes insights by providing a reliable and secure platform that can be integrated into a variety of reporting tools, including Excel and Looker, and most of the BI tools such as PowerBI.

Master Data and Metadata Data Catalogue is scalable data discovery and metadata management solution that uses a simple but effective faceted-search interface to find usable data. It is also used to automatically synchronize technical information and generate schematized tags for business metadata. It uses Cloud Data Loss Prevention integration to automatically flag sensitive data and allows users to get access and grow without having to set up or maintain infrastructure [40].

For data analytics products, Dataproc Metastore is a fully managed, highly available within a region, auto healing serverless Apache Hive Metastore (HMS) on Google Cloud [36]. It facilitates interoperability across data processing programmed in the open-source data ecosystem, as well as supporting HMS and serving as a vital component for managing the metadata of relational entities. Its uses include the following:

- A centralized metadata store that can be shared among numerous ephemeral Dataproc clusters that use open-source engines like Apache Hive, Apache Spark, and Presto.
- Interoperability between cloud-native services like Dataproc and other open source-based partner offerings on Google Cloud thanks to a single view of tables [41].

Data Classification, Encryption, and Access Data Loss Prevention (DLP) on the Cloud aids in the understanding, management, and protection of sensitive data. It can quickly categorize and redact sensitive data in text-based information and images, including content stored in Google Cloud storage repositories. Google Cloud [36], without needing any action from the customer, encrypts all client material kept at rest using one or more encryption algorithms.

- For storage, data is separated into chunks, and each chunk is encrypted with its own data encryption key. These data encryption keys are encrypted (or "wrapped") with key encryption keys that are only kept and used inside Google's central Key Management Service (GCP KMS) [39].
- IAM provides technologies that make managing resource permissions simple and automated. For data platform administrators, a complete audit trail of permission authorization, removal, and delegation is instantly displayed.

Data Storage Cloud Storage in Google [18] is the Google Cloud object storage service. It comes with several useful capabilities out of the box, including as object versioning and fine-grained permissions (by object or bucket), which may simplify development and save operating costs. Several Google services are built on top of Google Cloud Storage.

- *Data Archival, Backup, and Recovery* – Google Nearline [18] is a public cloud storage service for archiving, backup, and disaster recovery. It is one of the Google Cloud Storage platform's four cloud storage classes.
- *Data Visualization* – Looker [15] is used to serve real-time dashboards (updated every few seconds) for more in-depth, consistent analysis. It provides insights easily (slide and dive into data dimensions within the visualization layer). Data Visualization (Looker) is a highly customizable semantic layer to make custom apps that give data experiences that are as unique to the business. Data Studio is a free offering from Google that provides community-based connectors to various databases. It is customizable data transformation and visualization layer and gives lesser real-time in a sense (updated every few tens of minutes).
- *AI/ML Workbench* – There are three options for ML modelling. In either case model results are stored back in Big Query:
- Big Query [36] can be used to execute and create machine learning models in Big Query using SQL queries. This ML workbench can be implemented through Looker cloud service.
- Build Custom ML models using Python libraries on Jupyter Notebook and deploy models on Vertex AI. Vertex AI can be used to manage several stages in the ML workflow applicable to the business use cases such as building a dataset and uploading data, training an ML model on the data, evaluating the model's accuracy, and tuning hyper parameters (custom training only). It can also be used to save and upload models to Vertex AI. The trained model is deployed to an endpoint to serve predictions and send prediction queries to the endpoints.
- Cloud AutoML makes it possible to train models without writing code on picture, tabular, text, and video datasets. It automates data pre-processing, parameter extraction, selection, and engineering, algorithm selection, and hyperparameter tweaking. Vertex AI [39] unifies the AutoML and AI Platform APIs, client libraries, and user interface. Both AutoML training and bespoke training are accessible with Vertex AI.

The Google Cloud Healthcare and Life Sciences team has announced a collection of APIs and datastores for healthcare data analysis and machine learning applications, data-level healthcare system integration, and safe storage and retrieval of many forms of electronic patient healthcare data (ePHI). The Google Healthcare API, which focuses on healthcare apps, provides developers a robust toolbox for converting ideas into practical solutions. Google's machine learning has accurate recognition and can be taught to execute a range of high-accuracy prediction tasks.

3 Comparative Analysis of Services Required from AWS and GCP

Table 3 shows the various stages required for building a cloud-based big data analytics platform and the corresponding services required from AWS and GCP product suites.

Other than Amazon Web Services (AWS) and Google Cloud Platform (GCP), big data analytics cloud platform provider such as IBM Cloud [42] and Microsoft Azure [16, 43–46] also works in similar manner to provide data analytics for predictive and prescriptive insights.

4 Challenges

Data collection is one of the primary challenges when considering any cloud-based big data application for business analytics or IT analytics, the data sources may be incomplete, contain interferences and errors, the consent needs to be taken from organizations or individuals before considering any analytics application which uses personal identifiable data [47]. It also becomes very important to integrate data generated from external/internal system in order to arrive at meaningful analytics. It also becomes very difficult to process such heterogeneous data, which varies widely [48]. When integrating data in numerous formats, ensuring accuracy and consistency in decision-making becomes a concern. In order to have processing of such varied data and keeping the infrastructure costs in check, there needs to be a lot of thought which needs to go in at the time of inception which is time taking and extensive [49]. The volume of data poses another challenge, live feeds of data across globe related to organizations or individuals are challenging storage tasks and need highly efficient computational capabilities with high input/output speeds for efficient storage management [50].

Data analysis of such multi-dimensional data also poses a significant challenge, as it is not interpreted in appropriate graphical representations, often it may require more than two to three charts to derive significant inferences from the data sets across visualizations. Confidential data is very sensitive and thus data security is a crucial part of building the cloud data platform for data analytics [48]. It is necessary to have proper controls, authentication mechanisms, and encryption of data done to enhance the security of data, which is available in cloud.

5 Conclusion

This chapter reviews the cloud computing services which are used to build an architecture of a cloud-enabled data platform which is used in business analytics

Table 3 Services required from AWS and GCP

Sr No.	Stage	Requirement	GCP Services	AWS Services
A	Data Ingestion and Processing	Integration with existing source systems (internal and external) To plan data transformation, you need have a workflow management and scheduling system Data acquisition or data delivery jobs Should have audit and error logs for auditing and troubleshooting GUI interface for checking the errors, scheduling, and restarting of ETL jobs	Dataflow and Dataproc (using Apache Hadoop framework)	AWS Glue and AWS EMR (using Apache Hadoop framework)
B	Data Preparation			
	Data Modeling	Develop data models based on data flowing in from various data sources	Big Query	AWS Redshift
	Master Data and Metadata	Ensure data lineage Data quality, integrity, compliance, auditing Data Governance Framework	Cloud Data Catalogue	AWS MDM
	Data Classification, Encryption, and Access	Use proper encryption keys for customer datasets Individual user-level access Role-based access Group-based access (e.g., cardiology, medicine, cancer) Access for the applications which intend to use the data present in the data platform	DLP API, KMS, IAM	AWS KMS and IAM
	Data Storage	Where and how data is being stored? How is data going to be accessed by different user groups? Methodology for reconciliation of data from various source systems	Google Cloud Storage (Landing Zone) Big Query (Data Lake)	AWS S3 and AWS Redshift
C	Data Archival, Backup, and Recovery	Define data backup, archival, and recovery policies based on industry requirement	GCS	AWS Glacier
D	Data Visualization	Visualizations and advanced analytics dashboards	Looker	AWS Quick sight
E	AI/ML Workbench	Experimentation and Model building Model evaluation, deployment, and visualization	Vertex AI	AWS Sage maker

using two popular cloud service providers, that is, AWS and GCP. During the study, it was concluded that both these cloud computing service providers offer their own set of individual services for accessing and collating data from different input data sources used in business industry. These services can be easily procured and employed for studying and analyzing the data and then deriving valuable inputs from it, which is used in devising the machine learning model. While AWS is more stable, reliable, and offers a more global reach owing to early adoption by the business industry, GCP is a container-based model with flexible pricing models and enhanced computing capabilities. The mix of services or the evaluation of a particular platform can be done based on the actual use cases, which are to be implemented for advanced analytics use cases.

As discussed earlier, both AWS and GCP have come up with their targeted solutions for different industries and the competition will grow more over time. The ultimate objective of using a cloud platform for data analytics in industries is to achieve the creative vision necessary to advance IT analytics to the next level of innovation, so enabling it to both drive and support substantial improvements in building quality measures and outcomes.

References

1. Sestino, A., Prete, M. I., Piper, L., & Guido, G. (2020). Internet of Things and Big Data as enablers for business digitalization strategies. *Technovation, 98*, 102173, ISSN 0166-4972. https://doi.org/10.1016/j.technovation.2020.102173
2. Srinivasan, A. Cloud computing. Pearson India ISBN: 9789332537439. *Cloud computing: Concepts, technology & architecture*. Prentice Hall Service Technology Series (1st ed.), ISBN-10: 9780133387520
3. Kavis, J. *Architecting the cloud: Design decisions for cloud computing service models (SaaS, PaaS, and IaaS)* (1st ed.). Wiley, ISBN-10: 1118617614.
4. Alreshidi, E. (2019). Comparative review of well known cloud service providers. *Science International (Lahore), 31*(8), 65–170, ISSN-1013-5316.
5. Stankovic, J. A. (2016). Research directions for cyber physical systems in wireless and mobile healthcare. *ACM Transactions on Cyber-Physical Systems, 1*(1), 1–12.
6. Kune, R., Konugurthi, P. K., Agarwal, A., Chillarige, R. R., & Buyya, R. (2016). The anatomy of big data computing. *Software: Practice and Experience, 46*, 79–105. https://doi.org/10.1002/spe.2374
7. Rizwan, A., Zoha, A., Zhang, R., et al. (2018). A review on the role of Nano communication in future healthcare systems: A big data analytics perspective. *IEEE Access, 6*, 41903–41920.
8. Khan, S., Shakil, K. A., & Alam, M. (2018). *Cloud-based big data analytics—A survey of current research and future directions*. ©Springer Nature Singapore Pte Ltd. Aggarwal, B., et al. (Eds.), Big data analytics. *Advances in Intelligent Systems and Computing* 654. https://doi.org/10.1007/978-981-10-6620-7_57
9. https://www.gartner.com/. Accessed 31 Jan 2020.
10. Rajendran, V. V., & Swamynathan, S. *Parameters for comparing cloud service providers: A comparative analysis*. https://doi.org/10.1109/CESYS.2016.7889826, IEEE Xplore: 30 March 2017. 11. Dutta, P., & Dutta, P. (2019). Comparative study of cloud services offered by Amazon, Microsoft, and Google. *International Journal of Trends in scientific Research and Development (IJTSRD), 3*(3), 981–985.

11. Zhing, L. C. (2020). Cloud computing: State of Art and research challenges. *Journal of International Services and Applications, 1*(1), 7–18.
12. Practical Amazon EC2, SQS, Kinesis, and S3. eBook. SpringerDoi: https://doi.org/10.1007/978-1-4842-2841-8
13. Pradhananga, Y., Karande, S., & Karande, C. *High performance analytics of big data with dynamic and optimized Hadoop cluster*. IEEE. https://doi.org/10.1109/ICACCCT.2016.7831733
14. Dawelbeit, O., & McCrindle, R. A novel cloud based elastic framework for big data preprocessing. In *IEEE Conference Publications*. https://doi.org/10.1109/CEEC.2014.6958549
15. Gonzales, J. U., & Krishnan, S. P. T. *Building your next big thing with Google Cloud Platform*. SpringerDOI: https://doi.org/10.1007/978-1-4842-1004-8
16. Singh, M. P., Hoque, M. A., & Tarkoma, S. A survey of systems for massive stream analytics, arXiv1605.09021v2
17. Ambeth Kumar, V. D., Ashok Kumar, V. D., Divakar, H., & Gokul, R. *Cloud enabled media streaming using Amazon Web Services*. IEEE. https://doi.org/10.1109/ICSTM.2017.8089150
18. Subia, S. (2018). Data Storage SpringerDOI: 978-3-319-21569-3_7 10, *Procedia Computer Science*.
19. Nakhimovsky, A., & Myers, T. Google, Amazon, and beyond: Creating and consuming Web services. SpringerDOI: 9781590591314.
20. Mohanty, H., Bhuyan, P., & Chenthati, D. Chapter 2: Big data architecture. In *Big data: A primer*. Springer DOI: 9788132224938
21. Begam, S. S., Selvachandran, G., Ngan, T. T., & Sharma, R. (2020). Similarity measure of lattice ordered multi-fuzzy soft sets based on set theoretic approach and its application in decision making. *Mathematics, 8*, 1255.
22. Thanh, V., Rohit, S., Raghvendra, K., Le Hoang, S., Thai, P. B., Dieu, T. B., Ishaani, P., Manash, S., & Tuong, L. (2020). Crime rate detection using social media of different crime locations and Twitter part-of-speech tagger with Brown clustering. *Journal of Intelligent & Fuzzy Systems, 38*, 4287–4299.
23. The Old Bailey and OCR: Benchmarking AWS, Azure, and GCP with 180,000 Page Images DocEng '20: In *Proceedings of the ACM Symposium on Document Engineering*, September 2020. Article No.: 19, pp. 1–4. https://doi.org/10.1145/3395027.3419595
24. Ta, V.-D., Liu, C.-M., & Nkabinde, G. W. (2016). Big data stream computing in healthcare real-time analytics. In *2016 IEEE International Conference on Cloud Computing and Big Data Analysis (ICCCBDA)*, pp. 37–42, https://doi.org/10.1109/ICCCBDA.2016.7529531
25. Saraswat, M., & Tripathi, R. C. (2020). Cloud computing: Comparison and analysis of cloud service providers-AWS, Microsoft and Google. In *2020 9th International Conference System Modeling and Advancement in Research Trends (SMART)*, pp. 281–285. https://doi.org/10.1109/SMART50582.2020.9337100
26. Daniels, M., Rose, J., & Farkas, C. (2018). Protecting patients' data: An efficient method for health data privacy. In *Proceedings of the 13th International Conference on Availability, Reliability and Security*, ACM, p. 9.
27. He, Z., Cai, Z., Sun, Y., et al. (2017). Customized privacy preserving for inherent data and latent data. *Personal and Ubiquitous Computing, 21*(1), 43–54.
28. Mahmoud, M. M. E., Rodrigues, J. J. P. C., Ahmed, S. H., et al. (2018). Enabling technologies on cloud of things for smart healthcare. *IEEE Access, 6*, 31950–31967.
29. Nguyen, P. T., Ha, D. H., Avand, M., Jaafari, A., Nguyen, H. D., Al-Ansari, N., Van Phong, T., Sharma, R., Kumar, R., Le, H. V., Ho, L. S., Prakash, I., & Pham, B. T. (2020). Soft computing ensemble models based on logistic regression for groundwater potential mapping. *Applied Sciences, 10*, 2469.
30. Jha, S., et al. (2019). Deep learning approach for software maintainability metrics prediction. *IEEE Access, 7*, 61840–61855.
31. Emara, K. (2017). Safety-aware location privacy in VANET: Evaluation and comparison. *IEEE Transactions on Vehicular Technology, 66*(12), 10718–10731.

32. Sharma, R., Kumar, R., Sharma, D. K., Son, L. H., Priyadarshini, I., Pham, B. T., Bui, D. T., & Rai, S. (2019). Inferring air pollution from air quality index by different geographical areas: Case study in India. *Air Quality, Atmosphere & Health, 12*, 1347–1357.
33. Sharma, R., Kumar, R., Singh, P. K., Raboaca, M. S., & Felseghi, R.-A. (2020). A systematic study on the analysis of the emission of CO, CO2 and HC for four-wheelers and its impact on the sustainable ecosystem. *Sustainability, 12*, 6707.
34. Xiao, W., Miao, Y., Fortino, G., Wu, D., Chen, M., & Hwang, K. (2022). Collaborative cloud-edge service cognition framework for DNN configuration toward smart IIoT. *IEEE Transactions on Industrial Informatics, 18*(10), 7038–7047. https://doi.org/10.1109/TII.2021.3105399
35. Erhan, L., Ndubuaku, M.U., Mauro, M.D., Song, W., Chen, M., Fortino, G., Bagdasar, O., & Liotta, A. (2020). Smart anomaly detection in sensor systems: A multi-perspective review. arXiv: Learning.
36. Krishnan, S. P. T., & Ugia Gonzalez, J. L. Google BigQuery. SpringerDOI: https://doi.org/10.1007/978-1-4842-1004-8_10
37. Dansana, D., Kumar, R., Das Adhikari, J., Mohapatra, M., Sharma, R., Priyadarshini, I., & Le, D.-N. (2020). Global forecasting confirmed and fatal cases of COVID-19 outbreak using autoregressive integrated moving average model. *Frontiers in Public Health, 8*, 580327. https://doi.org/10.3389/fpubh.2020.580327
38. Malik, P. K., Sharma, R., Singh, R., Gehlot, A., Satapathy, S. C., Alnumay, W. S., Pelusi, D., Ghosh, U., & Nayak, J. (2021). Industrial Internet of Things and its applications in industry 4.0: State of the art. *Computer Communications, 166*, 125–139, ISSN 0140-3664. https://doi.org/10.1016/j.comcom.2020.11.016
39. Gupta, Y. K., & Mittal, T. (2020). Comparative study of Apache Pig & Apache Cassandra in Hadoop Distributed Environment. In *2020 4th International Conference on Electronics, Communication and Aerospace Technology (ICECA)*, pp. 1562–1567. https://doi.org/10.1109/ICECA49313.2020.9297532
40. Sharma, R., Kumar, R., Satapathy, S. C., Al-Ansari, N., Singh, K. K., Mahapatra, R. P., Agarwal, A. K., Le, H. V., & Pham, B. T. (2020). Analysis of water pollution using different physicochemical parameters: A Study of Yamuna River. *Frontiers in Environmental Science, 8*, 581591. https://doi.org/10.3389/fenvs.2020.581591
41. Dansana, D., Kumar, R., Parida, A., Sharma, R., Adhikari, J. D., et al. (2021). Using susceptible-exposed-infectious-recovered model to forecast coronavirus outbreak. *Computers, Materials & Continua, 67*(2), 1595–1612.
42. Patil, A., Rangarao, D., Seipp, H., Lasota, M., dos Santos, R. M., Markovic, R., Casey, S., Bollers, S., Gucer, V., Lin, A., Richardson, C., Rios, R., VanAlstine, R., & Medlin, T. *Cloud Object Storage as a Service IBM Redbooks*. https://www.redbooks.ibm.com/redbooks/pdfs/sg248385.pdf
43. Klein, S. *IoT Solutions in Microsoft's Azure IoT Suite*. SpringerDOI: 9781484221426.
44. Copeland, M., Soh, J., Puca, A., Manning, M., & Gollob, D. *Microsoft Azure*. SpringerDOI: 9781484210444.
45. Moemeka, E. *Azure in the enterprise*. Springer DOI: 9781484230862.
46. Reagan, R. *Web applications on Azure*. Springer DOI: 9781484229750.
47. Morshed, M. G., & Yuan, L. (2017). Big data in cloud computing: An analysis of issues and challenges. *International Journal of Advanced Studies in Computer Science and Engineering, 6*(4), 345–350.
48. Abouelmehdi, K., Beni-Hssane, A., Khaloufi, H., & Saadi, M. (2017). Big data security and privacy in healthcare: A review. *Procedia Computer Science, 113*, 73–80, ISSN 1877-0509. https://doi.org/10.1016/j.procs.2017.08.292
49. Google Cloud Platform Products. https://cloud.google.com/gcp/
50. Fortino, G., Messina, F., Rosaci, D., & Sarné, G. M. L. (2020). Using blockchain in a reputation-based model for grouping agents in the Internet of Things. *IEEE Transactions on Engineering Management, 67*(4), 1231–1243. https://doi.org/10.1109/TEM.2019.2918162

An Intelligent Model for Optimizing Sparsity Problem Toward Movie Recommendation Paradigm Using Machine Learning

Manash Sarkar, Shiwangi Singh, V. L. Soundarya, Maroi Agrebi, and Ahmed Alkhayyat

1 Introduction

As we all know, the world is expanding at an unprecedented rate. Everyone is hurrying to achieve their ultimate objectives. This thirst leads to the growth of, practically, every industry. One of them is online business. We do not have time to go to the market, and this is not the end of it. We do not even have enough time to select an item from the collection. This sparked the birth of internet shopping, which has now grown into a massive tree with hundreds of branches. As the online market expands at an exponential rate, it is inevitable that competition will spread to other areas as well. Owners of separate websites must now entice their readers by giving appealing features. One of the features available to users is recommender engines. Recommender engines are the most well-known machine learning approach now in effect. We have all encountered services and websites that try to suggest books, movies, and articles based on our previous behavior. They attempt to infer tastes and

M. Sarkar (✉)
Department of Computer Science & Engineering, Atria Institute of Technology, Bengaluru, Karnataka, India

Expert Data Scientist, Limendo, Bolzano, Italy
e-mail: manash.sarkar@atria.edu; manash.sarkar@limendo.com

S. Singh · V. L. Soundarya
Department of Computer Science & Engineering, Atria Institute of Technology, Bengaluru, Karnataka, India

M. Agrebi
LAMIH UMR CNRS 8201, Department of Computer Science, Université Polytechnique Hauts-de-France, Valenciennes, France

A. Alkhayyat
College of technical engineering, The Islamic University, Najaf, Iraq
e-mail: ahmedalkhayyat85@iunajaf.edu.iq

R. Sharma et al. (eds.), *Data Analytics for Internet of Things Infrastructure*, Internet of Things, https://doi.org/10.1007/978-3-031-33808-3_10

preferences as well as recognize undiscovered interesting items. Recommendations are so commonplace in our environment that we sometimes fail to notice how effortlessly these systems created to improve our purchasing choices have been integrated into virtually every device and platform. Although we are accustomed to getting suggestions all the time, their integration into our everyday lives has a long and intriguing history that has been shaped by ups and downs, the ambition of a few pioneering firms, and the efforts of other dreamers. You may already be aware that a recommender system is a system that typically tries to forecast the value a user would attribute to a certain piece of material. These engines provide predictions by merging data models from several natural sources.

Sama et al. [1] have described in their research that using artificial intelligence technologies to create a movie recommendation system is difficult. Finding satisfying results is challenging because not all information obtained from the internet is trustworthy and useful. Experts advise using a recommendation system that gives pertinent information rather than continuously searching for the matching information. Collaborative filtering methods based on user information (such as gender, geography, or preference) are often effective. However, in today's world, everyone is concerned with advice or principles in order to make a choice. Our study focuses on developing a system for suggesting movies that may provide users with a list of movies that are comparable to the one they have chosen. Recommendation system, which falls under the category of model-based filtering approach, is explained in the research performed by Srikanth et al. [2]. It is built using cutting-edge deep autoencoders. NLP and computer vision applications employ deep autoencoders, an artificial neural network. Here, using an unsupervised learning technique, deep autoencoders are utilized to obtain the Top N movie recommendations.

Netflix's 3-month implicit value dataset with patterns discovered through collaborative filtering was utilized to assess the performance of the suggested recommender system. This chapter can successfully forecast the ratings that are absent from the sparse matrix that are first provided for input. The top model out of 20 created models is connected to the frontend.

By proposing movies that match the viewer's preferences, several online movie or video streaming companies may keep their users interested. Making more tailored suggestions is a major area of study for recommender systems. Two filtering strategies are covered in the study by Salmani et al. [3], content-based filtering and collaborative filtering. The former recommends items (movies) to users (viewers) based on their past viewing behavior and preferences, while the latter does so by considering the opinions and actions of other users (viewers) who have similar preferences (movies). Collaborative filtering has produced user-based, item-based, SVD++, and SVD algorithms, and their performance has been evaluated. A hybrid recommendation engine that combines both the content-based and SVD filtering models offers the greatest performance and improved movie selections for preserving active viewer engagement with the service. In order to suggest movies based on several criteria, the study done by Sunandana et al. [4] created a movie recommendation system. The project's main goal is to provide a system for movie recommendation that would allow users to be prescribed movies. There are several algorithms that may be used to create a recommendation system. The content-based algorithm has been used to propose movies based on their similarities to other movies by looking at the movie's content. The degree of resemblance has been assessed using the cosine similarity method. The parameters in this case were taken from the TF-IDF vectorization output, and a linear kernel was used to get the cosine similarity. The most comparable movies are then recommended. The recommendation system uses things based on user preferences to help with decision-making. There are many different sectors where recommendation systems are employed, similar to social networking, advertising, and e-commerce. Collaborative filtering is a common approach for creating recommendation systems; however, it becomes less reliable if the amount of data is too little. As a result, using the weighted average technique can increase the precision of suggestions. Today, recommender system is also used for restaurant [5]. Restaurant recommendation is very useful for tourism industries.

The most common types of recommendation systems are as follows:

- *Content-based recommender systems*: To recommend things such as movies or music, content-based algorithms use meta data such as genre, producer, actor, and musician. A good example of such a recommendation would be recommending *Infinity War*, which stars Vin Diesel, because someone enjoyed *The Fate of the Furious*. Content-based systems are founded on the assumption that if you like one thing, you will probably like another similar thing.
- *Collaborative filtering recommender systems*: The behavior of a group of users is used to offer suggestions to other users in collaborative filtering. The preference of other users is used to provide recommendations.

There are two types of collaborative filtering systems:

- *Collaborative filtering based on users*: User-based collaborative filtering: In this paradigm, products are recommended to a user based on how popular they are among users who are similar between users.
- *Item-based collaborative filtering*: Related items are identified by these systems based on the users' earlier ratings. *Hybrid Recommender System*: Hybrid Recommender Systems merge any of the two interconnected systems in a customized sector. It is the most sought-after recommender system, as it merges the virtues of many recommender systems while also removing any potential flaws that might arise from using just one recommender system.
- *Weighted Hybrid Recommender*: The score of a recommended item is calculated using the results of all of the system's various recommendation strategies in this method. P-Tango, for example, combines content-based and collaborative recommendation systems, initially giving both equal weight but gradually changing the weighting as user ratings predictions are validated or disproved. Pazzani's hybrid does not use numerical scores, instead treating the output of each recommender as a series of votes that are then combined in a consensus-building procedure.

- *Switching Hybrid Recommender*: Switching Hybrid Recommender is a feature that allows you to switch between recommendation algorithms based on certain criteria. The switching hybrid recommender can deploy a content-based recommender system initially, and if that one fails, it will deploy a collaborative-based recommender system. This is possible if we combine content- and collaboration-based recommender systems.
- *Mixed Hybrid Recommender*: If a large number of recommendations can be made at the same time, mixed recommender systems should be used. Because recommendations from multiple techniques are given at the same time, the user can choose from a wide range of options. The majority of media and entertainment organizations use Smyth and Cotter's PTV system, which is mostly a recommended program, to recommend television viewing to customers.

Today, recommender system is the most powerful technique for intelligent business model to recommend the users' based on their choice and budget. There is an issue in recommender system called sparsity problem. In this problem, many values are not available in the dataset. This creates trouble at the time of analysis of historical data. In this research, the sparsity problem is optimized by proposing an intelligent model. An optimized decision to develop a collaborative filtering system that forecasts movie ratings for a user based on a large database of user ratings. A statistical analysis is carried out in order to create a pilot survey model for analyzing the real-time dataset. Ant Colony Optimization (ACO) is used to determine the group members' ratings for future recommendation. Finally, fourfold cross-validation is used to assess the performance of the proposed model.

The following sections comprise the remainder of this chapter: Section 2 reviews related works. Section 3 describes a fundamental of Big Data. Section 4 contains the proposed model, followed by Sect. 5's Results and Discussion. Section 6 contains the conclusion and future scope.

2 Similar Works Done

This section reviews earlier studies in the topic of recommendation systems, particularly those that use methods based on systems for recommending movies. The numerous studies are described in the section that follows.

According to this study undertaken by Omega et al. [6], user weighted average and movie weighted average have an impact on how users are recommended films. Additionally, it demonstrates that the weighted average recommendation system outperforms the collaborative recommendation system in terms of accuracy.

In their publication, Ajith et al. [7] go into great detail about the goal of their study, which is to provide a knowledge graph and particle filtering-based movie recommendation system. Every method of movie recommendation that is currently available makes use of clustering and machine learning technology. The groundbreaking movie recommendation system that is suggested in this chapter interacts directly with the database, is very effective thanks to the expressive power of knowledge graphs, and offers the user the most pertinent suggestions. This movie recommendation tool uses particle filtering and a knowledge graph database to show the user the best movies based on their preferences for genres, directors, and other elements.

In contrast to previous systems, this recommendation system does not keep track of the browser history for forthcoming suggestions. The large amount of data makes it challenging for consumers to locate relevant movie resources fast; however, the movie recommendation model is a great way to deal with this problem. It is advised to apply GRU (Gate Recurrent Unit) to mine the rules and characteristics of their presence from the text data of movies in addition to deep neural networks (DNN) and factor decomposition machines FM to process the features of customers' wishes for various movie genres. By combining the link between users and movies, Xiong et al. [8] offer a movie recommendation model based on GRU, DNN, and FM in order to improve the accuracy of the suggested outcome. The Movielens-1m dataset is then used for a significant number of tests. The experimental findings demonstrate that the model outperforms CNN (Convolutional Neural Network) and LSTM (Long-term and Short-term Memory Network) recommendation models in terms of performance and recommendation accuracy.

Wang et al. [9] have discussed achieving lower marketing costs and higher revenues, traditional recommender models make use of technologies including collaborative filtering, matrix factorization, learning to rank, and deep learning models. Moviegoers, however, give the same film varying evaluations depending on the situation. The audience's mood, the setting, the weather, etc., are all significant movie-watching contexts. It is quite advantageous for recommender system builders to be able to use contextual data. However, widely used methods like tensor factorization require unreasonably large amounts of storage, which significantly affects their viability in actual environments. The MatMat framework is used in this study to create a context-aware movie recommendation system that is superior to conventional matrix factorization and similar in terms of fairness. By matrix fitting, MatMat factors matrices.

Due to their numerous uses, recommendation systems today play a significant role in society. Many businesses utilize it frequently in order to provide better customer experiences and accomplish growth by employing it properly. Information on the strategies and techniques utilized in recommendation systems is presented in this publication. Collaborative filtering approach, content-based approach, and hybrid approach are the three categories into which recommendation systems are divided. The study performed by Kukreja et al. [10] will outline the content-based method, its benefits and drawbacks, and how to use it to provide movie recommendations based on this concept. The report will also detail all other approaches, their methods, and their restrictions. Although we have chosen a content-based strategy, there may be various approaches depending on the business. Users can get their preferred movie programs from a wide range of films thanks to the user-oriented recommendation system. Qiu et al. [11] have discussed a type of heterogeneous network information called a knowledge graph that offers a movie recommendation system a wealth of structural knowledge. It aids in resolving the cold start and data sparseness issues. We developed a knowledge graph-based movie recommendation model using this information. The knowledge graph is first integrated into the recommendation model, the vector representation is then updated using the engaging attention network represented by the user or object, rich local knowledge is then acquired through cross-interactive information dissemination, and predictions are finally produced. We experiment with open data sets, and the results show that our strategy is superior to the accepted practices. Lavanya et al. [12] have discussed recommendations that are so commonplace in our environment that we sometimes fail to notice how effortlessly these systems created to improve our purchasing choices have been integrated into virtually every device and platform. Although we are accustomed to getting suggestions all the time, their integration into our everyday lives has a long and intriguing history that has been shaped by ups and downs, the ambition of a few pioneering firms, and the efforts of other dreamers. You may already be aware that a recommender system is a system that typically tries to forecast the value a user would attribute to a certain piece of material. These engines provide predictions by merging data models from several natural sources. The dataset that was abandoned for the evaluation of qualities with CF includes nominal ratings for a wide range of films. A complex method of collaborative filtering is effective for the motivation of recommendations during evaluation and to link the performance of the arrangement with the inclusion of characteristics from user reviews and deprived of the attributes from user assessments. This method uses an enhanced methodology that is based solely on nominal ratings and movie characteristics like genres.

Agrawal et al. and Roy et al. [13, 14] have proposed a methodology that presents a hybrid approach that combines content-based filtering, collaborative filtering, SVM as a classifier, and genetic algorithm. Comparative results show that the proposed approach improves the accuracy, quality, and scalability of the movie recommendation system over pure approaches in three areas: accuracy, quality, and scalability. The benefits of both techniques are combined in a hybrid strategy that also seeks to minimize their drawbacks. This could happen when looking for things like books, music, films, and job listings. Therefore, it is necessary for a recommendation system to assist in recommending products to consumers that are more pertinent, accurate, and meet their wants and requirements. A movie recommendation system is used to make suggestions for films that are compatible with the user's interests and tastes. There are several ways to implement this system, including basic recommendation systems, collaborative recommendation systems, content-based recommendation systems, metadata-based recommendation systems, and demographic-based recommendation systems. In this study, they have employed a straightforward recommendation system, a content-based filtering strategy, and a collaborative filtering strategy. Python, the B/S framework, and the MySQL database have been used in the study performed by Xu et al. [15] to develop a system for recommending movies. The movie data for Douban is first crawled using crawler technology, after which the movie data is stored in MySQL. Users of the system can browse, look up, and collect movie-related data. The user-based collaborative filtering algorithm is used to suggest movies that may be of interest to users based on their search history. The entire system of movie suggestion is made possible by the design of these useful components. In order to investigate user interests and offer the best products to users, collaborative filtering algorithms are frequently employed in e-commerce websites' recommendation systems. These systems are based on the analysis of a huge number of past behavior data from users. In this chapter, we concentrate on the development of a highly accurate and dependable algorithm for movie recommendation. It is important to note that the algorithm can be used in many different e-commerce domains and is not only restricted to movie recommendation. In the study undertaken by Zhou et al. [16], they have constructed a movie recommendation system in Ubuntu using Java language. The system can manage massive data sets thanks to the MapReduce framework and the item-based recommendation algorithm. Zhang et al. [17] have discuseed these two problems in their study. First, a straightforward but very effective recommendation algorithm is suggested, which makes use of user profile information to divide users into various clusters. The size of the original user-item matrix is drastically reduced by creating a virtual opinion leader to represent each cluster as a whole. The weighted slope one-VU approach is then applied to the virtual opinion leader-item matrix to produce the recommendation results. When compared to traditional clustering-based CF recommendation schemes, our approach can significantly reduce the time complexity while still achieving comparable recommendation performance. By analyzing the user's profile to suggest the most relevant content, a recommendation system saves the user the time of looking for the information. For making suggestions, a number of strategies have been put forth by Hwang et al. [18] that included content-based, collaborative, and knowledge-based methodologies in their study. Recommendations for content like books, music, and videos are made using recommendation systems. They are also frequently utilized in online shopping. A collaborative filtering method based on genres, which is frequently used in movie recommendation systems, is used in particular by the South Korean film industry to make movie recommendations. This tactic might not be as effective when customers first discover movie suggestion services or have specific movie interests, like preferences for actors or directors.

One of the best and most appropriate techniques utilized in suggestion is collaborative filtering. Providing a prediction of the many goods that a user might be interested in based on their preferences is the core goal of the recommendation. Kharita et al. [19] have discussed in their study that when there is enough data, recommendation systems based on collaborative filtering approaches can produce predictions that are roughly correct. In the past, user-based collaborative filtering techniques have been quite effective at recommending products based on user preferences. However, there are also additional difficulties, such as data sparsity and scalability issues, which get worse as user and item counts rise. Finding the relevant information quickly on a big website is challenging. Selecting a movie that makes use of current technology to satisfy user needs has become more difficult as a result of the growth in the number of movies available. The growing usage of online services has led to a rise in the prevalence of recommendation systems. Today, the goal of all recommendation systems is to use filtering and clustering algorithms to suggest content that people will find interesting. Users' preferences are initially discovered to allow them to evaluate movies of their choice by finding user profiles of people with similar interests.

To obtain precise recommendations based on customers' preferences, a hybrid recommendation model built on crowd search optimization is suggested. Collaborative and content-based filtering are combined to create the hybrid recommendation as described by Sarkar et al. [20].

The study undertaken by Joseph et al. [21] provides suggestions for a media commodity like movies to customers. After some usage, the recommender algorithm gets to know the user and suggests movies that are more likely to get good reviews. It is crucial for every area to implement recommender systems to give people advice on all technological elements. Different types of recommendation systems, including content-based, hybrid-based, and collaborative filtering-based systems, are available. Based on collaborative filtering, collaborative filtering for recommendations is separated into user-based and item-based filters. The study performed by Shrivastava et al. [22] aims to assess the collaborative filtering recommendation approach for a dataset of movies. The findings of the user-based and item-based recommendation methods compare how well each performed and which strategy produced the best outcomes.

3 Fundamentals of Big Data

Information that is present in large quantities and numbers is referred to as Big Data. Data refers to the various ways in which the computer has saved and recorded information as well as activities. This includes data that was provided by both people and machines. Big Data is a body of information that has grown significantly over time. Big Data is extremely large and complicated in terms of quantity and storage, making it impossible for conventional data management solutions to analyze and interpret it.

Big Data [23, 24] is collected in a number of methods that are critical. People's experiences and purchasing histories can be tracked through feedback and user reviews. This is also accomplished by conducting surveys that are presented to clients in the shape of a survey, which asks them different questions dependent on the data collecting goal. Every transaction a customer makes can earn them credits, and those points can then be redeemed for rewards based on specific standards. This makes it possible for the client to keep up a profile on the portal that details interests and behaviors; businesses use this kind of data.

There are some social media programs that monitor the information and activity on the concerned parties' profiles. In actuality, one of the most often used retargeting methods is social media. Other data collection techniques and resources exist as well, including cookies, email tracking, and satellite pictures, all of which operate similarly and are self-explanatory.

3.1 Properties of Big Data

Big Data, a large amount of data that is not handled by standard data storage or processing units makes up Big Data. Many multinational organizations utilize it to handle data and conduct business. The data flow would exceed 150 exabytes per day prior to replication. The properties of Big Data are explained by its 5 Vs (Fig. 1).

The 5 Vs of Big Data are as follows:

- Volume
- Veracity
- Variety
- Value
- Velocity
- *Volume*: Big Data is a term used to describe enormous amounts of data. Big Data is a term used to describe the enormous "volumes" of data that are produced every day from several sources, such as corporate processes, machines, human interactions, social media platforms, networks, and many more. About a billion messages, 4.5 billion "Like" button hits, and more than 350 million new postings are uploaded to Facebook every day. Big Data technology can manage large amounts of data.

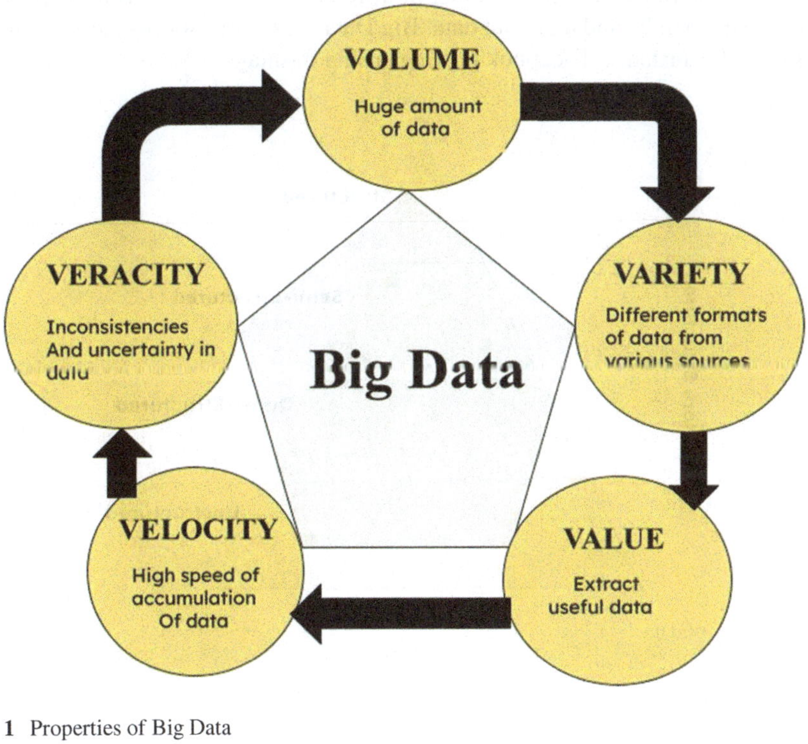

Fig. 1 Properties of Big Data

- *Variety*: Big Data can be gathered from a variety of sources and can be structured, semi-structured, and unstructured. Data was once only gathered through databases and sheets, but today it can also be found in PDF files, emails, audio files, social media posts, pictures, and videos. The data is arranged in the following way:
- *Structured data*: The structured schema contains each and every required column. It is arranged on a table. Structured data is kept in a relational database management system.
  - *Semi-structured Data*: Semi-structured data lacks a properly defined schema for technologies like XML, JSON, CSV, TSV, and email. The OLTP (Online Transaction Processing) systems handle semi-structured data. In order to store the data, tables or relations are used.
  - *Unstructured Data*: Unstructured data includes all unstructured files, including audio, log, and image files. Some businesses have a lot of data, but they are unsure how to value it because the data has not been processed.
  - *Quasi-structured Data*: The data format, which was created with effort and patience using a limited number of technologies, contains textual data with variations in data formats For example, a server creates and maintains a log file that contains a record of actions and is used by web servers (Fig. 2).
- *Veracity*: Veracity is a concept used to describe how reliable the data is. You may filter or manipulate data in a number of different ways. Veracity is the capacity to effectively handle and maintain data. Big Data is essential for business expansion as well, for instance, Facebook postings using hashtags.

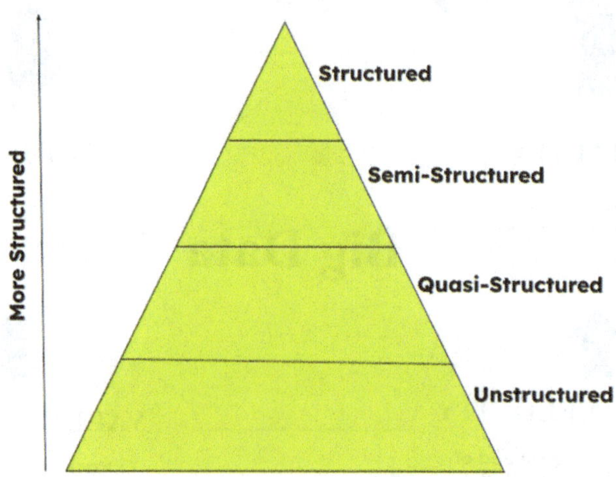

Fig. 2 Variety of data

- *Value*: The importance of value in large data cannot be overstated. The information we handle or store is not the issue. We store, process, and analyze useful and trustworthy information.
- *Velocity*: In contrast to other variables, velocity has a big impact. The rate at which data is produced in real time is known as velocity. It incorporates the incoming data sets' speed, rate of change, and activity bursts. Big Data's primary objective is to promptly supply highly demanded data. Big Data velocity is the rate at which data moves from sources including corporate processes, networks, social media sites, sensors, and mobile devices.

3.2 Big Data in Entertainment Industry

Nowadays, people are highly enthusiastic about discovering new material in terms of what they watch and choose to watch since media and entertainment have permeated every aspect of their life. The days of single-channel viewing, with no choice, integration, or consultation with viewers, are long gone. These factors are altering, though, since there are now millions of watching options to select from, and they can all be streamed across many platforms, making them much more accessible. Unquestionably, the use of Big Data, which is a mix of existing data, time series data, network data, and other types of data, has considerably assisted in the creation and successful application of such ideas and options in the media and entertainment sector. The only way to understand the behavior of consumers (who are the viewers here) is to analyze the different consumer data that is accessible throughout and as a product of multiple platforms. Big Data is aiding from a distance, but its influence is particularly obvious and felt in the sector. Let us examine Big Data's definition, applications in the media and entertainment industry, and sources. The recommendation system is very useful in entertainment industries [25]. Nowadays, OTT platform used the concept of recommender system to serve their customer smoothly.

3.2.1 Uses of Big Data in Media and Entertainment

The media and entertainment industry likewise integrates and gathers similar data from numerous sources in order to better understand viewer behavior and improve in a way that would allow them to thrive and be the viewers' favorite between each of them. The more you know your customer, the better you can predict their preferences and modify content, pricing, and user interface appropriately. This is a well-known marketing and financial reality (Fig. 3).

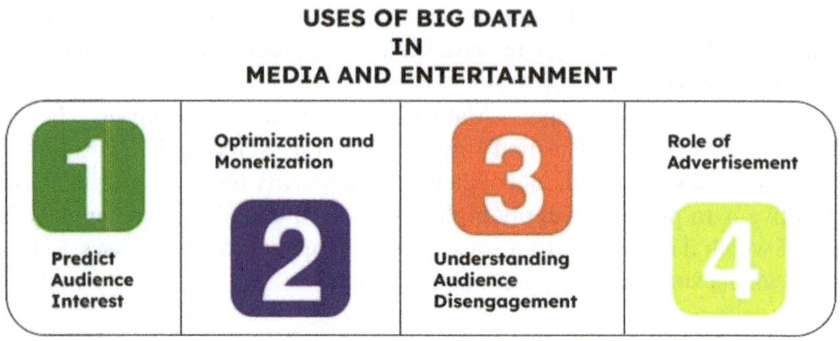

Fig. 3 Uses of Big Data in entertainment. (Courtesy: www.google.com)

- *Predict Audience Interest*: Media firms and entertainment channels, particularly those that are focused on online streaming, can benefit from Big Data analysis. Because satisfying their customers' needs is so crucial to them, such channels want to be prepared with the content and categories necessary to accommodate practically any type of viewer and endeavor to ensure that content variety reigns supreme. Media and entertainment goliaths benefit from a plethora of classified data provided by Big Data. Search histories, genre ratings, followed social media trends, language, age, and other crucial pieces of this data can help businesses predict a particular viewer's interests, enabling them to not only customize the viewer experience but also come up with the most practical and popular program initiatives based on their audience's preferences. Some media organizations are also able to monitor how long viewers watch a video or movie for as well as their responses to it on other social media platforms. Additionally, there is a mechanism for tracking trending hashtags, and in a situation like this, user interaction makes it simpler to control and remove content, like a movie, from the board. They may occasionally provide the user a list of suggested readings to keep them interested in the site.
- *Optimization and Monetization*: Based on current market releases and trends, businesses may decide to add a specific movie to their list of available material just because it is popular and people would want to watch it. Because they often monetize such unconventional content, which keeps viewers interested and draws in new users seeking for the same content, this might bring in more money for the businesses than usual. Such discussion boards are more frequently used by them for amusement. There is also material when companies release a program or a film depending on how viewers respond to its trailers and elect to reserve a certain film or television program for a membership-only set of viewers, who are often charged requiring subscribers to pay a bundle to access these. These businesses further attempt to pique user interest by making only a small portion of a television show or movie available to viewers without a subscription. There are several types of material available to members and non-members of many entertainment firms, including Netflix, Hotstar, and Amazon Prime.

- *Understanding Audience Disengagement*: In any industry, losing a customer is still a frustrating problem for any company. The same is true for media and entertainment businesses; they provide memberships for access to certain material, which is a crucial step in the customer conversion procedure. Any company's membership duration in relation to its pricing is undoubtedly subjective, and it is continuously revised and updated. Big Data can also reveal information about returning clients and devoted fans. In certain cases, despite receiving several push-up reminders and calls to action, members choose to opt-out of the membership program and do not renew their memberships. This needs to be taken into consideration by the media and entertainment industries so that inaccurate information and stale, boring content may be seen. Big Data aids in acquiring up-to-date insights into customer behavior and might lead a company to modify content in response to the platform's widespread demand. As an illustration, the business may add multilingual material if the majority of customers choose not to renew their membership speech problems in the content introduced in the form of feedback.
- *Role of Advertisements*: The key factor that continues to determine a company's market worth and income is advertising. Inquiries from advertisers on a company's analysis of viewing behavior are essential. This may facilitate the timely and accurate distribution of relevant, tailored, and related adverts. The understanding of user behavior and what they are likely to buy as a consequence of targeted marketing is also aided by Big Data applications. Advertisements are common and seem to be an essential part of every entertainment sector. This gives companies the ability to act as retargeting agents, which means that if a customer is watching a show or a movie about one of their goods, an advertising will pop up more appropriately. The audience is more likely to purchase 3D glasses if they are marketed in the middle of a science fiction or technological film. An example of this is a fashion movie that mimics online shopping and encourages viewers to purchase items like clothing and freshly declared trends. Big Data enables businesses to construct successful advertising campaigns based on a variety of criteria, including weather, timing, and the use of second screens, in addition to enabling customers and businesses to generate content-related advertisements.

4 Proposed Model

An intelligent recommender model is proposed to minimize the sparsity problem. At the beginning, a statistical analysis using Brownian process is performed. A metaheuristic-based optimization technique called Ant Colony Optimization (ACO) is applied to optimize the sparsity problem.

4.1 Mathematical Background

At the beginning, a statistical analysis [26, 27] is performed to understand the nature of the data and its distribution. As per the real dataset, all the required features are not in scaled. Normalization is performed first to set the all features value with a certain range. After data visualization, it is found that some outlier data are there which are not ignorable. To find out the distance between the values, Brownian process [28] is deployed.

Assume that random distance time for approximating the sum of independent random values is considered.

Let $x_1 + x_2 + \ldots. + x_n$ where x_i is a random variable $\forall i \geq 1$

As per the Brownian motion theory, it was a continuous time (t) where $0 \leq t \leq \infty$ and μ as a mean of the random data point $V(t)$ and

$$V(t) = 0 \quad (1)$$

$$P_\mu \{V(t) - V(s) \leq x\} = \phi \left[{(x - \mu(t - s))} \Big/ {(t - s)^{\frac{1}{2}}} \right]$$

where

$$\forall\ 0 \leq s < t < \infty \quad (2)$$

$$\text{and } 0 \leq s_1 < t_1 \leq s_2 < t_2 \leq \ldots\cdots \leq s_n < t < \infty \text{ where } n = 2, 3 \ldots\ldots \quad (3)$$

The random data $V(t_i) - V(S_i)$ where $i = 1, 2, 3 \ldots\ldots n$ are stochastically independent. $Z(t)$ is a continuous function of t. The standard normal distribution is denoted as ϕ. If a random data x_i will be independent and distributed randomly with mean value μ and variance 1, then it will be as Eq. (4).

$$S_n = x_1 + x_2 + \ldots. + x_n \text{ where } n = 0, 1, 2, \ldots \text{ and } V(n) = 0, 1, 2, \ldots \forall \\ \{\ V(t), 0 \leq t < \infty\} \quad (4)$$

To implement the time in a discrete way, a random selection S_n, $n = 0, 1, 2, \ldots$ is considered. In this chapter, a random data selection is performed to avoid the biasness of the proposed model toward the particular class of data. To achieve this target, data sample with discrete time interval, the likelihood ratio of the $\{V(S), S \leq t\}$ under P_{μ_1} relative to P_{μ_2} will be as Eq. (5)

$$l\,(t, V(t); \mu_1, \mu_2) = \exp\left[(\mu_1 - \mu_2)\ V(t) - \tfrac{t}{2}\ (\mu_1^2 - \mu_2^2)\ \right] \\ \text{for any } -\infty < \mu_i < \infty \text{and } (i = 1, 2 \ldots) \text{ and } t > 0 \quad (5)$$

4.2 *Ant Colony Optimization (ACO)*

To achieve the desired output, a bio-inspired intelligent modll based on Ant Colony Optimization technique is deployed. At the beginning, a statistical analysis is performed to understand the data pattern and to visualize the data.

The Ant Colony Optimization metaheuristic's limited history is primarily one of experimental research. All early researchers were directed by trial and error, and most ongoing research efforts are still guided by trial and error. This is the normal scenario for almost all known metaheuristics: when experimental work has demonstrated the usefulness of the metaheuristic, researchers attempt to develop a theory to explain how it functions in addition to doing ever-more complex tests. Convergence is typically the first theoretically motivated topic to be addressed. If sufficient resources are available, will the metaheuristic discover the optimal solution? Numerous issues are frequently looked into, including the speed of convergence, ethical methods for choosing the metaheuristic's parameters, connections to preexisting strategies, identification of problem traits that increase the metaheuristic's likelihood of success, and comprehension of the significance of the various metaheuristic components. Here, we discuss the problems for which we currently have an answer. We focus on how ACO compares to other well-known techniques like stochastic gradient ascent as well as how various ACO algorithms converge to the best result.

4.2.1 Theoretical Considerations on ACO

The relatively broad concept of when attempting to demonstrate the theoretical merits of the ACO metaheuristic, ACO [29] presents the researcher with their first significant challenge. Generality is a desirable property that makes theoretical analysis much more difficult, if not impossible, because it enables ant-based algorithms to be applied to discrete optimization problems ranging from static problems like the traveling salesman problem to time-varying problems like routing in telecommunications networks. The convergence problem is the first theoretical aspect of ACO that we will look at in this chapter. Is it conceivable that the algorithm will eventually identify the ideal solution? ACO algorithms are stochastic search processes, and they may never find the best answer due of the bias provided by pheromone trails. This makes the topic intriguing. It is crucial to bear in mind that there are at least two different forms of convergence to take into account when analyzing a stochastic optimization algorithm: value convergence and solution convergence. Assuming indirectly that in the event of a problem with several optimum solutions, we are interested in assessing the likelihood that the algorithm will produce an ideal answer at least once while investigating convergence in value. We want to determine the likelihood that the algorithm will provide an ideal result at least once. Convergence of values for ACO algorithms in ACObs; convergence of tmin and solutions for ACO algorithms in ACObs; small tmin parameter may vary during an algorithm run, as shown by tminy, and the iteration counter for the ACO algorithm is y. We next demonstrate that these findings hold even when common ACO components like search engine and heuristic information are taken into account. Finally, we discuss the consequences of these discoveries and demonstrate that the MMAS and ACS ACO algorithms, two of the most successful ACO algorithms now in use, directly benefit from the evidence of value convergence. Sadly, there are currently no findings on the rate of convergence of any ACO algorithm. Even if convergence may be demonstrated, there is presently no alternative to conducting large experimental procedures in order to assess algorithmic performance. The formal relationship between ACO and other techniques is another theoretical subject explored in this chapter. We embed ACO in the broader model-based search framework, to better comprehend the relationships between ACO, stochastic gradient ascent, and the more modern cross-entropy method [30, 31]. To find out the optimize solution of the users' rating, ACO, a metaheuristic algorithm have been applied. In this chapter, the users are considered ants and their corresponding rating are considered as pheromone. The users give their rating based on the choice. Similarity index is determine based on the rating value. The users too change their choice from movie I to movie j with probability.

$$P_{i,j} = \frac{\left(\tau_{i,j}^{\alpha}\right)\left(\eta_{i,j}^{\beta}\right)}{\sum\left(\tau_{i,j}^{\alpha}\right)\left(\eta_{i,j}^{\beta}\right)} \tag{6}$$

where

$\tau_{i,j}$ → The amount of pheromone (Rating) on the edge i, j

α → Parameter to control the influence of $\tau_{i,j}$

$\eta_{i,j}$ → The desirability of edge i, j

β → A parameter to control the influence of $\eta_{i,j}$

Algorithm

A well-known bio-inspired algorithm called Ant Colony Optimization is deployed for recommending movies

Algorithm: Ant Colony optimisation for movie recommendation
Begin
Input: Movielens dataset
Output: Optimistic decision making for movie recommendation
Initialization: A set of data from movie lens dataset is used for pilot survey.
Step 1: Randomly select the data from the dataset.
Step 2: User U with m numbers of features. $U = \{f_1, f_2, f_3, \ldots\ldots\ldots, f_m\}$
Step 3: Consider the total number of users is n. $U = \{U_1, U_2, U_3, \ldots\ldots\ldots, U_n\}$
Step 4: Determine the co-variance of the users' with their features.
Step 5: Pheromone update User rating "r," will rate the next rating "s" based on the following formula. $s = \begin{cases} \max_{j \in N_i^k} \left\{ \tau_{ij}(t).\eta_{ij}^{\beta} \right\} & \text{if } q \prec Q_0 \\ \text{according to AS equation} & \text{otherwise} \end{cases}$ N_i^k are the unrated movies of Kth user $\tau_{ij}(t)$ is rating in (i,j) edge during t turn and $\eta_{i,j}$ is the distance between user i to user j and "q" is the stochastic variable with uniform distribution. Q_0 is the threshold value in which initializing procedure begins.
End

4.3 *Data Preparation*

To validate the proposed model, a real-time dataset named "movies.csv" [32] is used as a test-bed.

5 Results and Discussion

A real-time dataset is used as a test-bed to validate the proposed model. In this research, 5000 data are considered. The users' ratings are visualized in Fig. 4.

The rating of the users' is varying based on their choice. Figure 5 shows the distribution of users' ratings. Some users' ratings are outliers. In this chapter, 5000 data is considered from Movielens and the outlier data is also considered for better results.

Figure 6 shows the difference in ratings between the user 20 and other users. It displays the distance between User Id 20 and other User Ids (Fig. 7).

5.1 *Performance Analysis*

Performance analysis based on various parameters of the confusion matrix is developed to assess the effectiveness of the proposed model. The fourfold cross-validation method is used on the provided data set in this chapter. Table 1 displays the various ratios that were chosen for the training and testing data sets over four

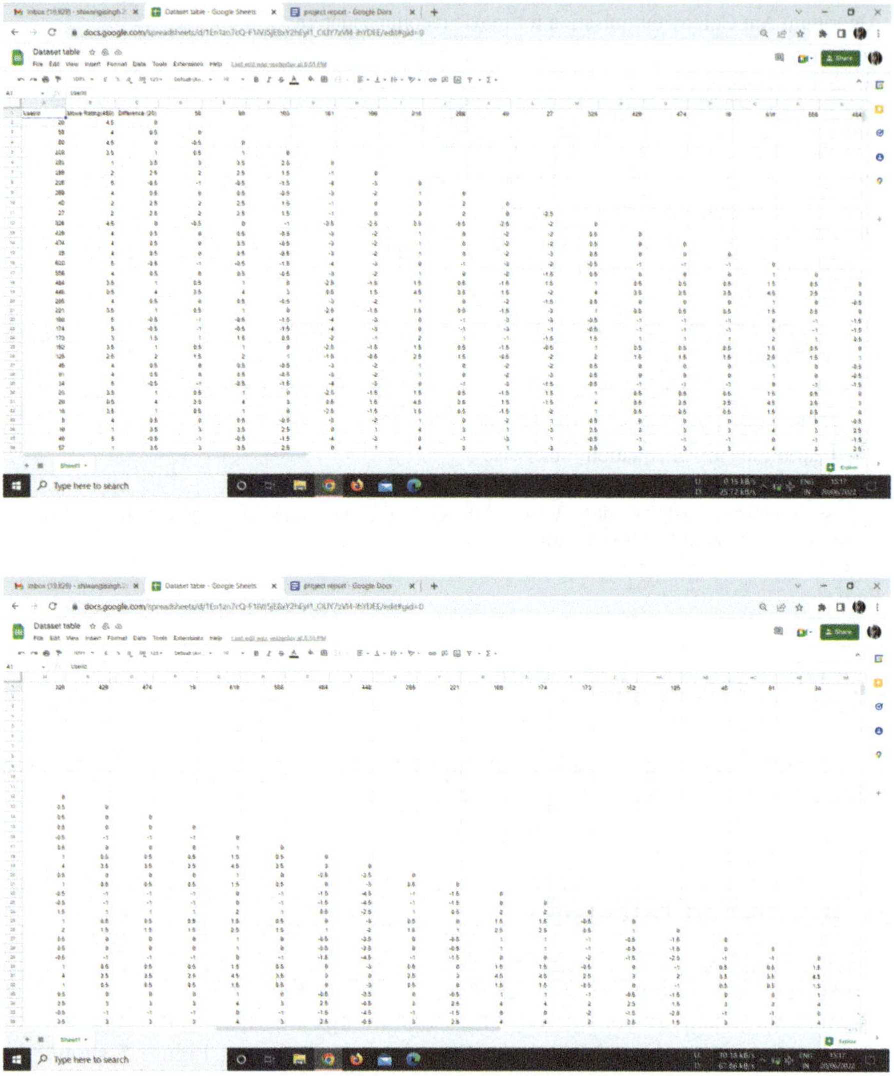

Fig. 4 Snapshot of difference in users and their ratings

iterations. The error is fixed after each iteration to make the model more accurate. The suggested model is used to provide users with the best recommendations based on their preferences.

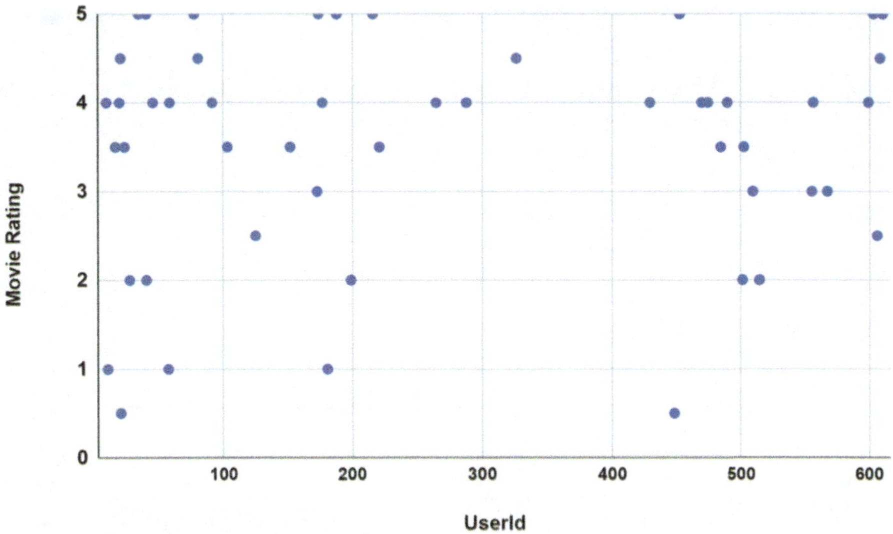

Fig. 5 Distribution of the users' rating

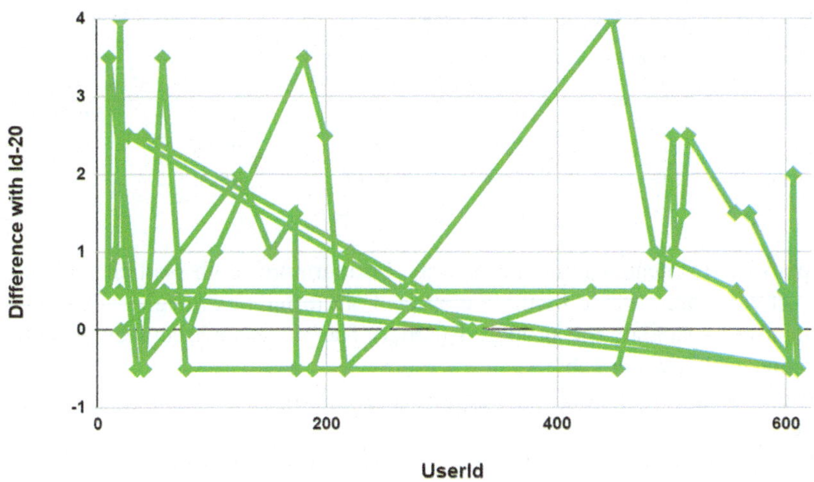

Fig. 6 Distance between user Id 20 and other users

5.1.1 Evaluation Metrics

Evaluation metrics are frequently used to gauge how accurate a recommendation system is. In areas like security, image processing, and information retrieval, the evaluation metrics are very helpful. In general, evaluation metrics make it simple to verify the effectiveness and efficiency of recommenders. Thus, fictitious frameworks could become actual applications. On the other hand, it identifies shortcomings,

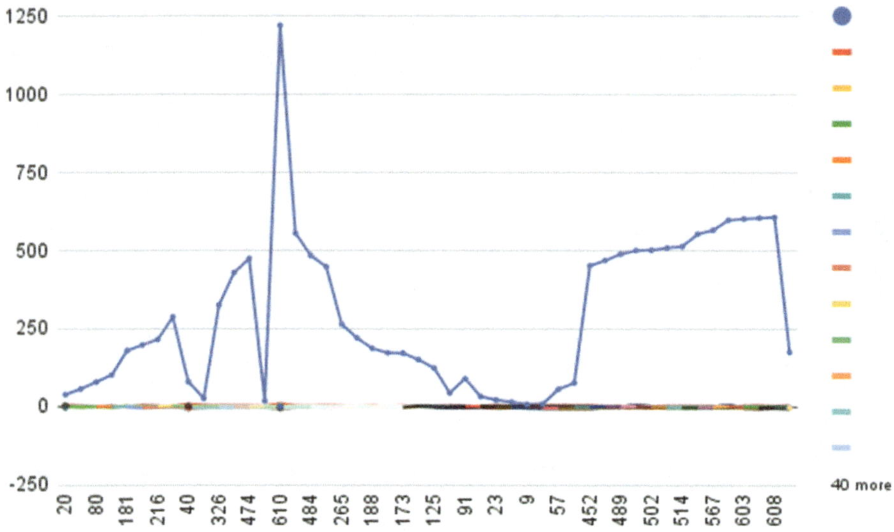

Fig. 7 Distribution of random user's ratings

Table 1 Performance at four different strategies

Training and testing strategy	Strategy 1 (50%–50%)	Strategy 2 (60%–40%)	Strategy 3 (70%–30%)	Strategy 4 (80%–20%)	Average
Accuracy	91.56	90.75	95.88	98.25	94.11
Recall	88.68	92.45	94.84	96.68	93.16
Precision	90.34	91.64	92.42	95.24	92.41

accuracy, and potential remedial actions. Furthermore, both explicit and implicit user activities are covered by these metrics. Four different strategies are applied for the dataset to validate the proposed model. The values of Accuracy, Recall, and Precision are shown in Table 1.

The average value of the Accuracy, Recall, and Precision, from Table 1, is shown in Fig. 8.

The average values of the Accuracy, Recall, and Precision are 94.11, 93.16, and 92.41, respectively.

6 Conclusion

The media sector has seen a significant amount of research work focused on various aspects of big data analysis, including the indexing, capture, storage, mining, and retrieval of multimedia big data. The entertainment industry now does a huge amount of business online in this digital age. During the pandemic, when social isolation is required, the idea of the entertainment through online become popular.

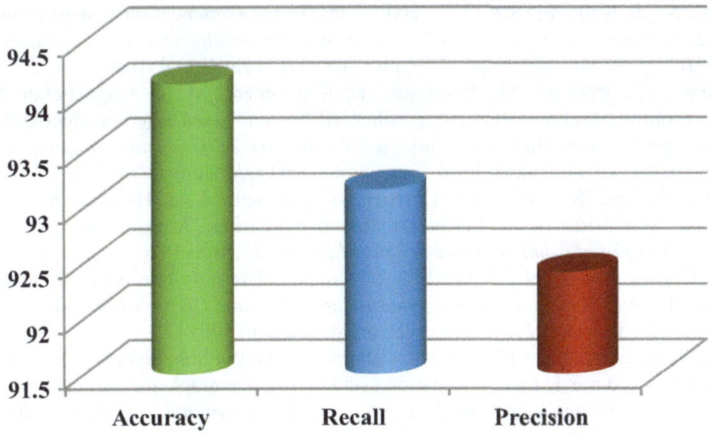

Fig. 8 Average value of the evaluation metrics

Industries use the idea of big data to represent customer feedback, comments, and choices in order to provide better service. In this chapter, an intelligent algorithm that makes an optimistic decision to develop a collaborative filtering system that predicts a user's movie rating based on a sizable user rating database. It recommends movies that are the best fit for users based on the genres of media they enjoy. The list of recommended movies is generated by the combined impact of user ratings and reviews. In order to create a pilot survey model to analyze the real-time dataset, a statistical analysis is carried out. Ant Colony Optimization (ACO) is used to rate the group members in order to make recommendations in the future. In a recommender system, sparsity problems will be optimized in this manner. As a result, the suggested model could be used to successfully train a recommender. The experimental findings show that there are very few false positive and false negative recommendations. Additionally, the abundance of data does not affect the model's accuracy. The results show that the proposed model's Accuracy, Recall, and Precision are 94.11, 93.16, and 92.41, respectively. Future research will focus on using deep learning-based intelligent models to reduce sparsity and enhance Precision and Recall values for recommendation systems in the context of big data.

References

1. Sama, L., Wang, H, & Makkar, A. (2021). Movie recommendation system using deep learning. In *9th International Conference on Orange Technology (ICOT)*, Tainan. https://ieeexplore.ieee.org/document/9680609
2. Purushothaman Srikanth, E. Ushitaasree, S. M. Bhargav Bhattaram, G., & Anand, P. (2021). Movie recommendation system using deep autoencoder, In *5th International Conference on Electronics, Communication and Aerospace Technology (ICECA)*, Coimbatore. https://ieeexplore.ieee.org/document/9675960

3. Salmani, S., & Kulkarni, S. (2021). Hybrid movie recommendation system using machine learning. In *International Conference on Communication information and Computing Technology (ICCICT)*, Mumbai. https://ieeexplore.ieee.org/document/9510058
4. Sunandana, G., Reshma, M., Pratyusha, Y., Kommineni, M., & Gogulamudi, S. (2021). Movie recommendation system using enhanced content-based filtering algorithm based on user demographic data. In *6th International Conference on Communication and Electronics Systems (ICCES)*, Coimbatore. https://ieeexplore.ieee.org/document/9489125
5. Roy, A., Banerjee, S., Sarkar, M., Darwish, A., Elhosen, M., & Hassanieen, A. E. (2018). Exploring New Vista of intelligent collaborative filtering: A restaurant recommendation paradigm. *Journal of Computational Science, Elsevier, 27*, 168–182.
6. Christ Zefanya Omega, H. (2021). Movie recommendation system using weighted average approach. In *2nd International Conference on Innovative and Creative Information Technology (ICITech)*, Salatiga. https://ieeexplore.ieee.org/document/9590147
7. Ajith, T. T., Ajay Krishnan, C. V., Nandakishore, J., Ananthu Subramanian, M. S., Siji Rani, S. (2021). Enhanced movie recommendation using knowledge graph and particle filtering. In *2nd International Conference on Smart Electronics and Communication (ICOSEC)*, Trichy. https://ieeexplore.ieee.org/document/9591834
8. Xiong, W., & He, C. (2021). Personalized movie hybrid recommendation model based on GRU. In *4th International Conference on Robotics, Control and Automation Engineering (RCAE)*, Wuhan. https://ieeexplore.ieee.org/document/9638949
9. Wang, H. (2021). MovieMat: Context-aware movie recommendation with matrix factorization by matrix fitting. In *7th International Conference on Computer and Communications (ICCC)*, Chengdu. https://ieeexplore.ieee.org/document/9674549
10. Soni, N., Kumar, K., Sharma, A., Kukreja, S., & Yadav, A. (2021). Machine learning based movie recommendation system. In *IEEE International Conference on Parallel & Distributed Processing with Applications, Big Data & Cloud Computing, Sustainable Computing & Communications, Social Computing & Networking*, Dehradun. https://ieeexplore.ieee.org/document/9667602
11. Qiu, G., & Guo, Y. (2021). Movie big data intelligent recommendation system based on knowledge graph. In *IEEE International Conference on Parallel & Distributed Processing with Applications, Big Data & Cloud Computing, Sustainable Computing & Communications, Social Computing & Networking*, New York City. https://ieeexplore.ieee.org/document/9644835
12. Lavanya, R., & Bharathi, B. (2021). Systematic analysis of Movie Recommendation System through Sentiment Analysis. In *International Conference on Artificial Intelligence and Smart Systems (ICAIS)*, Coimbatore. https://ieeexplore.ieee.org/document/9395854/authors#authors
13. Agrawal, S., & Jain, P. (2017). An improved approach for movie recommendation system. In *International Conference on I-SMAC (IoT in Social, Mobile, Analytics and Cloud) (I-SMAC)*, Palladam. https://ieeexplore.ieee.org/document/8058367
14. Roy, S., Sharma, M., & Singh, S. K. (2019). Movie recommendation system using semi-supervised learning. In *Global Conference for Advancement in Technology (GCAT)*, Bangalore. https://ieeexplore.ieee.org/document/8978353
15. Xu, Q., & Han, J. (2021). The construction of movie recommendation system based on Python. In *IEEE 2nd International Conference on Information Technology, Big Data and Artificial Intelligence (ICIBA)*, Chongqing. https://ieeexplore.ieee.org/document/9687872
16. Zhou, T., Chen, L., & Shen, J. (2017). Movie recommendation system employing the user-based CF in cloud computing. In *IEEE International Conference on Computational Science and Engineering (CSE) and IEEE International Conference on Embedded and Ubiquitous Computing (EUC)*, Guangzhou. https://ieeexplore.ieee.org/document/8005971
17. Zhang, J., Wang, Y., & Yuan, Z. *Personalized real-time movie recommendation system: Practical prototype and evaluation.* Tsinghua Science and Technology. https://ieeexplore.ieee.org/document/8821512
18. Hwang, S., & Park, E. (2021). Movie recommendation systems using actor-based matrix computations in South Korea. In *IEEE Transactions on Computational Social Systems*.https://ieeexplore.ieee.org/document/9566476

19. Kharita, M. K., Kumar, A., & Singh, P. Item-based collaborative filtering in movie recommendation in real time. In *2018 First International Conference on Secure Cyber Computing and Communication (ICSCCC)*, Jalandhar. https://ieeexplore.ieee.org/document/8703362
20. Sarkar, M., Roy, A., Agrebi, M., & AlQaheri, H. (2022). Exploring new vista of intelligent recommendation framework for tourism industries: An itinerary through big data paradigm. *Information, 13*(2), 70. https://doi.org/10.3390/info13020070
21. Anandkumar, R., Dinesh, K., Obaid, A. J., Malik, P., Sharma, R., Dumka, A., Singh, R., & Khatak, S. (2022). Securing e-Health application of cloud computing using hyperchaotic image encryption framework. *Computers & Electrical Engineering, 100*, 107860, ISSN 0045-7906. https://doi.org/10.1016/j.compeleceng.2022.107860
22. Sharma, R., Xin, Q., Siarry, P., & Hong, W.-C. (2022). Guest editorial: Deep learning-based intelligent communication systems: Using big data analytics. *IET Communications*. https://doi.org/10.1049/cmu2.12374
23. Sharma, R., & Arya, R. (2022). UAV based long range environment monitoring system with Industry 5.0 perspectives for smart city infrastructure, 108066, ISSN 0360-8352. *Computers & Industrial Engineering, 168*. https://doi.org/10.1016/j.cie.2022.108066
24. Rai, M., Maity, T., Sharma, R., et al. (2022). Early detection of foot ulceration in type II diabetic patient using registration method in infrared images and descriptive comparison with deep learning methods. *The Journal of Supercomputing*. https://doi.org/10.1007/s11227-022-04380-z
25. Sharma, R., Gupta, D., Maseleno, A., & Peng, S.-L. (2022). Introduction to the special issue on big data analytics with internet of things-oriented infrastructures for future smart cities. *Expert Systems, 39*, e12969. https://doi.org/10.1111/exsy.12969
26. Sharma, R., Gavalas, D., & Peng, S.-L. (2022). Smart and future applications of Internet of Multimedia Things (IoMT) using big data analytics. *Sensors, 22*, 4146. https://doi.org/10.3390/s22114146
27. Sharma, R., & Arya, R. (2022). Security threats and measures in the internet of things for smart city infrastructure: A state of art. *Transactions on Emerging Telecommunications Technologies*, e4571. https://doi.org/10.1002/ett.4571
28. Zheng, J., Wu, Z., Sharma, R., & Lv, H. (2022). Adaptive decision model of product team organization pattern for extracting new energy from agricultural waste, 102352, ISSN 2213-1388. *Sustainable Energy Technologies and Assessments, 53*(Part A). https://doi.org/10.1016/j.seta.2022.102352
29. Mou, J., Gao, K., Duan, P., Li, J., Garg, A., & Sharma, R. (2022). A machine learning approach for energy-efficient intelligent transportation scheduling problem in a real-world dynamic circumstances. In *IEEE Transactions on Intelligent Transportation Systems*.https://doi.org/10.1109/TITS.2022.3183215
30. Priyadarshini, I., Sharma, R., Bhatt, D., et al. (2022). Human activity recognition in cyber-physical systems using optimized machine learning techniques. *Cluster Computing*. https://doi.org/10.1007/s10586-022-03662-8
31. Priyadarshini, I., Alkhayyat, A., Obaid, A. J., & Sharma, R. (2022, 103970, ISSN 0264-2751). Water pollution reduction for sustainable urban development using machine learning techniques. *Cities, 130*. https://doi.org/10.1016/j.cities.2022.103970
32. Pandya, S., Gadekallu, T. R., Maddikunta, P. K. R., & Sharma, R. (2022). A study of the impacts of air pollution on the agricultural community and yield crops (Indian context). *Sustainability, 14*, 13098. https://doi.org/10.3390/su142013098

Techniques to Identify Image Objects Under Adverse Environmental Conditions: A Systematic Literature Review

Navjot Kaur, Kalpana Sharma, and Anuj Jain

1 Introduction

Numerous resources are available from various digital and nondigital sources. Over time, scholars from different areas have started exploring various fields of study to gather information from different fields. For traffic surveillance and security surveillance, moving object detection and segmentation are critical. Detecting moving objects in dynamic environments is more difficult than it is in static environments. The inclination of research scholars and industries to transform the quality and quality of unstructured data has increased over time. Many data are available in the field of object identification and classification for researchers dig into to investigate various techniques for the identification of objects.

Furthermore, thanks to the immense growth and availability of online resources, users have a lot of exposure to various ideas, approaches, opinions, and recommendations on various methods. Such ample numbers of data open up new opportunities for scholars to analyze the existing techniques in their area of interest. Image processing has various applications in different fields of study. Among all these applications, the contributions of image-processing techniques and computer vision in the field of security and surveillance are remarkable. Moving object surveillance is an active area of research that detects, identifies, and tracks objects from a moving sequence of images. Objects in the video frame sequence are identified for video

N. Kaur (✉)
Bhagwant University, Ajmer, India

Lovely Professional University, Phagwara, Punjab, India
e-mail: navjot.20506@lpu.co.in

K. Sharma
Bhagwant University, Ajmer, India

A. Jain
Lovely Professional University, Phagwara, Punjab, India

R. Sharma et al. (eds.), *Data Analytics for Internet of Things Infrastructure*,
Internet of Things, https://doi.org/10.1007/978-3-031-33808-3_11

sequencing [1]. In order to track objects from moving vehicles or video sequencing of moving vehicles, it is important to detect an object that appears first in the video sequence or detect objects from every image frame of a video sequence. Using a good-quality, high-intensity video camera is required to capture and acquire inputs for high effectiveness and clear object detection. The identification of objects in videos follows three important steps:

1. Detection of object of interest
2. Tracking objects from each image frame
3. Identification of object behavior in an image

Morphology provides the operations for analyzing objects of different forms and shapes. This aids in object analysis and recognition. The required information is extracted from the image for analysis to help yield an improved image. Morphological procedures are contingent on the comparative assembling of pixel values [2]. The structuring component supports determining the method into which the structuring element fits to identify an image.

1.1 Morphological Operations on Image

1. *Erosion*: In this process, boundaries are eroded away. This shrinks the object. The mathematical erosion of image I can be defined as follows:

$$I + T = \{G \,||\, [\,(T)A \in I \tag{1}$$

Erosion reduces the size of objects by etching the object borders. Structural elements pass through all the pixels of the image.

2. *Dilation*: This operation allows objects to expand, filling in small holes and connecting disjointed objects.

$$I + T = \{G \mid [(T)\ A \cap I] \in I\} \tag{2}$$

For descriptor extraction, grayscale image representations are employed. Such representations simplify and minimize the computing requirements [21–23] (Fig. 1).

Fig. 1 Image scale

Fig. 2 Moving vehicle in (**a1**) dust wind, (**b1**) rain, (**c1**) hail storm, and (**d1**) fog

1.2 Impact of the Environment on Objects

Humans infer object-to-object interactions, part-to-whole object hierarchies, object properties, and 3D scene structures in addition to identifying and finding items in a scene. Having a better understanding of situations would help with applications such as robotic interactions, which often need information beyond item identification and position. This requires not only scene perception but also knowledge of the physical surroundings. An image has pixels, and every pixel of an image has a numerical intensity value. In order to classify images, it's very important to identify objects that are static or moving, and they must be clearly visible. Many factors influence the visibility of the image. One such key factor includes environmental conditions such as fog, rain, snow, dust, and so on (Fig. 2).

The environment also significant affects the identification of images. Images taken in hazy or foggy weather are difficult to identify because of their unclarity. Fog and dust, in particular, drastically diminish visibility distance. Because of light dispersion and attenuation, the color of nearby objects seems to be quite similar, with poor saturation. Under such circumstances, it is difficult to distinguish between objects because the edges between the background and the item in the foreground become blurred. Many cameras are installed in modern automobiles, and they are used for a wide range of purposes. The detection of fog from images captured by a photographic camera mounted on a vehicle is a difficult task that could be useful in a variety of situations [3]. So far, methods have focused on the attributes of nearby items in the image, such as lane markings, traffic signs, vehicle backlights, and approaching vehicle headlights. In contrast to all these previous studies, some researchers suggest adopting methods that utilize image descriptors and take a cataloging approach to distinguish between images with fog and those without fog.

2 Methodology and Research Description

Image-processing techniques are used to visibly improve image appearance. These techniques help to improve the interpretation of an image by a human or an automated system. Image-processing techniques may also be used to identify

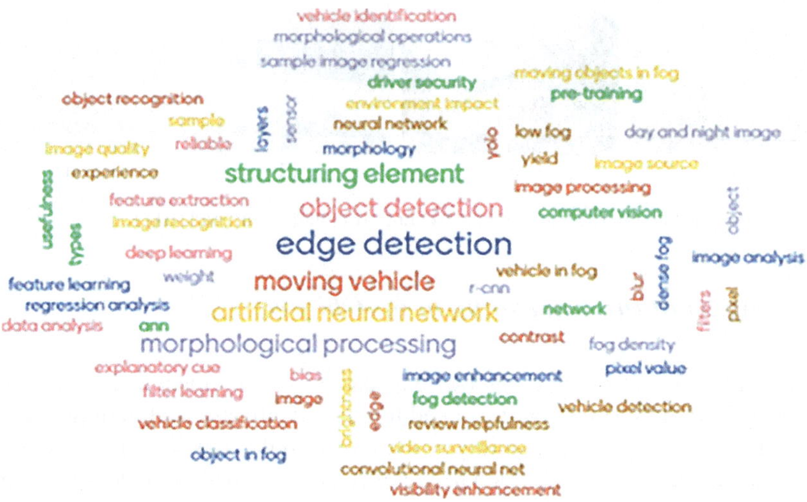

Fig. 3 Keyword cloud of the selected research papers

images and vehicles under adverse weather circumstances, such as fog, rain, hail, etc. Table 2 presents research conducted in similar fields.

Some researchers have proposed conducting a literature review on studies that aimed to identify objects. For the analysis and identification of objects, the Scopus database has been selected because this database indexes a wide range of engineering literature from conferences and journals. The scholars also can explore a wide range of research articles from the Scopus database. For this study, papers from 2011 to 2022 are included for analysis. For the selection of research papers, we focused on keywords such as "object detection," "object recognition," "moving vehicles in the fog," "vehicle in the fog," "morphology," "neural network," and "deep learning." We also used Boolean connectors "AND," "OR," and the symbol "+" to influence the search so that more specific and meaningful data could be gathered. In order to retrieve the desired research papers, a search using keywords such as "object identification," "fog," and "moving vehicle" was conducted; the search yielded 7300 research papers in the first attempt. To refine the results, the second phase of the search for studies published in journals was limited to 1100 publications. Further, in the third phase, we focused primarily on object identification in fog, because of which the number of research papers dropped to 423 (Figs. 3 and 4).

3 Findings and Results

This section is divided into three sections: year-by-year statistics, journal-by-journal statistics, and theme-by-theme reporting. In the first subsection, the study displays

Fig. 4 Keywords used for paper searching

the year-by-year distributions of publications and a list of periodicals. In the second subsection, the study shows the high-frequency words from the title and author keywords of the publications studied. In the third subsection, the study shows diverse themes across several image-processing approaches. Various functional factors, such as contrast-sensitive images, brightness, transparency, texture gradient, and light, affect the visibility of an image display. Further, a few more key performance indicators, also known as image-quality factors, influence image quality and visibility, such as image sharpness, noise, dynamic range, color accuracy, alterations, homogeneity, blaze, artifacts, compression, and links. In addition to these factors, a few environmental factors also influence image visibility. Such factors include rain, hail, dusty wind, and fog. Various researchers have carried out significant research on all these factors and their effects, and they have proposed suggestive and corrective techniques to overcome all these problems, except for fog.

Even though a few researchers have conducted research on object identification in foggy weather, they have not put any significant weight on the identification of objects in moving vehicles in fog or video sequencing of the vehicles moving in fog. This leaves a huge gap to fill. We have searched through research papers in this field and have identified and focused on their techniques for identifying objects in fog.

This section is divided into three subsections: year-by-year statistics, journal-by-journal statistics, and theme-by-theme reporting. The first subsection displays the year-by-year distributions of research articles and a list of periodicals. The study shows high-frequency words from the title and author keywords of the articles studied in the second subsection. The study shows the diverse themes across several image-processing approaches in the third subsection. Various researchers have proposed different methods for identifying objects in moving vehicles so that the vehicles themselves can also be identified (Figs. 5 and 6).

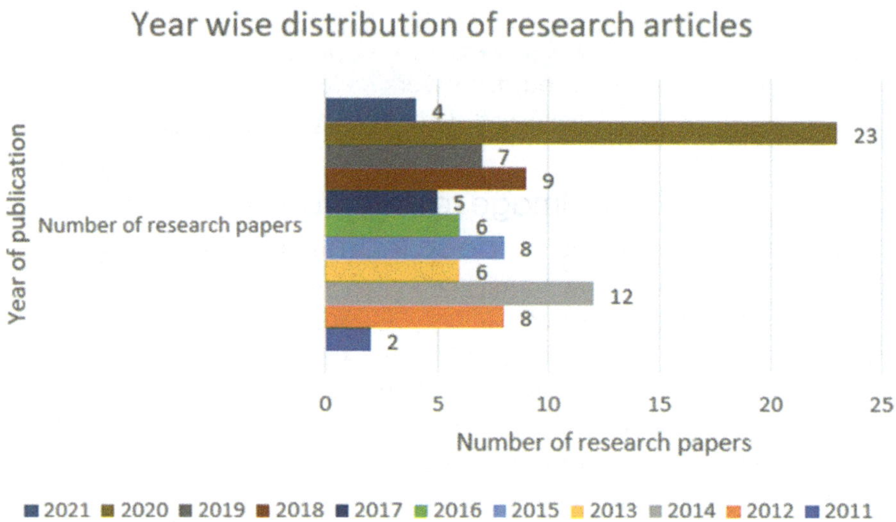

Fig. 5 Year-by-year distributions of research articles

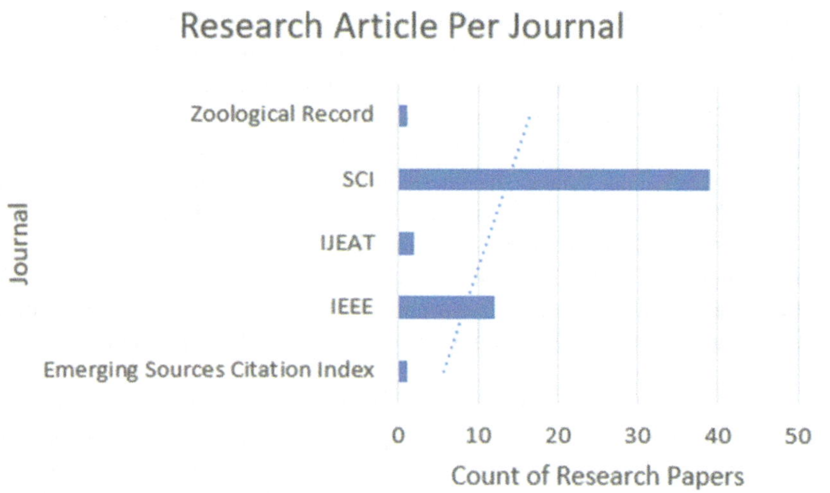

Fig. 6 Research articles per journal

All the referenced papers were analyzed, read, and categorized according to vehicle identification and fog estimation. Various researchers have proposed different techniques for vehicle identification under adverse environmental conditions, such as rain, haze, and fog. Researchers classify fog density as a visibility feature of a vehicle preceding another vehicle and the distance to that other vehicle. Researchers proposed a method to recognize daytime images in dense fog and applied Fourier transformation with global features. The proposed method was applied to many

Table 1 Visibility distance range

Visibility distance (meters)			
1000	No fog	100–300	Fog
300–1000	Low fog	100	Dense fog

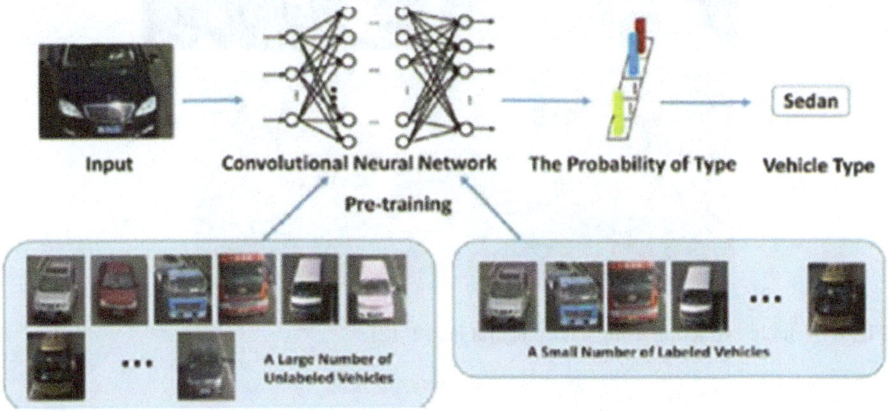

Fig. 7 Representation of the analyzed works that use local features [1–4]

images. Here, 96% of fog-free photographs were categorized as fog-free, whereas 93% of fog images were classified as fog images. The researchers used annotated photos with fog values, as shown in Table 1. The data did not include profiles of roads [4]. To categorize fog, various calculations have been performed on the basis of visibility distance.

While following the proposed methodology, researchers assessed fog images and fog-free images. Afterwards, they proposed using a semisupervised convolutional neural network. A sparse Laplacian filter was applied to gather information about the vehicles. The SoftMax classifier layer was used as the output layer on the labeled vehicles. We worked with the BIT-Vehicle data set. This data set includes 9850 images to test the proposed method. Only 10% of the images were nighttime images [3, 4] (Fig. 7).

First, foreground objects are extracted from the video. After this, the hierarchical multi-SVMs (support vector machines) method is applied for vehicle classification. For final precision, a voting-based correction scheme is used. Vehicle classification is achieved in complicated traffic scenes by taking the proposed approach. Singh et al., Zhuo et al., Murugan et al., Chowdhury et al. presented a vehicle classification technique that is based on neural networks. In the pretraining stage, GoogLeNet is applied to the ILSVRC-2012 data set, and after fine-tuning, a vehicle data set of 13,700 images is used for the final classification [5–8]. Liu et al. have proposed techniques to detect the number of vehicles on the road in real time. They applied image subtraction on foreground and background images and used image binarization, a method of counting objects that results in a faster computing technique. Wang et al. have proposed classifying images on the basis of deep learning and generative

Fig. 8 Vehicle extraction using a deep neural net [9–12]

adversarial networks (GANs) [9, 10]. Jyothi et al. have proposed using a high-accuracy convolutional neural network (CNN) technique for vehicle classification. The vehicles in focus were cars and trucks [11]. Morphological operations were applied in vehicle detection and identification. Video sequencing of vehicles was reviewed by Chandrika et al., and frames were extracted from it [12] (Fig. 8).

Kalyan et al., Shyamala et al., Şentaş et al., Hedeya et al., Zahra et al., Jagannathan et al., Miclea et al. have applied image binarization and Sobel edge detection. Morphological operations were applied to the image to identify objects. Researchers used their own video data set. They tested the SVMs classifier and tiny-YOLO on a data set, and their paper proposed an adaptive histogram equalization and the Gaussian mixture model. Researchers have implemented a technique to enhance vehicle image quality and detected vehicles from denoised images [13–19]. Kim et al. have proposed an efficient technique for visibility measurement under foggy weather conditions [20]. International Commission on Illumination (CIE) defined *meteorological visibility distance* as the distance beyond which a black object of adequate size with a contrast of less than 5% can be seen. Jiang et al. have applied the Canny–Deriche filter [21]. The purpose of this research was to propose a framework for recovering the contrast of photographs captured from a moving vehicle. Nam and Nam first computed the weather conditions, then inferred the scene structure, which was refined during the restoration process [22]. The proposed visibility enhancement algorithm was not designed for road photos [23]. Hautière et al. proposed techniques in which the fog region is segmented using the calculation from the direction charts. They also computed fog density and used a method that restored the contrast and assumed that the road was flat, to detect vertical objects [24]. Abbaspour et al. used a technique on a single in-vehicle camera. This method has better performance in terms of accuracy and speed to detect fog density from images [25]. Hautière et al. proposed image improvement techniques appropriate under daytime fog conditions in differential geometries, where the partition of unity

Fig. 9 Image identification by applying image enhancement [24–27]

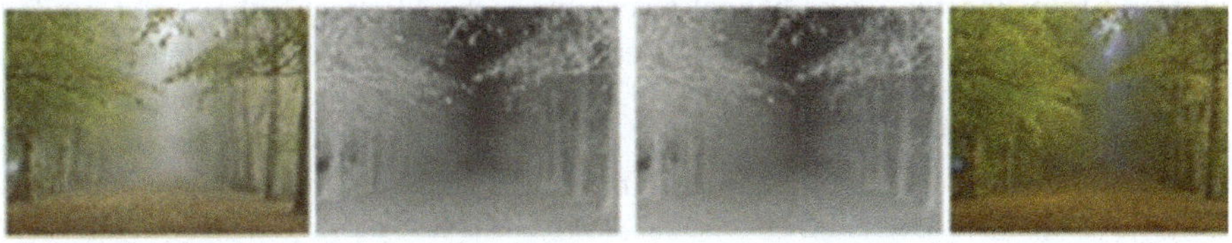

Fig. 10 Image identification by removing haze from an image [28–30]

was the base of this proposed method. This model was the most suitable for contrast restoration under foggy conditions [26]. These researchers suggested an algorithm for reducing the turbidity of an image. In their method, they assumed that an image has a reference intensity level and a characteristic intensity level. A low-pass filter was used to obtain the reference intensity level: intensity level = original intensity level – reference intensity level [27] (Fig. 9).

Negru et al. conducted an experiment to contrast an image by providing quantifiable proof that road safety is increased by Advanced Driver Assistance Systems (ADAS). Next, using a modified Piéron's rule as a foundation, a quantitative model was developed for target visibility (Vt), which is calculated from onboard camera images [28]. Other researchers presented a pixel-based technique to eradicate haze artifacts from images by using a single image-based dehazing framework. Halmaoui et al. recommended conducting a haze density analysis to determine the level of atmospheric light. A transmission map can then be estimated and refined by using a bilateral filter [29] (Fig. 10).

For hazy photos and videos, Kim et al., Su et al. have proposed a fast and optimal dehazing method. The contract term and the data loss term were combined into a cost function by these researchers. The suggested approach improves the contrast and maintains information by minimizing the cost function. This method is expanded to real-time video dehazing from the static-image-dehazing algorithm [31, 32]. Sharma and Arya provided a technique for retrieving image data from a single blurry image. The dark-channel-prior (DCP) algorithm has a tendency to underestimate bright region transmission. The predicted value of the shady network of a hazy image was utilized as an estimation of this offset to adjust the transmission because the intensity in a dark channel that was influenced by haze created the same offset [33].

For the real-time processing of haze removal from high-definition videos, a GPU-accelerated parallel computing solution was proposed [34]. For real-time processing, Rai et al. proposed a single image haze reduction approach that implements hardware. The suggested method uses computationally efficient image-processing techniques [35]. This research investigated aureole pieces generated by scattering and nonuniformly dispersing lighting in low-light foggy circumstances. To reduce the influence of multiple scatterings, an image was subdivided into a halo and scene layer. Next, following the Retinex theory, they calculated the spatially variable ambient light [36]. In the transmission map, they employed the mean shift segmentation technique to separate the sky areas from the foreground, which were obtained by using the dark-channel-prior approach. Afterward, they used guided image filtering to smooth the map by separately increasing the brightness of each sky region in the transmission map [37]. Yuan et al. proposed dark channels prior to masking the sky regions focused on road edge recognition, and dehazing was similarly focused—together resulting in improved visibility under foggy conditions [38].

In their study, Mou et al. proposed an algorithm consisting of sky segmentation and area-wise medium-transmission-based image dehazing [39]. Another approach was proposed that first detects the sky and divides the image into sky and nonsky regions and then independently estimates the transmissions of the two sections, followed by a refining step [40]. This is an efficient strategy for improving fog-degraded traffic images. The fog-degraded image is divided into blocks. The block with the least amount of sparsity is chosen to compute the local transfer function [41]. To simulate the mathematical model for the fog, Hu et al. used a deep neural network [42].

The edge-sharpening cycle-consistent adversarial network was proposed as a generative adversarial network, namely ESCCGAN [43]. To solve the performance limitations of using an atmospheric disintegrating archetypal-based method, a residual-based dehazing net was developed [44]. The generative adversarial network (GAN) dehazing method has been used to dehaze image areas. This technique considers the many degrees of haze concentration that need to be adjusted while preserving the original image's details [45]. The suggested algorithm employs a supervised machine-learning technique to approximate the transmission medium's pixel-wise extermination factors and uses a unique compensation scheme to correct the erroneous enlargement of white objects after dehazing.

Feng et al. used an edge-preserving maximum reflectance prior (MRP) method to reduce the color effect of hazy photos taken at night. The transmission map was then obtained by feeding the hazy image with no color effect into the self-encoder net [46]. Feng et al. continued their research by considering the following parameters: contrast, intensity, image noise, image resolution, illumination, heavy occlusions, visibility distance, classification accuracy, speed, changes in illumination, aspect ratio, compactness, the accuracy of a crossing vehicle, hue, precision, accuracy, light angle, size variety, relative width, length and area, Histogram of Oriented Gradients

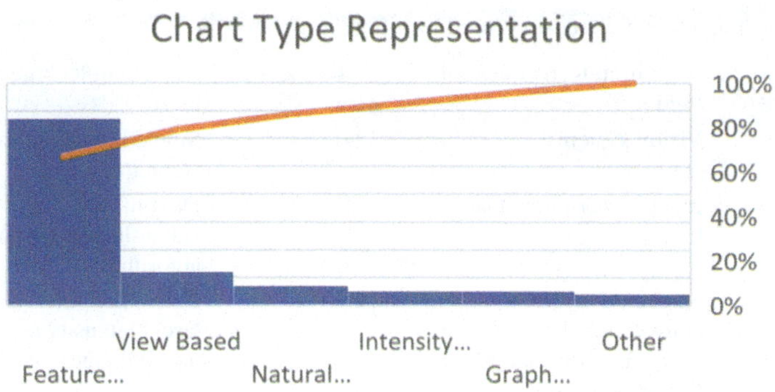

Fig. 11 Representations based on various feature types

(HOG), precision, vehicle class, classification accuracy, computation complexity, processing speed, reliability, standard deviation, entropy, visibility distance, saturation value, chroma, visible threshold, attenuation, meteorological visibility distance, object distance, camera response, pitch angle sensitivity, inflection point, processing time, reaction time, and radiance. Our proposed review is carried out on the basis of the effect that the parameters have on the input image under consideration.

Table 2 highlights the implemented techniques and the key points extracted from all the referenced research papers (Fig. 11).

4 Conclusion

According to the key findings of all the research papers referenced in the survey, a flexible fog-estimating method is required. A hybrid algorithm is not used/proposed by any of the researchers to detect the edges of moving/dynamic objects. In addition to this, none of the researchers focused on identifying moving vehicles in foggy weather, which leaves a huge gap to fill. According to the research conducted to date, researchers have proposed many techniques and methods to identify objects but have not focused much on the identification of *moving* objects. The few researchers who considered moving object identification did not include the impact of the environment on moving vehicles, especially the impact of foggy weather on the visibility and identification of moving vehicles. Therefore, an algorithm that can efficiently identify moving objects in fog needs to be developed.

Table 2 Techniques and dimensions used by different researchers

Author and year	Algorithm/technique/tool used	Author and year	Algorithm/technique/tool used
[1]	Vision system	[24]	Single in-vehicle camera technique
[2]	Image descriptors, Gabor filters	[25]	Daytime fog single-image-enhancement algorithm
[3]	Semisupervised convolutional neural network, Laplacian filter, SoftMax classifier	[26]	Calculated reference and target intensity to degrade a turbid image
[4]	Hierarchical multi-SVMs classifier	[27]	Piéron's rule-based quantitative model
[5]	Adaptive Gaussian thresholding technique, LiDAR with camera	[28]	Histogram of oriented gradient-based pedestrian detector
[6]	Convolutional neural network (CNN) model	[29]	Visibility enhancement algorithm
[7]	Artificial neural fuzzy inference system	[30]	Real-time video dehazing
[8]	Image processing, object counting methodology	[31]	Transposed filter based on a memory access pattern
[9]	Generative adversarial nets and convolutional neural network	[32]	Unified nighttime hazy-image-enhancement framework
[10]	R-CNN technique	[33]	Mean shift segmentation technique
[11]	Object identification	[34]	Dehazing technique
[12]	Erosion and dilation	[35]	Sky segmentation
[13]	Edge detection, morphological processing, segmentation using background subtraction, Sobel operator	[36]	Convolutional neural network
[14]	Neural fuzzy inference system, Gabor and log-Gabor transforms	[37]	Deep neural net
[15]	Used the tiny-YOLO with the SVMs classifier	[38]	Image-dehazing method
[16]	Deep-network-based vehicle-type classification	[39]	Fog-removal technique
[17]	Conditional generative adversarial network	[40]	Sky segmentation
[18]	Histogram equalization with Gaussian mixture model	[41]	Image improvement technique

(continued)

Table 2 (continued)

Author and year	Algorithm/technique/tool used	Author and year	Algorithm/technique/tool used
[19]	Fog detection and visibility enhancement based on sensors	[42]	Deep neural network
[20]	Canny–Deriche filter	[43]	Edge-sharpening cycle-consistent adversarial network
[21]	Contrast restoration	[44]	Gate fusion network
[22]	Visibility enhancement algorithms	[45]	Generative adversarial network (GAN) dehazing method
[23]	Fog-density computation	[46]	Supervised machine-learning technique

References

1. Pavlic, M., Belzner, H., Rigoll, G., & Ili, S. (2011). *Image based fog detection in vehicles*. IEEE.
2. Pavlic, M., Belzner, H., Rigoll, G., & Ili, S. 2012. Image based fog detection in vehicles. In *Intelligent Vehicles Symposium Alcalá de Henares*, SCI indexed.
3. Dong, Z., Wu, Y., Pei, M., & Jia, Y. (2015). Vehicle type classification using a semisupervised convolutional neural network. *IEEE Transactions on Intelligent Transportation Systems, SCI Indexed, 16*(4), 2247–2256.
4. Fu, H., Ma, H., Liu, Y., & Lu, D. (2016). A vehicle classification system based on hierarchical multi-SVMs in crowded traffic scenes. *Neurocomputing, SCI indexed, 211*, 182–190.
5. Singh, R., Singh, S., & Kaur, N. (2016). A review: Techniques of vehicle detection in fog. *Indian Journal of Science and Technology, Zoological Record, 9*(45). https://doi.org/10.17485/ijst/2016/v9i45/106793
6. Zhuo, L., Jiang, L., Zhu, Z., Li, J., Zhang, J., & Long, H. (2017). Vehicle classification for large-scale traffic surveillance videos using convolutional neural networks. *Machine Vision and Applications, SCI, 28*(7), 793–802.
7. Murugan, V., & Kumar, V. R. (2018). Automatic moving vehicle detection and classification based on artificial neural fuzzy inference system. *Wireless Personal Communications, SCI, Springer, 100*(3), 745–766.
8. Chowdhury, P. N., & Ray, T. C. (2018). A vehicle detection technique for traffic management using image processing. In *International Conference on Computer, Communication, Chemical, Material and Electronic Engineering (IC4ME2)*, SCI.
9. Liu, W., Luo, Z., & Li, S. (2018). Improving deep ensemble vehicle classification by using selected adversarial samples. *Knowledge-Based Systems, SCI, 160*, 167–175.
10. Wang, X., Zhang, W., Wu, X., Xiao, L., Qian, Y., & Fang, Z. (2019). Real-time vehicle type classification with deep convolutional neural networks. *Journal of Real-Time Image Processing, SCI, 16*(1), 5–14.
11. Jyothi, R. A., Babu, R. K., & Bachu, S. (2019). Moving object detection using the genetic algorithm for real times transportation. *International Journal of Engineering and Advanced Technology (IJEAT), 8*(6).
12. Chandrika, R. R., Ganesh, G. N. S., & Raghunath, K. M. K. (2020). Vehicle detection and classification using image processing. *IEEE Xplore*, SCI.

13. Kalyan, S. S., Pratyusha, V., Nishitha, N., & Ramesh, T. K. (2020). Vehicle detection using image processing. In *IEEE International Conference for Innovation in Technology*, SCI.
14. Shyamala, A. (2020). Certain investigations on moving vehicle detection and classification using soft computing techniques. *shodhganga*.
15. Şentaş, A., Tashiev, İ., & Küçükayvaz, F. (2020). Performance evaluation of support vector machine and convolutional neural network algorithms in real-time vehicle type and color classification. *Evolutionary Intelligence, SCI, 13*(1), 83–91.
16. Hedeya, M. A., Eid, A. H., & Abdel-Kadar, R. F. (2020). A super learner ensemble of deep networks for vehicle-type classification. *IEEE Access, SCI, 8*, 98266–98280.
17. Zahra, G., Imran, M., Qahtani, A. M., Alsufyani, A., Almutiry, O., Mahmood, A., & Alazemi, F. E. (2021). Visibility enhancement of scene images degraded by foggy weather condition: An application to video surveillance. *Computers, Materials & Continua Tech Science Press, SCI*. https://doi.org/10.32604/cmc.2021.017454
18. Jagannathan, P., Kumar, S. R., Frnda, J., Divakarachari, P. V., & Subramani, P. (2021). Moving vehicle detection and classification using Gaussian mixture model and ensemble deep learning technique. *Hindawi Wireless Communications and Mobile Computing, SCI*. https://doi.org/10.1155/2021/5590894
19. Miclea, R. C., Ungureanu, V. I., Sandru, F. D., & Silea, I. (2021). Visibility enhancement and fog detection: Solutions presented in recent scientific papers with potential for application to Mobile systems. *Sensors, SCI, 21*, 3370. https://doi.org/10.3390/s21103370
20. Kim, K., Kim, S., & Kim, K. S. (2018). Effective image enhancement techniques for fog-affected indoor and outdoor images. *IET Image Processing Research Article, SCI*. https://doi.org/10.1049/iet-ipr.2016.0819
21. Jiang, Y., Sun, C., Zhao, Y., & Yang, L. (2017). Fog density estimation and image defogging based on surrogate modeling for optical depth. *IEEE Transactions on Image Processing, SCI*. https://doi.org/10.1109/TIP.2017.2700720
22. Nam, Y., & Nam, Y. C. (2018). Vehicle classification based on images from visible light and thermal cameras. *Journal on Image and Video Processing, SCI*. https://doi.org/10.1186/s13640-018-0245-2
23. Pesek, J., & Fiser, O. (2013). Automatically low clouds or fog detection, based on two visibility meters and FSO. In *13th Conference on Microwave Techniques COMITE*.
24. Hautière, N., Tarel, J. P., & D. Aubert (2007). Towards fog-free in-vehicle vision systems through contrast restoration. In *Proc. IEEE Comput. Soc. Conf. Comput. Vis. Pattern Recognit.*, pp. 0–7. https://doi.org/10.1109/CVPR.2007.383259
25. Abbaspour, M. J., Yazdi, M., & Masnadi-Shirazi, M. (2016). A new fast method for foggy image enhancement. In *2016 24th Iranian Conference on Electrical Engineering (ICEE) 2016*, pp. 1855–1859. https://doi.org/10.1109/IranianCEE.2016.7585823
26. Hautière, N., Tarel, J. P., Halmaoui, H., Brémond, R., & Aubert, D. (2014). Enhanced fog detection and free-space segmentation for car navigation. *Machine Vision and Applications, 25*(3), 667–679. https://doi.org/10.1007/s00138-011-0383-3
27. Negru, M., & Nedevschi, S. (2013). Image based fog detection and visibility estimation for driving assistance systems. In *Proceedings, 2013 IEEE 9th International Conference on Intelligent Computer Communication and Processing 2013*, pp. 163–168, https://doi.org/10.1109/ICCP.2013.6646102
28. Negru, M., Nedevschi, S., & Peter, R. I. (2015). Exponential contrast restoration in fog conditions for driving assistance. *IEEE Transactions on Intelligent Transportation Systems, 16*(4), 2257–2268. https://doi.org/10.1109/TITS.2015.2405013
29. Halmaoui, H., Joulan, K., Hautière, N., Cord, A., & Brémond, R. (2015). Quantitative model of the driver's reaction time during daytime fog-application to a head up display-based advanced driver assistance system. *IET Intelligent Transport Systems, 9*(4), 375–381. https://doi.org/10.1049/iet-its.2014.0101
30. Yuan, H., Liu, C., Guo, Z., & Sun, Z. (2017). A region-wised medium transmission based image dehazing method. *IEEE Access, 5*(c), 1735–1742. https://doi.org/10.1109/ACCESS.2017.2660302

31. Anandkumar, R., Dinesh, K., Obaid, A. J., Malik, P., Sharma, R., Dumka, A., Singh, R., Khatak, S., & Securing e-Health application of cloud computing using hyperchaotic image encryption framework. (2022). 107860, ISSN 0045-7906. *Computers & Electrical Engineering, 100*. https://doi.org/10.1016/j.compeleceng.2022.107860
32. Sharma, R., Xin, Q., Siarry, P., & Hong, W.-C. (2022). Guest editorial: Deep learning-based intelligent communication systems: Using big data analytics. *IET Communications*. https://doi.org/10.1049/cmu2.12374
33. Sharma, R., & Arya, R. (2022). UAV based long range environment monitoring system with Industry 5.0 perspectives for smart city infrastructure, 108066, ISSN 0360-8352. *Computers & Industrial Engineering, 168*. https://doi.org/10.1016/j.cie.2022.108066
34. Rai, M., Maity, T., Sharma, R., et al. (2022). Early detection of foot ulceration in type II diabetic patient using registration method in infrared images and descriptive comparison with deep learning methods. *The Journal of Supercomputing*. https://doi.org/10.1007/s11227-022-04380-z
35. Sharma, R., Gupta, D., Maseleno, A., & Peng, S.-L. (2022). Introduction to the special issue on big data analytics with internet of things-oriented infrastructures for future smart cities. *Expert Systems, 39*, e12969. https://doi.org/10.1111/exsy.12969
36. Sharma, R., Gavalas, D., & Peng, S.-L. (2022). Smart and future applications of Internet of Multimedia Things (IoMT) using big data analytics. *Sensors, 22*, 4146. https://doi.org/10.3390/s22114146
37. Sharma, R., & Arya, R. (2022). Security threats and measures in the internet of things for smart city infrastructure: A state of art. *Transactions on Emerging Telecommunications Technologies*, e4571. https://doi.org/10.1002/ett.4571
38. Zheng, J., Wu, Z., Sharma, R., & Lv, H. (2022). Adaptive decision model of product team organization pattern for extracting new energy from agricultural waste, 102352, ISSN 2213-1388. *Sustainable Energy Technologies and Assessments, 53*(Part A). https://doi.org/10.1016/j.seta.2022.102352
39. Mou, J., Gao, K., Duan, P., Li, J., Garg, A., & Sharma, R. (2022). A machine learning approach for energy-efficient intelligent transportation scheduling problem in a real-world dynamic circumstances. *IEEE Transactions on Intelligent Transportation Systems*. https://doi.org/10.1109/TITS.2022.3183215
40. Priyadarshini, I., Sharma, R., Bhatt, D., et al. (2022). Human activity recognition in cyber-physical systems using optimized machine learning techniques. *Cluster Computing*. https://doi.org/10.1007/s10586-022-03662-8
41. Hussain, F., & Jeong, J. (2016). Visibility enhancement of scene images degraded by foggy weather conditions with deep neural networks. *Journal of Sensors, 2016*. https://doi.org/10.1155/2016/3894832
42. Hu, A., Xie, Z., Xu, Y., Xie, M., Wu, L., & Qiu, Q. (2020). Unsupervised haze removal for high-resolution optical remote-sensing images based on improved generative adversarial networks. *Remote Sensing, 12*(24), 1–20. https://doi.org/10.3390/rs12244162
43. Ha, E., Shin, J., & Paik, J. (2020). Gated dehazing network via least square adversarial learning. *Sensors (Switzerland), 20*(21), 1–15. https://doi.org/10.3390/s20216311
44. Chen, J., Wu, C., Chen, H., & Cheng, P. (2020). Unsupervised dark-channel attention-guided cyclegan for single-image dehazing. *Sensors (Switzerland), 20*(21), 1–15. https://doi.org/10.3390/s20216000
45. Ngo, D., Lee, S., Lee, G. D., & Kang, B. (2020). Single-image visibility restoration: A machine learning approach and its 4K-capable hardware accelerator. *Sensors (Switzerland), 20*(20), 1–27. https://doi.org/10.3390/s20205795
46. Feng, M., Yu, T., Jing, M., & Yang, G. (2020). Learning a convolutional autoencoder for nighttime image dehazing. *Information, 11*(9), 1–13. https://doi.org/10.3390/info11090424

Technology-Enhanced Teaching and Learning During the COVID-19 Pandemic

Hung Phu Bui, Tra Thu Dao, Thuy Thanh Dao, and Van Huong Vi

1 Introduction

The Covid-19 outbreak has affected the landscape of education worldwide. Many educational institutions have adopted technology-assisted teaching and learning as a temporary essential measure to avoid face-to-face interaction in the traditional classroom and prevent the spread of coronavirus. This teaching modality is applied in a synchronous or asynchronous mode or a combination of both, assisted by a computer or mobile device connected to the Internet.

The unprecedented shift from traditional classroom teaching and learning to remote teaching during the crisis without sufficient training and preparation has raised doubts about the effectiveness of this teaching and learning modality, aroused concerns about its effects on teachers' and learners' well-being, and showed potential challenges for both teachers and students [1]. However, technology-assisted language teaching and learning should be well planned to be successful. Schools should have established careful instructional design [2]. Relevant online resources should be available to support the teaching and learning activities [3, 4]. Teachers need pedagogical and technical skills to deliver effective instruction and facilitate the learning process [5–7]. Students need relevant skills to achieve their expected outcomes and overcome challenges emerging from the technological

H. P. Bui (✉)
University of Economics Ho Chi Minh City (UEH University), Ho Chi Minh City, Vietnam
e-mail: hungbp@ueh.edu.vn

T. T. Dao · T. T. Dao
Hong Duc University, Thanh Hoa University, Vietnam
e-mail: daothutra@hdu.edu.vn; daothanhthuy@hdu.edu.vn

V. H. Vi
Tay Bac University, Son La, Vietnam
e-mail: mrvihuong@utb.edu.vn

R. Sharma et al. (eds.), *Data Analytics for Internet of Things Infrastructure*,
Internet of Things, https://doi.org/10.1007/978-3-031-33808-3_12

divide [8]. The ultimate goal of education should be the sustainable development of individuals and society instead of "to provide a temporary access to instruction and instructional supports in a manner that is quick to set up [. . .] during an emergency or crisis" [9].

There is a paucity of investigations into the use of technology-assisted language teaching during the Covid-19 pandemic. Although a vast amount of research has been conducted on online teaching and learning, most of these studies were conducted in Western countries. This chapter first reviews the literature on how to effectively provide a distance education course and instruction assisted by technology. It then reviews the inherent challenges of remote teaching encountered by instructors and students and the strategies they use to teach and learn effectively during the Covid-19 pandemic. A framework for effective technology-assisted language teaching is introduced and discussed. Finally, the chapter recommends necessary training for teachers and students for effective teaching and learning on the Covid-19 pandemic. Given the possibility of unexpected similar emergent situations and school closure in the future, the education system might need to look back at relevant literature to prepare for better language teaching in crisis when the Covid-19 pandemic is over.

2 Current Perspectives on Technology-Enhanced Language Teaching and Learning

Online learning has become a popular and global trend in education. It affects education considerably [10] and tests hypotheses about the integration of technologies into language teaching and learning [11]. It offers technology-assisted teaching and learning opportunities that support distance education [12]. The term *technology-assisted teaching* mainly describes the use of technologies to deliver a course to students in a distance program. It has motivated the employment of online resources [13] and devices in teaching and learning [14], from which educators have developed different learning modalities, such as online, blended, and hybrid learning.

Technology-assisted learning is widely conceptualized as an alternative learning modality that is student-centered, inventive, and adaptable. Recent research has put forward the use of computer-mediated communication (CMC) as "computer-mediated learner-learner interaction provides unique opportunities for L2 learners for the active control of the topic selection and management and provides rich opportunities for learners to understand and adapt to diverse interactional patterns through collaboration among the interactants" [15]. CMC can also be used to facilitate the teacher's live instruction in online teaching and learning. Teachers can take advantage of CMC and immerse learners in virtual environments of asynchronous and synchronous communication in which particular linguistic features emerge through interaction between users [16].

Asynchronous communication has been predominantly applied in e-learning. Accordingly, users do not need to get online simultaneously for interaction and educational outcomes [17], which allows learners to get involved in interactions with different users and do different tasks at the same time [18]. Asynchronous communication is deemed to align with the hypotheses of social constructivism in which learners receive and cognitively process the input provided by the teacher and/or other resources to develop their own knowledge; therefore, the teacher is considered a collaborator rather than an input provider [19].

Several studies identified the benefits of asynchronous learning. The study by Meyer [20] found that online learning can enhance learner autonomy as students can manage their time and consider the content of their message carefully before they send it. Regarding learner differences, different learners may spend more time doing a task than another, depending on their competencies. Also, some research has found that learners' discussions in asynchronous environments have in-depth content because they have extra time to revise their messages [21]. The study by Tu and Corry [22] confirmed this result that learners' negotiations have interesting ideas, showing their cognitive and metacognitive processes in composing and editing their messages.

However, asynchronous learning does not always engage students in the learning process effectively. Research by Rourke and Kanuka [23] analyzed students' reports and showed evidence of students' low cognitive engagement. Results showed that asynchronous communication might fail to engage students socially in group work as cohesion between group members is insufficient, failing to achieve the group's goals. Also, in traditional teaching, teachers may employ various tools to motivate students, facilitate their learning, and collaborate with them in person. Considering the limitations of courses without a teacher available to give live instruction, immediate feedback, and instant advice, some researchers argue for the teacher's presence in engaging students in the learning process [24, 25].

Synchronous computer-mediated communication (SCMC) provides opportunities for direct communication between teachers and students and/or between peers through text chat and/or video calls. The nature of SCMC is that it allows for communication in discourse in which chat sequences require learners' concentration to follow the threads of messages and participate in discussions meaningfully. Regarding social presence, synchronous chat is widely considered to be simpler to detect social presence than in asynchronous discussion [26].

Several studies have attempted to compare the impacts of these synchronous and asynchronous learning modalities. Research by Schwier and Balbar [26] compared the effects of these two models in a graduate course and found that synchronous communication was effective for content in discussions. Also, synchronous computer-mediated communication assists live instruction to students and encourages live interaction between students, while the use of the asynchronous mode is prioritized for providing more time, concentration, and deeper thinking. Previous research identified an insignificant difference in student achievements in an educational psychology course in which half of the students adopted asynchronous interaction and the other half used synchronous text chat [27]. In an attempt to compare the effects of the learning modes in content learning, the study by Kuyath [28] found that the students who synchronously interacted significantly outperformed those using emails to discuss the course content in the posttest. Given the importance of communication in the classroom, the endeavor by Moradi and Farvardin [29] found insignificant difference in the quantity of negotiations between a face-to-face class and a synchronous class. However, it is important to note that the quality of negotiation contributes greatly to academic achievements.

Regarding the benefits and drawbacks of synchronous and asynchronous computer-mediated communication, several studies have suggested the combination of these two modes. Some research has indicated that students used these two modes for different purposes [30]. They used SCMC when they needed urgent responses and negotiations and asynchronous mode when they wanted to leave the recipient sufficient time to think or make an important decision. Researchers also identified the efficacy of SCMC [31]; meanwhile, they argued for the importance of asynchronous CMC. Some researchers [32–35] recommend a combination of face-to-face, synchronous, and asynchronous learning to develop the learning outcomes to the most.

3 Computer-Mediated Communication and Interaction Approach

Given that interaction is a driving force in the language learning process, the interaction approach hypothesizes the relationships between input, output, and corrective feedback [36]. Accordingly, input is defined as learners' exposure to the target language, output as language use opportunities, and corrective feedback as a modification to learners' linguistic knowledge. Second language development takes place in the interactional processes in which interlocutors negotiate meanings, and their language is modified from relevant corrective feedback [37].

The interaction approach has motivated strings of research. The first string is on negotiation [38, 39] and students' perceptions of communication [40, 41]. Another string is on the effects of interactional feedback on language development [42, 43]. Some other studies investigated the impacts of recasts on L2 learning [44]. The final string is about the efficacy of feedback [45, 46]. Overall, these studies have provided positive results that interaction and linguistic features used in interaction influence learners' L2 development.

The emergence of technology-assisted language teaching, particularly computer-mediated communication, has motivated a vast amount of research testing the hypotheses of the interaction approach. Although computer-mediated communication was originally used to describe communication synchronously or asynchronously mediated by the computer, advances in technology have extended the boundaries of CMC with a growing interest in using mobile devices (e.g., mobile phones). This research strand has manifested itself with confirmed evidence about how learners' second language develops from social synchronous and asynchronous interaction with others and how technology assists the L2 learning process.

A literature review also reveals the benefits and challenges of task-based learning and teaching in technology-mediated environments [47]. Accordingly, L2 learners perform significantly better in face-to-face communication than in CMC regarding the quality of their interaction. This claim provides further research on comparing the quality of L2 learners' interaction in the two environments.

4 Research on Remote Teaching in Crisis Situations

During the Covid-19 pandemic, schools worldwide have been suddenly compelled to adopt online teaching to prevent the spread of the coronavirus. This unprecedented transition has aroused concerns about technology-assisted teaching and learning. Potential challenges from a lack of social interaction, students' and teachers' insufficient training, and miscommunication between teachers and students on video calls provoke the concerns of teachers, parents, and students.

A study on teachers' and students' concerns about the use of technology-assisted language teaching during the Covid-19 pandemic in Saudi Arabia enrolled 735 participants (61 language teachers and 674 language students) who had to adopt the online teaching and learning mode [48]. While the teachers identified most of the students' concerns, there were some differences in the students' and teachers' ratings. First, the teachers did not recognize how boring the virtual classes were to the students. The teachers also raised their concerns about the reliability of online test results and students' concentration in virtual learning. The teachers and students were concerned about two main limits of virtual education: lack of interpersonal relationships and a sense of group work. The researchers concluded that virtual students need support and resources to achieve their goals and expected outcomes. Orientation in academic honesty may be necessary because cheating can be of different forms, which some students may not recognize. The researchers argued that a better understanding of technology acceptance could change teachers' and students' beliefs.

Another study investigated Indonesian high school students' willingness to participate in virtual L2 English classes and their familiarity with the technology needed for virtual learning during the Covid-19 pandemic [49]. The study employed Google Form to administer a 1–5 Likert-scale survey to 85 students. The results, mainly based on descriptive statistics, showed that many students could not use and were unsure about their ability to use the technology needed for e-learning. Also, students expressed their high level of willingness (agree and strongly agree) to participate in e-learning. The researchers argued that readiness to participate in online learning mainly stems from students' willingness to participate and technical skills. Therefore, it is advised that schools offer students a training course for e-learning.

Concerned about students' engagement in learning, teachers can use some strategies to engage students in remote learning during the Covid-19 pandemic [50, 51]. The three main types of engagement are learner-learner, learner-teacher, and learner-content [50]. Teachers can first break the ice by letting students introduce themselves, followed by collaborative tasks. These activities make students feel that they belong to a class and interact with their classmates rather than with the teacher personally. Groupwork activities may improve students' interaction and foster idea-exchanging opportunities [52]. Teachers can ask informal and follow-up questions, give scaffolding feedback, and use personalized emails. Learner-learner engagement and teacher-learner engagement can contribute to the success of remote teaching because it may foster students' uptake of knowledge and skills [53] and enhance students' sense of social presence [50]. Students might be required to work with provided structured texts and interact with content in social media, e.g., watching video clips. Teachers should use authentic materials to make them feel part of the real world.

The study by Iglesias-Pradas and associates [54] investigated the impacts of the sudden shift from face-to-face education to distance education on teachers and students in a Spanish context. The results showed that most teachers and students preferred using videoconferencing by using the technological tools they were familiar with to deliver synchronous instruction. Amazingly, students' academic results were significantly better than the achievements of those students in face-to-face learning. However, it is unlikely to conclude that the teachers' adoption of technology-assisted teaching is a better alternative. The researchers argued that the students' learning strategies and self-regulation were the main driving force behind their academic achievements. The researchers conclude that the digital tools used by teachers, delivery methods they adopted, and class size do not have significant impacts on the learning outcomes.

The study by Misirli and Ergulec [55] explored Turkish parents' experiences and perspectives on technology-assisted teaching. The study involved 983 parents with children taking an online course at different education level during the lockdown time. The qualitative and quantitative findings indicated that children in preschools and primary schools received more parent support than did middle school students. Some parents even engaged in their children's learning. Parents were also concerned about a wide range of problems, including, for instance, infrastructure (e.g., Internet access), social interaction, course length, motivation, and evaluation. The researchers suggested six areas to improve remote teaching in the crisis: (1) content, (2) assistance to students in need, (3) interaction, (4) live class time, (5) communication between teachers/ administrators and parents, and (6) parents' guidance.

The study by Shamir-Inbal and Blau [56] made insights into teachers' experiences in the challenges and benefits of technology-assisted teaching in terms of pedagogy, technology, and organization. The study administered an online questionnaire with open-ended questions to 133 primary and secondary teachers at many schools in Israel. The findings showed that pedagogical challenges mainly derived from teachers' adaptability to the new teaching modality, understanding students, and interpersonal communication. Also, insufficient support from authorities and institutions' slow response. Teachers also reported teachers' and students' inadequate training and unpreparedness. However, all the participants admitted that technology-assisted teaching was an alternative to face-to-face teaching in terms of learning continuity.

In short, the studies surveyed above have investigated different aspects of remote teaching and learning. Some attempted to explore aspects of language learning and learners [48, 49]. Others, although investigating educational aspects, gave implications for improvements of the application during the sudden shift to online instruction [51, 54–56]. As teachers' and students' experiences mainly reflect the context, further studies can raise the voice of teachers and students in other contexts in the world.

5 Technology Acceptance Model

To situate technology-assisted teaching in emergency, the Technology Acceptance Model [57] framework should be considered. Technology Acceptance Model posits that the acceptance of technology should depend on three main dimensions: perceived usefulness, perceived ease of use, and user acceptance of the technology. During the Covid-19 pandemic, a digital or technological tool is a medium for teachers to deliver instruction to students. The unprecedented use of and sudden adoption of this teaching mode exposes inherent uncertainties, such as the length of remote teaching, users' familiarity with the tool, available resources for teaching and learning, and assessment resources. Therefore, without prior experience with digital tools to be accepted [58], some teachers and learners may not have sufficient time to assess the tool(s) that they have to use. The acceptance of technologies is mainly based on their perceptions and familiarity with the digital tool.

Technology Acceptance Model is used to understand "predictors of human behaviors towards potential acceptance or rejection of the technology" [59]. It was first developed by Davis (1989) [57] with three dimensions and then modified with two main dimensions: perceived usefulness and perceived ease of use [60]. These two primary dimensions can be predicted by "previous experience with technology, enjoyment, self-efficacy, computer anxiety, and social norm" [48, 58].

Some researchers argue that Technology Acceptance Model does not particularize educational technology [61]. Current criticisms against this model have mainly focused on its lack of essential factors. First, it mainly takes into account internal factors. In fact, in deciding on a technology to use, group, social, and cultural factors should be cautiously considered because the usefulness and ease of technology can be perceived differently by different users, societies, and cultures [62]. Accordingly, educational technology should consider two other predictors: subjective norm and perceived playfulness.

This model has been recently modified. With recently added factors, the model becomes nonlinear. The choice of technology is not confined to the concepts of usefulness and ease to use, but it makes an environment a playful place. A learning environment is not only a place to study, but it should also satisfy the learner's enjoyment; therefore, a choice of technology for education should especially take into account the newly added factors.

6 Responsive Online Teaching in Crises

Another framework that should be considered in technology-assisted language teaching is responsive online teaching in crises [63]. This conceptual framework describes the essential elements to consider in distance education as an alternative in an emergency. Although this framework was developed from teachers and designers' opinions about technology-assisted teaching, it is proposed by the researchers to be used in future crises.

Based on the four components in a learning environment, technologies, users, configurations of the platform, and environment, researchers propose that responsive emergent online teaching should be assessed regularly [63]. The evaluation should take into account three main phases: inquiry, classifying resources, and designing.

First, we should consider two main elements for effective teaching: teachers and students. Inquiry into teachers' knowledge, skills, and time should be initiated because teachers may need time to prepare for their teaching. Some teachers may need time to familiarize themselves with the technology to use. Also, the pedagogies that they will apply should be workable and compatible with the technology. As students are the main participants in the learning process, teachers should ensure that the technology to be used and resources to satisfy students' needs should be accessible to students. Students' health and safety should also be considered. These concerns should be revisited regularly by examining teachers' and students' experiences.

After considering the factors in the inquiry phase, it is necessary to classify them. Teachers and students may have some shared characteristics and variables. For instance, in a wealthy district, students may afford and easily access technologies needed, such as computers and the Internet; however, in another district, some students may not have the required technical devices. Classifying students into groups can help administrators and teachers better understand students' needs, expectations, and difficulties to maximize the effects of technology-assisted teaching in emergencies.

The work by Means and associates [64] proposed three main considerations in online teaching: whether, when, and how. The study by Whittle and associates [63] referenced this proposal and introduced eight dimensions of course design for online teaching in an emergency. In this final phase, the teachers should consider the eight dimensions in which the course design reflects the factors in the classification phrase. Accordingly, constants might serve as the course foundation, and teachers can view variables as means to maximize individual learning.

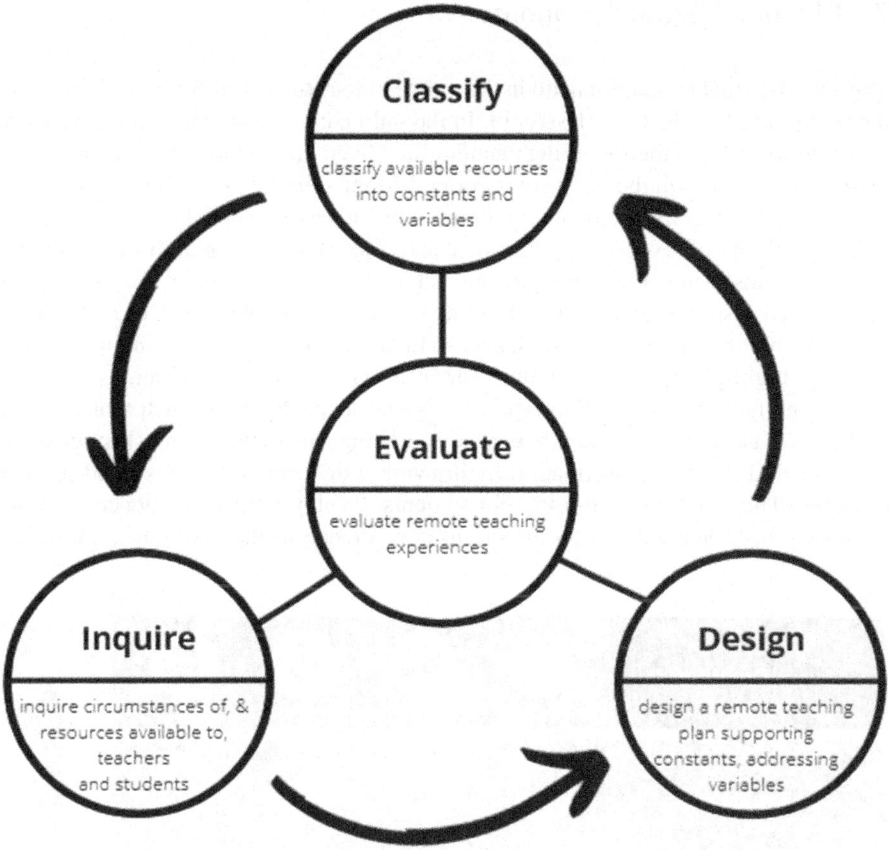

Fig. 1 A three-phase process of remote teaching in emergency [63]

The eight dimensions of the design phase include (1) critical learning goals, (2) teacher student ratio, (3) communication method, (4) building agency, (5) assessment, (6) social role of the instructor, (7) pedagogy and the student role, and (8) feedback. Some of these dimensions (e.g., teacher-student ratio) may be predetermined, but others (e.g., feedback) occur during the teaching process and depend on students' performance and progress. Regarding the communication method, both synchronicity and asynchronicity can be used for different purposes (Fig. 1).

However, this framework may need to be modified [63]. First, the framework was developed from a small-scale study, with four teachers and five instructional designers in the US context. Second, although the study proposed to include students as agents in the framework, the study did not reflect students' opinions. Third, administrators should not be outsiders in evaluating an educational practice. Further studies may test this model by including all types of stakeholders in other contexts.

7 Bloom's Digital Taxonomy

The activities that students can do in the online classroom are different from those in the traditional face-to-face classroom. In the online classroom, students mainly use digital tools to make their learning meaningful. Given the importance of an authentic learning experience in the online classroom, Bloom's digital taxonomy proposed six groups of activities that students can do in online classrooms [65].

Blooms' digital taxonomy outlines the activities students can do by using digital tools that transform student thinking and achieve learning outcomes. Each group of activities corresponds to a level of thinking skills (Fig. 2). The lowest order thinking level is remembering in which students can memorize information by bookmarking, copying, highlighting, and searching for information. Understanding, as a higher order thinking level, requires interpretation of concepts by using such tools as blog writing, tweeting, tagging, and advanced searching. Instructors can ask students to calculate and illustrate information by drawing a diagram, edit a text, and present opinions. These activities aim to help students develop their competence to apply what they have learned. As an important competence in the taxonomy, analyzing

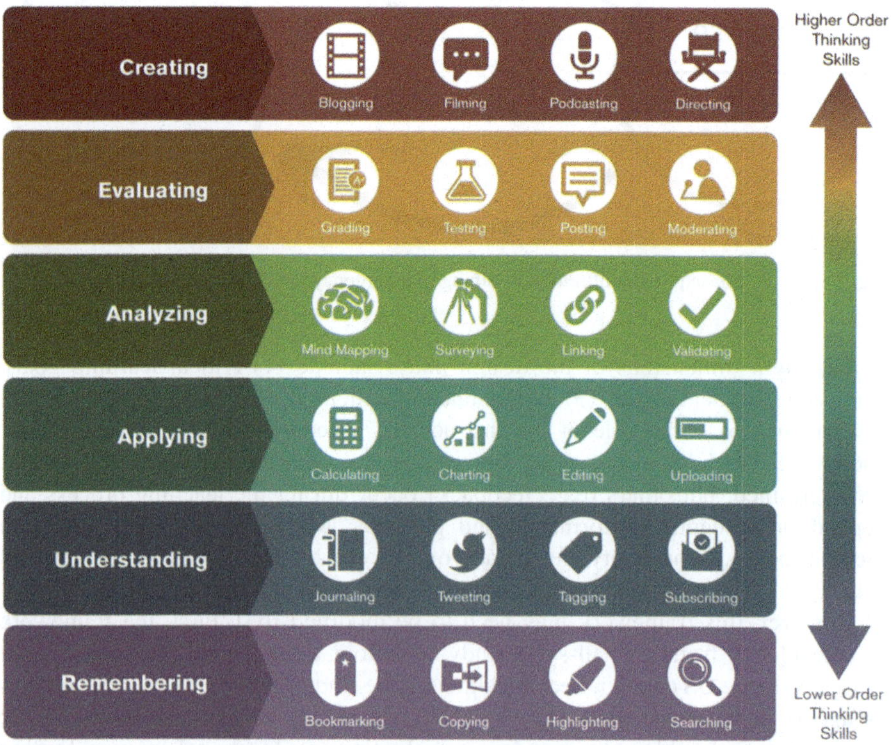

Fig. 2 Blooms' digital taxonomy [65]

is defined as establishing associations and interrelations between concepts. Such activities as mind mapping, conducting a survey, linking, and validating can be used to develop this competence. A higher order thinking level is evaluating which may be set through grading, testing, reviewing, and moderating. These activities aim to give students opportunities to judge and critique a perspective and decide which option(s) should be applied. Activities that require the highest order thinking level include blogging, filming, podcasting, and directing. These activities provide opportunities to create things.

Online classroom activities outlined by Bloom's digital taxonomy demonstrate different-order thinking levels. Teachers should be aware of this taxonomy to apply the right tool to obtain the right objective. The use of a particular activity should be based on the learners' needs, personal goals, and current abilities to help them achieve the learning outcomes.

8 Recommendations

The critical review above indicates some recommendations for remote teaching and remote language teaching during the Covid-19 pandemic. Accordingly, remote language teaching depends on sociocultural factors, characteristics of the stakeholders, available technologies, and course objectives.

First, the Technology Acceptance Model can be a reference for a technology choice. Investigations into perceived usefulness and perceived ease, which are based on stakeholders' prior experience, can help choose technology for first-time use in an emergency. Training should be offered to those in need. However, education should consider teachers' and students' emotions and well-being. A learning environment is not only a place for students to receive input and produce output but also provides learners with opportunities to enjoy their learning with joyful activities. The duration of a session, therefore, should be taken into account.

Second, school administrators should inquire about related issues. Evaluation of the circumstance might be the foremost thing to do. Administrators should consider the severeness of the circumstance. Is it necessary to apply remote teaching? Such school capacities as resources, available facilities, and technology accessibility should be examined cautiously to implement the subsequent steps. Teachers' experience, qualifications, and capabilities, e.g., technical skills and pedagogical abilities, should be investigated. Online resources should be available and accessible to students. Teachers and students must be informed of emergency remote teaching right after making a decision. Information should be sent to teachers and students via trustworthy and official channels to ensure they get updated. Regular evaluation is needed to improve the quality of teaching and learning.

Third, self-regulation is a crucial factor in online learning. As teachers' supervision of students' work might be impossible, students may not concentrate on their work or do not take the course seriously. Learner autonomy works best for those who are self-regulated. Parental guidance may be of help, especially for young kids. However, parents' excessive interference with coursework might undermine self-efficacy and the learning outcomes or even violate ethical issues. Parents are especially advised to minimize distractions from family activities during class hours and encourage their children's learning.

Fourth, teachers may need to reference Bloom's digital taxonomy to know what activities can be employed to assist learners in achieving the expected outcomes. Teachers might need to consider learners' current competencies to scaffold their uptake [53]. They can ascertain that the activities they require students to do aim at developing the right ability.

Finally, teachers and students should have the right to express their concerns because they are directly involved in teaching and learning. Opinions of parents, especially those of young learners, should be considered. Their perspectives should be collected on an iterative basis to modify and improve the teaching and learning quality, especially in contexts where prior preparation is marginal.

Be advised that different places may have different contextual factors, and different crises that occur at other times may have different characteristics. Although the current literature in the field provides us with references for our research and practice, practitioners should consider contextual factors to satisfy local needs and expectations, maximize the learning outcomes, and predict obstacles in the process.

9 Conclusion

Developments in technology have advanced language teaching and learning to a certain extent. Learners are given diverse education opportunities, such as online, blended, and hybrid learning. Teachers can use technology inside the classroom and use it to enhance out-of-class learning opportunities for students. Technology-enhanced language teaching and learning are generally affected by many factors. It might be crucial for people responsible to prudentially consider a technology to use, which should be accessible, easy to use, and useful. Given the importance of social interaction in language education, the accepted technology should have appropriate configurations to facilitate interaction between students and teachers, resulting in language learners' fostered competence. However, technology should not be accepted once and used forever. A historical account shows that technology has been developing, and software or platforms should be modified to meet stakeholders' needs. It might be necessary for school administrators to collect users' opinions. In language teaching and learning, consulting experts in the field may benefit the success of teaching and learning.

Nevertheless, teachers and learners are the ones who use the adopted technology and may know what should be modified. In teaching and learning, it is essential for teachers and learners to know the goals and expectations of the program. Setting goals and understanding the course objectives might benefit learners in terms of academic achievements. Teachers can consider the digital taxonomy to assist learners in making academic progress.

Acknowledgments The authors would like to sincerely thank the editor and reviewers for their constructive feedback.

Funding Source This work was funded by University of Economics Ho Chi Minh City (UEH University), Thanh Hoa University, and Tay Bac University, Vietnam.

References

1. Huber, S. G., & Helm, C. (2020). COVID-19 and schooling: Evaluation, assessment and accountability in times of crises—Reacting quickly to explore key issues for policy, practice and research with the school barometer. *Educational Assessment, Evaluation and Accountability, 32*, 237–270.
2. Ergulec, F. (2019). Instructional strategies for forming online collaborative teams. *International Journal on E-Learning, 18*(4), 349–372.
3. Bozkurt, A., & Sharma, R. C. (2020). Emergency remote teaching in a time of global crisis due to coronavirus pandemic. *Asian Journal of Distance Education, 15*(1), i–vi.
4. Zhang, X. (2020). Thoughts on large-scale long-distance web-based teaching in colleges and universities under novel coronavirus pneumonia epidemic: A case of Chengdu University. In *4th international conference on culture, education and economic development of modern society (ICCESE 2020)* (pp. 1222–1225). Atlantis Press.
5. Barnard, L., Lan, W. Y., To, Y. M., Paton, V. O., & Lai, S.-L. (2009). Measuring self-regulation in online and blended learning environments. *Internet and Higher Education, 12*, 1–6.
6. Hoven, D. (2007). The affordances of technology for student teachers to shape their teacher education experience. In *Preparing and developing technology-proficient L2 teachers* (CALICO monograph series) (Vol. 6, pp. 133–164). Computer Assisted Language Instruction Consortium (CALICO).
7. Peters, M. (2006). Developing computer competencies for pre-service language teachers: Is one course enough? In *Teacher education in CALL* (pp. 153–166). John Benjamins.
8. Dhawan, S. (2020). Online learning: A panacea in the time of COVID-19 crisis. *Journal of Educational Technology, 49*(1), 5–22.
9. Hodges, C., Moore, S., Lockee, B., Trust, T., & Bond, A. (2020). The difference between emergency remote teaching and online learning. *Educause Review, 27*, 1–12.
10. Blake, R. J. (2017). Technologies for teaching and learning L2 speaking. In *The handbook of technology and second language teaching and learning* (pp. 107–117). Wiley.
11. Kim, H. Y. (2020). More than tools: Emergence of meaning through technology enriched interactions in classrooms. *International Journal of Educational Research, 100, 101543*.
12. Battro, A. M., & Fischer, K. W. (2012). Mind, brain, and education in the digital era. *Mind, Brain, and Education, 6*(1), 49–50.
13. Vu, N. N., Hung, B. P., Van, N. T. T., & Lien, N. T. H. (2021). Theoretical and instructional aspects of using multimedia resources in language education: A cognitive view. In *Multimedia technologies in the Internet of things environment* (Vol. 2, pp. 165–194). Springer.
14. Gallardo, E., Marqués, L., & Bullen, M. (2015). Students in higher education: Social and academic uses of digital technology. RUSC. *Universities and Knowledge Society Journal, 12*(1), 25–37.
15. Abrams, Z. I. (2003). The effects of synchronous and asynchronous CMC on oral performance in German. *Modern Language Journal, 87*, 157–167.
16. Zeng, G. (2017). Collaborative dialogue in synchronous computer-mediated communication and face-to-face communication. *ReCALL, 29*(3), 257–275.
17. Johnson, S. D., & Aragon, S. R. (2003). An instructional strategy framework for online learning environments. *New Directions for Adult and Continuing Education, 100*, 31–43.

18. Hewitt, J. (2005). Toward an understanding of how threads die in asynchronous computer conferences. *Journal of the Learning Sciences, 14*, 567–589.
19. Cavana, M. (2009). Closing the circle: From Dewey to web 2.0. In *Information technology and constructivism in higher education: Progressive learning frameworks* (pp. 1–13). IGI Global.
20. Meyer, K. A. (2003). Face-to-face versus threaded discussions: The role of time and higher-order thinking. *Journal of Asynchronous Learning Networks, 7*(3), 55–65.
21. Branon, R. F., & Essex, C. (2001). Synchronous and asynchronous communication tools in distance education: A survey of instructors. *Technology Trends, 45*, 36–42.
22. Tu, C. H., & Corry, M. (2003). Designs, management tactics, and strategies in asynchronous learning discussions. *The Quarterly Review of Distance Education, 4*, 303–315.
23. Rourke, L., & Kanuka, H. (2009). Learning in communities of inquiry: A review of the literature. *Journal of Distance Education, 23*(1), 19–48.
24. Heckman, R., & Annabi, H. (2005). A content analytic comparison of learning processes in online and face-to-face case study discussions. *Journal of Computer-Mediated Communication, 10*.
25. Rovai, A. (2007). Facilitating online discussions effectively. *The Internet and Higher Education, 10*, 77–88.
26. Schwier, R. A., & Balbar, S. (2002). The interplay of content and community in synchronous and asynchronous communication: Virtual communication in a graduate seminar. *Canadian Journal of Learning and Technology, 28*(2).
27. Johnson, G. (2008). The relative learning benefits of synchronous and asynchronous text-based discussion. *British Journal of Educational Technology, 39*, 166–169.
28. Kuyath, S. (2008). *The social presence of instant messaging: Effects on student satisfaction, perceived learning, and performance in distance education* [Ph.D. Thesis]. University of North Carolina at Charlotte.
29. Moradi, A., & Farvardin, M. T. (2019). Negotiation of meaning by mixed-proficiency dyads in face-to-face and synchronous computer-mediated communication. *TESOL Journal, 11*(1).
30. Ligorio, M. B. (2001). Integrating communication formats: Synchronous versus asynchronous and text-based versus visual. *Computers & Education, 37*, 103–125.
31. Hines, R. A., & Pearl, C. E. (2004). Increasing interaction in web-based instruction: Using synchronous chats and asynchronous discussions. *Rural Special Education Quarterly, 23*, 33–36.
32. Davidson-Shivers, G. V., Muilenburg, L. Y., & Tanner, E. J. (2001). How do students participate in synchronous and asynchronous online discussions? *Journal of Educational Computing Research, 25*, 351–366.
33. Dudding, C., & Drulia, T. (2009). Analysis of synchronous and asynchronous discussion forums: A pilot study. In *Proceedings of world conference on educational multimedia, hypermedia and telecommunications 2009* (pp. 631–634). AACE.
34. Hrastinski, S. (2006). Introducing an informal synchronous medium in a distance learning course: How is participation affected? *Internet and Higher Education, 9*, 117–131.
35. Rockinson-Szapkiw, A. (2009). *The impact of asynchronous and synchronous instruction and discussion on cognitive presence, social presence, teaching presence, and learning* [Ph.D. thesis]. Regent University.
36. Gass, S. M., & Mackey, A. (2006). Input, interaction and output: An overview. *AILA Review, 19*, 3–17.
37. Mackey, A. (2012). *Input, interaction, and corrective feedback in L2 learning*. Oxford University Press.
38. Iwashita, N. (2003). Negative feedback and positive evidence in task-based interaction. *Studies in Second Language Acquisition, 25*, 1–36.
39. Peterson, M. (2006). Learner interaction management in an avatar and chat-based virtual world. *Computer Assisted Language Learning, 19*, 79–103.
40. Carpenter, H., Jeon, K. S., MacGregor, D., & Mackey, A. (2006). Learner's interpretations of recasts. *Studies in Second Language Acquisition, 28*, 209–236.

41. Mackey, A. (2002). Beyond production: Learners' perceptions about interactional processes. *International Journal of Educational Research, 37*, 379–394.
42. Mackey, A., & Sachs, R. (2012). Older learners in SLA research: A first look at working memory, feedback, and L2 development. *Language Learning, 62*, 704–740.
43. Mackey, A., & Silver, R. (2005). Interactional tasks and English L2 learning by immigrant children in Singapore. *System, 33*, 239–260.
44. Loewen, S., & Philp, J. (2006). Recasts in the adult English L2 classroom: Characteristics, explicitness, and effectiveness. *Modern Language Journal, 90*, 536–556.
45. Ha, X. V., Nguyen, L. T., & Hung, B. P. (2021). Oral corrective feedback in English as a foreign language classrooms: A teaching and learning perspective. *Heliyon, 7*(7), 1–8. [e07550].
46. Yilmaz, Y. (2012). The relative effects of explicit correction and recasts on two target structures via two communication modes. *Language Learning, 62*, 1134–1169.
47. Lai, C., & Li, G. (2011). Technology and task-based language teaching: A critical review. *CALICO Journal, 28*(2), 1–24.
48. Shlowiy, A. A., Al-Hoorie, A. H., & Alharbi, M. (2021). Discrepancy between language learners and teachers' concerns about emergency remote teaching. *Journal of Computer Assisted Learning, 37*(6), 1528–1538.
49. Mabrur, I. A. M., Suwartono, T., & Lutfiana. (2021). Junior high school students' readiness to participate in e-learning and online EFL classes during the COVID-19 pandemic. *International Social Science Journal, 71*(241–242), 153–161.
50. Bolliger, D. U., & Martin, F. (2018). Instructor and student perceptions of online student engagement strategies. *Distance Education, 39*, 568–583.
51. Brown, W. S. (2021). Successful strategies to engage students in a COVID-19 environment. *Frontiers in Communication, 6*, 641865.
52. Hung, B. P. (2019). Impacts of cooperative learning: A qualitative study with EFL students and teachers at Vietnamese colleges. *Issues in Educational Research, 29*(4), 1223–1240.
53. Hung, B. P., & Nguyen, L. T. (2022). Scaffolding language learning in the online classroom. In *Multimedia technologies in the Internet of things environment* (Vol. 2, pp. 165–194). Springer.
54. Iglesias-Pradas, S., Hernández-García, A., Chaparro-Peláez, J., & Prieto, J. L. (2021). Emergency remote teaching and students' academic performance in higher education during the COVID-19 pandemic: A case study. *Computers in Human Behavior, 119*, 106713.
55. Misirli, O., & Ergulec, F. (2021). Emergency remote teaching during the COVID-19 pandemic: Parents experiences and perspectives. *Education and Information Technologies, 26*, 6699–6718.
56. Shamir-Inbal, T., & Blau, I. (2021). Facilitating emergency remote K-12 teaching in computing-enhanced virtual learning environments during COVID-19 pandemic - blessing or curse? *Journal of Educational Computing Research, 59*(7), 1243–1271.
57. Davis, F. D. (1989). Perceived usefulness, perceived ease of use, and user acceptance of information technology. *MIS Quarterly, 13*(3), 319–340.
58. Abdullah, F., & Ward, R. (2016). Developing a general extended technology acceptance model for E-learning (GETAMEL) by analyzing commonly used external factors. *Computers in Human Behavior, 56*, 238–256.
59. Granić, A., & Marangunić, N. (2019). Technology acceptance model in educational context: A systematic literature review. *British Journal of Educational Technology, 50*, 2572–2593.
60. Lee, L. (2004). Learners' perspectives on networked collaborative interaction with native speakers of Spanish in the US. *Language Learning & Technology, 8*, 83–100.
61. Dumpit, D. Z., & Fernandez, C. J. (2017). Analysis of the use of social media in Higher Education Institutions (HEIs) using the Technology Acceptance Model. *International Journal of Educational Technology in Higher Education, 14*, 5.
62. Bagozzi, R. P. (2007). The legacy of the technology acceptance model and a proposal for a paradigm shift. *Journal of the Association for Information Systems, 8*(4), 244–254.
63. Whittle, C., Tiwari, S., Yan, S., & Williams, J. (2020). Emergency remote teaching environment: A conceptual framework for responsive online teaching in crises. *Information and Learning Sciences, 121*(5–6), 311–319.

64. Means, B., Toyama, Y., Murphy, R., & Baki, M. (2013). The effectiveness of online and blended learning: A meta-analysis of the empirical literature. *Teachers College Record, 115*(3), 1–47.
65. Sneed, O. (2016). *Integrating technology with Bloom's taxonomy*. Retrieved from https://teachonline.asu.edu/

The Symbiotic Relation of IoT and AI for Applications in Various Domains: Trends and Future Directions

Aman Jolly, Vikas Pandey, Praveen Kumar Malik, and Turki Alsuwian

1 Introduction

The network of devices that are linked to the Internet is referred to as the Internet of Things (IoT). IoT is an umbrella phrase that refers to any device that is able to transmit and receive data via a network. In most cases, these gadgets take the form of sensors that monitor various physical parameters, such as the amount of light, humidity, or temperature. IoT is continuing to proliferate across a variety of corporate sectors, which has resulted in the development of brand-new possibilities for cooperation and innovation. Devices are able to gather data, take actions depending on that data, and then report the results of those activities back to a centralized server. This technology has been around for quite some time, but only lately has there been an explosion of new inventions in this sector of the industry. Artificial intelligence (AI) and the Internet of Things (IoT) are the two primary motivating factors behind the recent uptick in interest about linked gadgets. When coupled, AI and IoT provide even more opportunities for analyzing data and discovering new insights into how we may enhance the quality of our lives. In this article, we will discuss AI and IoT, focusing on how these two technologies might

A. Jolly (✉)
Department of CSE, Delhi Technological University, Delhi, India
e-mail: amanjolly_2k20phdco07@dtu.ac.in

V. Pandey
School of Engineering, Babu Banarasi Das University, Lucknow, Uttar Pradesh, India

P. K. Malik
School of Electronics and Electrical Engineering, Lovely Professional University, Phagwara, Punjab, India

T. Alsuwian
Electrical engineering department, Najran University, Najran, Saudi Arabia
e-mail: tmalsuwian@nu.edu.sa

R. Sharma et al. (eds.), *Data Analytics for Internet of Things Infrastructure*,
Internet of Things, https://doi.org/10.1007/978-3-031-33808-3_13

collaborate to provide innovative answers to common challenges. Some of the core areas where IoT has been used at its peak are as follows:

1. 5G: The next generation of wireless communication, 5G, will increase Internet access and network connection. The innovative design of 5G allows it to simultaneously connect more devices at greater speeds with less delay. This cellular IoT application was developed for the advantage of users at minimal cost and with increased speed. Using this innovative program, we can fully automate several sectors with the aid of smart grids. In the future, 5G will expand to support larger devices across wider regions, bridging the gap between smart cities and wireless vehicle communication. The public and commercial sectors may both profit greatly from this potential use of IoT. With 5G, it will be possible to remotely manage even more types of devices via software. The infrastructure for smartphones, tablets, and other mobile devices, as well as for sensors, medical equipment, and automobiles, is being established by IoT and 5G.
2. The Internet of Things and Augmented Reality: The connection between IoT and augmented reality is becoming stronger all the time. While IoT bridges the gap between physical assets and digital infrastructure, augmented reality (AR) brings digital components into the physical world. The future of AR and IoT seems bright in the medical field. Surgeons may, for instance, use software designed to reconstruct a body part in three dimensions, together with equipment monitoring the required statistics in real time. This software can be found on mobile devices. All these things could make difficult operations simpler for surgeons and more rewarding for patients who are patient.
3. Smart Cities on the Rise: The popularity of smart city technologies is at an all-time high right now, and analysts predict that interest in and investment in smart city technology will continue to soar in the not-too-distant future. City councils and other local authorities believe that smart city solutions are effective ways of involving the general public in the day-to-day operations of municipal administration and maintenance. Providing more comfort and convenience for the populace is just a minor portion of the whole bargain. Interactivity is being given a significant amount of consideration in the development of smart city initiatives all around the globe. There is a growing consensus that engaging individuals in a manner that requires greater hands-on participation is beneficial in all matters pertaining to the day-to-day maintenance of an urban space, which pays huge dividends for everyone concerned. Safe, habitable, and environmentally sustainable cities will not be possible without IoT-based smart city technologies. As city populations are expected to grow dramatically in the future, these strategies will become even more important.
4. Blockchain Technologies and Information Security: One of the most important developments in IoT technology is blockchain. Put together, IoT and blockchain technology are living up to their hype. As of today, the reliable exchange of money and data between IoT devices become possible once the blockchain technology provides them with a simple infrastructure for doing so. When these two phenomena are put together, they live up to their hype. The decentralized

nature of blockchain is analogous to the dispersed nature of the Internet of Things. The latter provides anonymity and security to numerous networks and the owners of those networks, while the digital signatures and private keys that accompany each transaction assure that the environment of the Internet of Things will be secure. IoT devices are intended to simplify a person's or an operational unit's day-to-day activities in some way. This necessitates the ongoing creation of personal data and introduces significantly increased opportunities for cybercriminal activity. In addition, it is anticipated that the quantity of data will exponentially increase in the same way as the number of machine-to-machine interactions increase. As a consequence of this, the decentralized structure of blockchain will cause massive numbers of data to be accessible whenever they are required and with a minimal number of associated security issues.

All the aforementioned applications require the use of IoT in various domains because IoT acts as a bridge between hardware sensors and the digital infrastructure. This work will focus on all the applications that employ IoT device to obtain real-time data, which will be further fed into the AI-based prediction models. In this review analysis, seven domains have been identified where AI and IoT are being employed: healthcare, sustainability, information security, education, pollution monitoring, robotics, and autonomous vehicles.

Healthcare Applications The IoT of the future will make it much simpler for doctors and nurses to keep track of their patients' activities and vital signs. This will be made possible by advancements in 5G wireless technology, AI, and sensor technology. A smart glucose-monitoring system and smart insulin pens will also be of assistance since they will automatically transmit the patient's important information to a monitoring system. This will serve to direct assistance, particularly in the situation involving insulin. The pen will be able to evaluate the quantity of insulin that has to be administered on the basis of the data that is derived from the patient. Security and patient safety are of the utmost importance in the clinical setting, and IoT may assist in enhancing the monitoring and transfer of patient data. The whole world has been forced to reconsider the significance of remote healthcare. It won't be long until patients and physicians won't even have to physically interact with one another, which might be quite helpful in times of lockdown. IoT developments will soon have stronger effects on healthcare, which will result in the proliferation of increasingly intelligent medical equipment.

Better Way to Store Data and Perform Data Analytics We've just recently become used to storing information on the cloud before it became necessary to make a switch. Edge computing, in its simplest form, enables interconnected gadgets to carry out computations, store results, and locally view results. Edge computing is a hybrid method of data processing that is altering the trajectory of the Internet of Things. The actual value that IoT may bring about via data analysis is in managing and analyzing these data. Therefore, machine learning and artificial intelligence will play increasingly crucial roles. These developments will aid in making our lives simpler and more comfortable and will provide efficient methods for completing jobs.

2 Recent Works on IoT and AI in Various Domains

2.1 Healthcare

The healthcare industry has always relied on a small number of centralized agents freely disseminating raw data to the public. This system still faces substantial threats and weaknesses. With AI, however, the system would consist of several agents working together and effectively interacting with their preferred host. The most cutting-edge and fascinating innovations in the area of intelligent healthcare include federated learning (FL), AI, and explainable AI (XAI). FL operates in a decentralized way and keeps the communication based on a model in the favored system without transmitting raw data. Multiple healthcare constraints may be alleviated with the integration of FL, AI, and XAI methods. In [1], Rahman provides a comprehensive evaluation of FL as it relates to the use of AI in forward-thinking medical settings. They used FL-AI in several healthcare technologies and categorized the results.

Using multisensing, edge-based, and on-device AI components, T. Montanaro et al. [2] constructed a real-time IoT-aware healthcare system that comprises three layers: an edge computing layer, a data visualization layer, and an intelligent data-acquisition layer. (i) Three sublayers make up the intelligent data-acquisition layer. (1) Advanced sensors include motion, temperature, location, and electric charge buildup in skin sensors. These devices gather data for deductions. (2) Computation and data preprocessing devices control the selection and collection of sensor data. (3) AI modules contain microcontrollers linked with AI algorithms. These algorithms identify data irregularities, accurately differentiate between behaviors and between people, classify measured data, etc. (ii) The edge computing layer receives data from the preceding layer, provides a gateway to the top layer, handles multiple protocol communications, and performs additional analyses. This layer may accept data from the bottom layer, transmit changes to it, transfer data to the top layer, and receive notifications and updates from the upper layer. This layer allows communicates with employers, families, and intimate partners. (iii) The data visualization layer connects storage and user interactions. Web dashboards provide authorized users with local device warnings and historical occurrences. Healthcare providers may utilize this dashboard to give people ideas. Advanced data analysis is also possible. Finally, REST APIs connect with lower levels and contain a database to store data.

M. M. Kamruzzaman et al. [3] identified new difficulties, possibilities, case studies, and edge-AI applications for linking healthcare in smart cities. Relevant publications and journals were studied, analyzed, and appraised, and this review also included secondary data sources such as Google Scholar, Science Direct, etc. Only papers including AI, edge AI, IoT, and deep learning (DL) were reviewed. The study selection and data abstraction yielded 22 relevant articles/research papers, which were grouped into two subtopics: edge AI and healthcare in smart cities. They addressed how the machine-learning (ML), DL, and IoT models could be used in healthcare. The accuracy of the models implemented in various research papers was assessed. A. Alghamdi [4] developed a VGG-Net model for analyzing electrocardiogram (ECG) images. VGG-MI-1 showed sensitivity, specificity, and accuracy values of 98.76%, 99.17%, and 99.02%, respectively, and the VGG-MI2 model showed sensitivity, specificity, and accuracy values of 99.15%, 99.49%, and 99.22%, respectively, which were the best so far.

Zhao-xia Lu et al. [5] examined the technical aspects of IoT, cloud computing, big data analysis, and machine learning in clinical medicine. They highlighted the application of AI and IoT in various medical scenarios and discussed challenges and future prospects in this rapidly evolving field. COVID-19 exacerbated the global scarcity of nurses and physicians. Automated tools employing IoT, cloud computing, ML, etc. may be helpful. IoT devices monitor real-time data and send them to the cloud. 5G has increased medical staff's accuracy and speed. Cloud computing services include processing power, machine learning, and storage. They provide data exchange, remote consulting, etc. Large data sets are analyzed by using big data. They can cluster, classify, and visualize data and can mine text. Clustering sorts data by proximity. Classification mining maps data to labels by using decision trees, neural networks, etc. Text mining methods shape unstructured medical data through preprocessing, segmentation, and semantic analysis. Visualizing data uses charts and graphs. This may prevent, diagnose, and cure illnesses. Machine learning creates data-driven models. High-dimensional, high-variance data are analyzed. Medical analysis uses supervised and unsupervised learning. Preprocessing techniques express data by using piecewise linear representation and Fourier transform. By using the sliding window, top-down, bottom-up, and other methods, time series data are segmented, but sliding window is the best method. Diminution involves feature selection and change, k-clustering, SVMs, etc. Decision trees, k nearest neighbors, SVMs, naïve Bayes, etc. identify and classify data. Behavior detection monitors patient behavior by using CNNs or RNNs. Unsupervised learning detects abnormal values. Clustering is simpler. Patient data privacy is also vital. K anonymizes identifiable information but not attributes. L-diversity and T-closeness models reduce granularity. Many clinical uses of IoT and 5G are have been explored. IoT and 5G allow remote diagnostics. Thanks to using IoT patient data and ML graphs, medical practitioners can cooperate. Neonatology, cardiology, and skin cancer screening have been used with near-professional precision. ML may be used for supplementary diagnostics, triage, and alerts for patient vitals. Patients' histories and conditions may be used to produce exercise and food regimens. Cloud-based medical photos may be examined by ML systems to diagnose patients. IoT data, cloud sharing, and ML and AI analysis may assist in diagnosis, spot warning signals, notify emergency services, and provide remote services. This study examines IoT-assisted wearable sensor systems, AI, blockchain difficulties, and other concerns that must be addressed to improve their use in Health management system (HMS).

The study by Junaid et al. [6] aimed to explain the need for and uses of new technologies (sensor-IoT-AI-blockchain) in the healthcare sector by analyzing these technologies as well as past approaches and methodology. In the healthcare administration system, Junaid et al. [6] utilizes papers and survey research on topics such as sensors, IoT, AI, and blockchain. The majority of the data used in their study came from a network of intelligent wearable IoT devices. These devices monitor a patient's vital signs and other pertinent data and display them in real time. For instance, the LIFE Shirt is a multisensor extended HMS that collects and analyses a patient's health data. This information is obtained from the patient by means of the device. In addition to that, an AI-based data synthesis was carried out in order to provide data for testing and validation purposes. There are no well-known data sets that are available for purchase in a prepackaged manner. A smart health ecosystem that limits access to a patient's electronic health record (EHR) might be built with the use of smart contracts. As a result of this research, Junaid et al. [6] concluded that a single ledger that is maintained by healthcare stakeholders may record a patient's entitlements, and smart contracts automate information gathering and distribution and calculate benefits in real time. The acquisition and analysis of real-time patient data from hospital and home devices is facilitated by using sensing technology. It's possible that a real-time analysis may increase the accuracy of patient safety monitoring and incident prediction. When using ML, decisions are made on the basis of studying the data rather than on the basis of making intuitive assumptions. The proposed HMS is able to function by collecting data from its users via the use of smart wristbands and then feeding that information into an artificial neural network (ANN) for risk assessment. The findings of the studies demonstrate that the suggested HMS is capable of accurately assessing the health states of patients.

Ramasamy et al. [7] offered an AI-enabled combination of the Internet of Things with a cyber-physical system (IoT-CPS) for doctors to identify illnesses in patients. Human intelligence improves AI. Computers are better at arithmetic and numbers, whereas humans excel at logic and reasoning. AI might make things "thinkable." Two algorithms make up the AI-powered IoT-CPS algorithm. The first component of the algorithm generates classification rules by categorizing the training data set of patients' illnesses. The second part of the algorithm classifies patients according to their symptoms in order to make disease predictions for the disease-testing data set. The experimental findings show that the suggested AI-enabled IoT-CPS method outperforms state-of-the-art algorithms on accuracy, precision, recall, F-measure, and execution time when diagnosing patients' illnesses. Figures 1 and 2 display the IoT-CPS process.

The importance of the Internet of Things and artificial intelligence in COVID-19 was the primary focus of the research carried out by Praveen Kumar et al. in 2022 [8]. The Internet of Things was studied on a new level: how it manages the COVID-19 epidemic. In this research, long short-term memory (LSTM) with a recurrent neural network (RNN) was employed for diagnostic purposes because

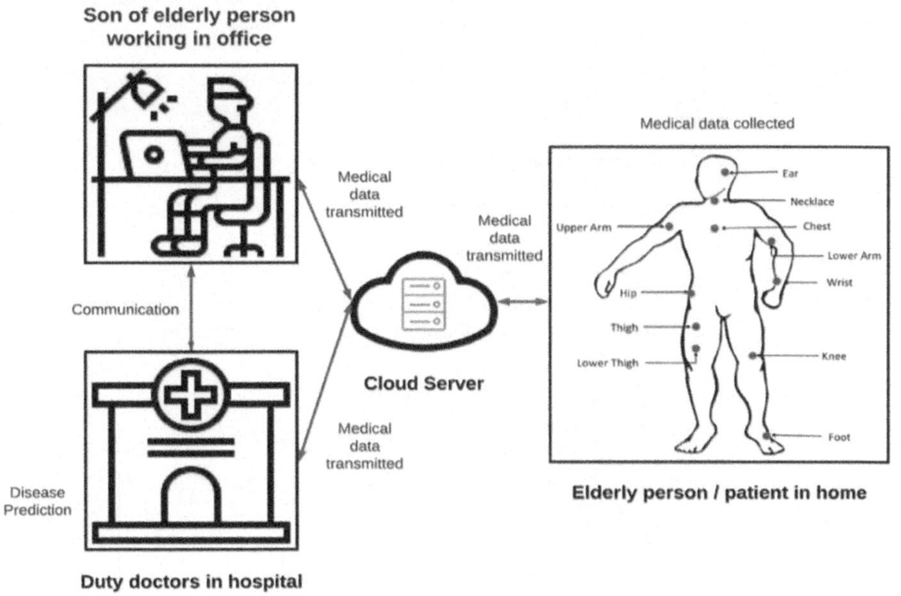

Fig. 1 IoT-CPS [7]

this architecture is particularly helpful in assessing the acoustic aspects of coughing and breathing. Real-time data on a patient's temperature and respiration levels are gathered via the use of a large number of IoT sensors.

The purpose of this chapter is to examine the use of various sensing devices to gather various types of information. The research focuses mostly on using AI systems on data sources. The major objective of this work is to gather data on the temperature and respiration rate of COVID-19 patients in real time by using a number of sensors.

Severe Acute Respiratory Syndrome Coronavirus (SARS-CoV), Middle East Respiratory Syndrome Coronavirus (MERS-CoV), and Severe Acute Respiratory Syndrome Coronavirus 2 (SARS-CoV-2) constitute serious public health hazards. These viruses risk countless lives and inflict economic damage. Recent IT and networking advances have led to IoT and AI applications in numerous sectors. The healthcare and diagnostic sectors are impacted by IoT and AI. By interfacing with smart gadgets and biometric sensors, they have expanded into telemedicine, healthcare, and disease prevention. Even though IoT and AI may improve disease diagnoses, surveillance, and quarantines, their influence is limited insofar as they aren't integrated or deployed quickly for a sudden outbreak. Conventional procedures fail to prevent large-scale illnesses and halt worldwide outbreaks via prediction, resulting in many deaths. Sungho Sim et al. [9] proposed a combined Internet of medical Things and AI (IoMT-AI) framework model to handle COVID-19 outbreaks. IoMT uses (1) remote monitoring, (2) prescriptions tracking, and (3) biometric sensors to provide health data to doctors. Image signal processing and

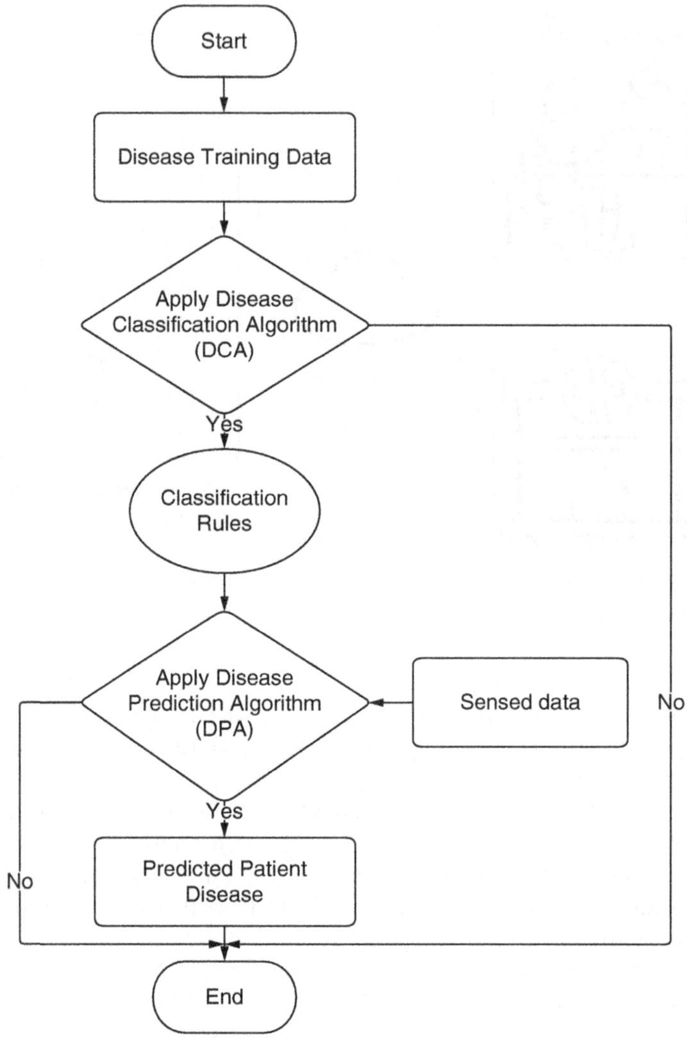

Fig. 2 Flowchart of an AI-enabled IoT-CPS [7]

virus infection detection using AI can carry out disease surveillance, risk prediction, medical diagnosis and screening, curative research, virus modeling and analysis, and the management of lockdown measures, as shown in Fig. 3 (Table 1).

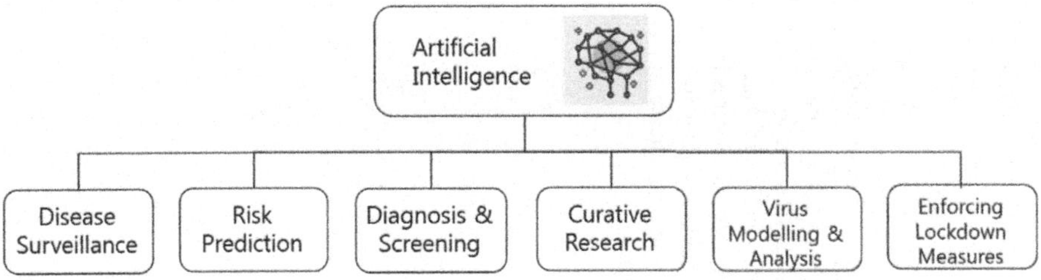

Fig. 3 Applications of AI in medical IoT [10]

2.2 *Sustainability*

Internet of things (IoT) makes it possible to digitalize a wide variety of tasks and procedures, including the distribution of water, preventive maintenance, and smart manufacturing. Although IoT technologies and ideas such as edge computing hold a great deal of potential for the digital transition to sustainability, they are not yet making contributions to the sustainable development of the IoT industry. This sector has a significant carbon footprint because it makes use of a limited supply of raw materials and a significant amount of energy in its production, operation, and recycling processes. However, its sustainable vision collides with edge artificial intelligence (edge AI), which demands more energy, which is why the green Internet of Things (GIoT) paradigm has emerged as a study topic to reduce carbon footprints. In article [11, 12], the authors investigate the process of designing and developing edge-AI GIoT systems. The concepts discussed in the paper are highlighted by using a real-world example of an Industry 5.0 application case. To enhance operator safety and operation tracking, a smart Industry 5.0 workshop should be held. This application takes advantage of an IoT mist architecture that is equipped with AI. After the application situation has been explained, the energy consumption and carbon footprint of the application are analyzed. This article offers guidelines for aspiring developers who want to design edge-AI GIoT systems using the aforementioned technologies (Table 2).

2.3 *Information Security*

The Internet of Things (IoT) has altered how humans live and has permeated all facets of human life, but it has also given rise to worries over data security, which in turn have led to a variety of technical ethical and security difficulties. For the information security of a company, these problems often center on the demand to obtain user data and maintain privacy when the IoT is deployed. This requirement becomes more sensitive when discussing the security of corporate information. The

Table 1 Recent works on IoT and AI in healthcare

Paper	Authors and year	Methodology	Data set	Advantages	Limitations/future scope
[1]	Rahman et al., 2022	This integrates the FL-AI into healthcare technologies and classifies them.	Arrhythmia database [13, 14] of ECG signals from the PTB-XL data set [15, 16]	Federated learning has more benefits than traditional machine-learning approaches do. Data security keeps locally training data on devices, so a data pool is not required. It contains heterogeneous data because it uses data from different users. Real-time continual learning is enabled by models that are continuously improved with client data. Hardware efficiency is improved by using less-complex hardware; federated learning does not need one complex central server for analysis.	The whole procedure runs over the Internet, so privacy and security are always major concerns in this field, especially in competent healthcare systems.

[3]	M. M. Kamruzzaman et al., 2022	This employs a systematic review of 425 articles between 2016 and 2021, using AMSTAR tools.	Multiple data sets [3]	Pros of healthcare AI and IoMT with smart cites include the following: (1) It will minimize processing delays and data traffic for essential analysis and patient monitoring. (2) Implementing a hierarchical edge-based computing system enables scalability, low latency, and the fortification of training model data, allowing researchers to evaluate COVID-19 with a reliable local edge server. (3) The SEG 3.0 strategy, which integrates, associates, and enables semantic interoperability, may be used everywhere in smart cities to enhance heterogeneous data.	Edge AI presents security and computational complexity problems.

(continued)

Table 1 (continued)

Paper	Authors and year	Methodology	Data set	Advantages	Limitations/future scope
[2]	T. Montanaro et al., 2022	This improves the precision and reliability of IoT in healthcare by using ultrathin, skin-compatible, flexible, high-precision piezoelectric sensors, low-cost communication technologies, on-device intelligence, edge intelligence, and edge computing.	ECG5000 [17, 18]	Some health conditions, such as a heart failure, are time critical, yet cloud-based solutions are slower. Onboard devices are quick and have minimal latency. The approach fosters hybrid solutions that combine advanced sensor technology with low-cost, low-power IoT technologies, and AI. Onboard intelligence further reduces latency.	The reliability of IoT devices and sensors can be further improved, and by using blockchain technologies, we can provide additional security and privacy to users.

[5]	Zhao-xia Lu et al., 2021	This summarizes AI and IoT in clinical care, analyses obstacles, and discusses current and potential advancements.	GHWA/WHO; Nursing Report 2020 [19, 20]; World Population Prospects 2019 [21, 22]; Tang X. data set [23, 24]	(1) Integrating AI with IoT gives IoT greater intelligence and is being utilized to improve health care. Remote diagnostic and treatment cooperation may advance service parity. Combined with medical IoT, real-time vital sign data may be acquired and utilized to detect when NICU and CCU patients are in early danger. (2) AI and IoT can deliver speedy and accurate diagnoses and treatments for emergency care, buying rescue time and opening the green life channel. (3) The early identification of high-risk VTE patients and appropriate management may minimize VTE incidence and enhance patient quality of life.	(1) The security of IoT devices is a concern. Because there are so many interconnected devices, one compromised device can compromise many others. (2) IoT devices lack standards. They are still new, and there are numerous vendors, each of which has a different standards and structures for data. (3) AI models focus on particular diseases, such as cancer or skin disease. Medical diagnoses sometimes need to have the entire picture because one part can affect another. (4) Medical ethics are concerning to some people. Patients have difficulty accepting diagnoses made by computers.

(continued)

Table 1 (continued)

Paper	Authors and year	Methodology	Data set	Advantages	Limitations/future scope
[6]	Junaid et al., 2022	This proposes an HMS that uses user-provided data gathered from connected smart wristbands to anticipate health status hazards by using an ANN. The experimental findings demonstrate the utility of the suggested HMS for assessing health conditions in humans.	Health data from wearable IoT devices [6]	Smart contract promotes restricted access to the patient's confidential medical data. Real-time analysis may improve patient safety monitoring and incident prediction accuracy. With ML, decision-making is based on data analysis rather than intuition.	The scarcity of cost-effective and accurate smart medical sensors, unstandardized IoT system architectures, the heterogeneity of linked wearable devices, multidimensional data, and interoperability are difficulties. Privacy and ethical difficulties and the lack of a legal framework to enable smart contracts are open research challenges in this sector. Healthcare data are diverse, incomplete, and ambiguous. Training an ML model with that much data is difficult.

[7]	Ramasamy et al., 2022	This proposes cyber-physical systems (CPSs). The smart device is programmed to help elderly people in emergencies. The sensors collect data from various parts/joints of the subject and process that data to draw conclusions.	Kaggle data set	(1) The experimental findings show that the suggested AI-enabled IoT-CPS method outperformed the state-of-the-art algorithms on accuracy, precision, recall, F-measure, and execution time when diagnosing patients' illnesses. (2) The accuracy when using IoT + CPS is the highest when compared to all the other techniques, i.e., naïve Bayes, SVMs, KNN, and ANN. Comparisons between accuracy and F-measure are as follows: naïve Bayes—83.5% (F-measure—83.8%), SVM—83.9% (F-measure—83.7%), KNN—75.0% (F-measure—75.7%), ANN—78.9% (F-measure—78.2%), and AI-enabled IoT-CPS—85.1% (F-measure - 86.4%).	(1) Disease propagation and contagiousness can be determined by mobile phone location, as a future scope. (2) Data encryption was not considered while writing this paper, but because health and condition data are personal, leaking it may embarrass individuals or swamp them with spam calls about treatments and drugs. (3) The large amount of information on IoT and CPS devices makes them targets for fraudsters, other criminals, and unethical users.

(continued)

Table 1 (continued)

Paper	Authors and year	Methodology	Data set	Advantages	Limitations/future scope
[8]	Praveen Kumar et al., 2022	This uses long short-term memory (LSTM) with a recurrent neural network (RNN) for the diagnosis of COVID-19 and collects data from IoT sensors.	Coughing and respiratory samples and thermal and breathing levels, obtained from various IoT devices [8]	The proposed approach shows the following: (1) IoT manages risky patients so that conditions such as hyperglycemia, pulse rate, and heart infections may be diagnosed. (2) IoT is inventive and powerful at fighting COVID-19 and can be used for strengthening patient data.	Taking evolutionary approaches may solve the problems of interference and complexity in migrating data. Prediction isn't perfect. Electronic enforcement, which incorporates secrecy and trust, is another difficulty.
[9]	Sungho Sim et al., 2021	This proposes a framework where IoMT handles monitoring the patient from a remote location and tracking medication prescriptions by using biometric sensors. And AI/ML handles image signal processing and virus infection detection by using AI.	WHO [25, 26]	While research on AI and IoT for health services has advanced, cost-effective and inexpensive smart healthcare services have received little attention. AI-driven IoT (AIIoT) for smart healthcare could change many facets of healthcare, but several technological hurdles must first be overcome.	Future research on a model for an international integrated viral defense system might build on existing analyses and make use of already-acquired data. To this end, it is important to collect detailed information on viral outbreaks on a global scale and in a timely fashion in order to create a system for the prevention of viral illness transmission and prevalence and for early warning signs.

Table 2 Recent works on IoT and AI for sustainability

	Authors and year	Methodology	Data set	Evaluation parameter	Advantages	Limitations/future scope
[11]	Paula Fraga-Lamas et al., 2021	This proposes an edge-AI GIoT system that improves operator safety and operation tracking.	Visual Wake Words Dataset, CIFAR-10, Google Speech Commands, and ToyADMOS (ToyCar)	Accuracy Visual Wake Words Dataset—80%, CIFAR-10—85%, Google Speech Commands—90%, and ToyADMOS(ToyCar)—85%	(1) Data may be captured and processed on the same device, reducing network connection difficulties. (2) When processing is carried out locally, all communication-related delays are eliminated, resulting in a latency that converges to the inference latency. (3) Reduced communication between IoT edge devices and the cloud reduces the risk of data loss, theft, or leakage. (4) Reducing IoT edge device connectivity with the cloud decreases bandwidth demands and costs, thus improving bandwidth efficiency.	Network, physical, software, and encryption threats require more protection. On-device learning must be protected against intrusions. The IoT industry is fragmented, so a fully standardized framework is needed to address edge-AI GIoT needs. Future developers must address how edge-AI GIoT devices affect CO_2 emissions. Such expansion varies from nation to nation depending on their available energy.

development of AI-based security solutions for IoT sensor networks has been a recent trend.

In [27, 28], the authors built a platform for the administration of information security that was comprised of four parts: the management of IoT data mining, management of equipment, management of keys, and management of databases. Testing for concurrency, stress, high data volume, and security were improved, together with the original architecture for physical security, which was also reorganized.

The purposes of this chapter are to provide some fresh suggestions for applying information security technology to IoT-based corporate management and to advocate for the use of IoT in industrial and commercial management in the near future. The following features are included in the functionality of the platform: authorization and revocation, staff scheduling, data storage and backup, rank role administration, and data encryption and mining (Table 3).

2.4 *Education*

In most cases, the services and administrations that rely on traditional libraries cannot function without them. A reader must follow a multistep procedure to borrow books, which includes entering a library to preserve books of interest, bringing books to a certain spot (i.e., the circulation desk), showing the librarian their identity for verification, and lastly confirming the books to borrow. The many occurrences of the targeted books being borrowed by others throughout the preceding method make it difficult to ignore them. If a reader doesn't know in advance which branch has the books they're looking for, they'll have to make many trips to different libraries. However, in the era of the "smart library," obtaining books requires only a few steps via a smart terminal device: confirm and make an appointment for the intended books and fetch the books under the smart guide and devices furnished by the library. This is the optimal schedule for the borrowing process. Thanks to the immense potential of AI and IoT, library circulation efficiency has greatly improved over more-conventional approaches.

One study [29] conducted an extensive literature review on the use of AI and IoT technologies in proposed future smart libraries in order to formally offer a systematic, organized, and comprehensive strategy for such a possible topic. When compared to librarianship that relies on only human effort, the smart library model, which incorporates AI and IoT, provides far superior service, as seen in the aforementioned smart circulation service. Smart service, smart sustainability, and smart security are the three primary foci of the author efforts. There is a wide variety of use cases for AI and IoT in a smart library. According to the findings of [29, 30], its authors have formally described the development of the smart library (Fig. 4).

Machine-learning and deep-learning techniques, including natural language preprocessing, deep-learning models, recurrent neural networks, and deep-learning-based recommendation systems, have been used throughout this study.

Table 3 Recent works on IoT and AI for information security

	Authors and year	Methodology	Data set	Evaluation parameter	Advantages	Limitations/future scope
[27, 28]	Hongbin Sun et al., 2022	With the current state of global informatization as a backdrop, this work is being carried out to maintain the information security of enterprises and provide modern upgrading means for enterprise management; to popularize cutting-edge IoT management technology for enterprises; and to address issues of information security in enterprise management.	The authors gathered 1000 groups of user data to create a data set, of which 800 groups were used for training and 200 for testing.	The response time of the system is measured according to four aspects: (1) concurrency testing: 0.136 s (information input); (2) stress testing: 0.137 s (event entry); (3) large data volume testing: 0.126 s (gap analysis); and (4) security testing: 0.124 s (gap analysis).	The proposed management system for IoT information security has outperformed other systems. The average response time to an event is less than 0.25 seconds; CPU consumption does not exceed 20%; and the memory requirements are not excessive. Thus, this product improves business information security management.	Limitations: (1) Different sectors' information security demands aren't considered. (2) The existing management platform is memory-intensive. Future scope: (1) It should address the information security demands of diverse sectors. (2) It should gather more corporate information security data and enhance the memory algorithm.

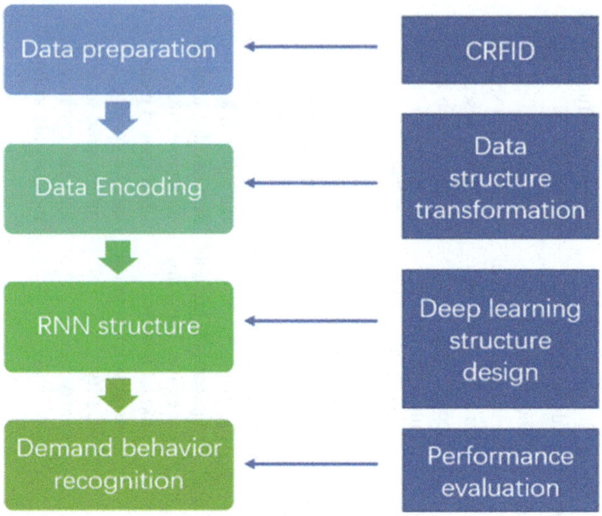

Fig. 4 The workflow of the personalized activity-learning system [31]

Using IoT with the help of artificial intelligence may significantly lessen the likelihood that sensitive data or valuables will be compromised. A new authentication system is being suggested to ensure the anonymity of readers. IoT and AI can work together to create a sustainable timetable that takes into account real-world requirements. The article proposes many AI-assisted IoT methods for controlling the library's lighting in order to maximize the efficiency of its use of natural light (Table 4).

2.5 Pollution Monitoring (Table 5)

2.6 Robotics (Table 6)

2.7 Other Related Works

Apart from the above-mentioned domain, Several papers have been identified that are relevant to the theme. Gültekin et al. [32] presented a deep learning approach based on multisensory data fusion for fault diagnosis in an industrial autonomous transfer vehicle. They focused on real-time fault detection and condition monitoring using edge artificial intelligence for industrial autonomous transfer vehicles [33]. Malik et al. [34] discussed the latest trends in the design and application of smart antennas. Rogers and Malik [25] explored the opportunities and challenges of planar and printed antennas for IoT-enabled environments. Abdul Rahim and Praveen Kumar Malik [36] analyzed and designed a fractal antenna for efficient communica-

Table 4 Recent works on IoT and AI for education

	Authors and year	Methodology	Advantages	Limitations/future scope
[29]	Siguo Bi et al., 2022	This tries to solve the problems with traditional libraries by (1) providing a consistent definition of smart libraries; (2) surveying IoT technology in use; (3) using AI to enhance IoT devices and user experiences; (4) comparing CRFID with mobile IoT technology; (5) comparing AI technologies, including deep learning, OCR, recommender systems, and KNN and SVMs; and (6) separating smart libraries according to their services, sustainability, and security. Thus, better and more-specific solutions may be built for each phase of the library book-borrowing process. to improve customer experiences.	AI-assisted IoT reduces the danger of exposing private information and objects. Novel authentications can preserve readers' privacy. IoT and AI can arrange sustainability on the basis of practical demands. AI-assisted IoT solutions are offered to regulate light shading in libraries to enhance natural light consumption.	Smart libraries have advanced AI-assisted IoT, but there remains room for growth. The performance values of smart services for cutting-edge recommendation-based situations, such as accuracy and recall, are low. When combined with IoT, strong natural language processing (NLP) technologies such as bidirectional encoder representations from transformers (BERT) and graph-based neural networks with side information improvement might power upstream or downstream data processing. Existing smart security systems deal with homogenous data from sensors such as radio-frequency identification (RFID) sensors. With the diversity of smart library situations, processing heterogeneous data from multipurpose sensors such as RFID, near-field communication (NFC), Bluetooth, and infrared sensors, among others, might be challenging.

Table 5 Recent works on IoT and AI for pollution control

	Authors and year	Methodology	Data set	Evaluation parameter	Advantages	Limitations/future scope
[37]	Chao-Tung Yang et al., 2021	This proposes a conceptual PM AIIoT framework that includes an approach for low-cost sensing technologies. IoT sensors and IoT industrial environmental quality sensors should follow a protocol.	PM 2.5 data set [37]	R2 compares each sensor's performance in FRM/FEM. R2 demonstrates near-perfect consistency around 1; near 0 shows no relationship. The Kaiterra/LE200 model used in plan tower sensor and reference instrument correlated well (R2 = 0.98). ITRI calibrates plan tower sensors by providing the most accurate and reliable data, according to the best-performing tester.	The paper examines the integration of AI and IoT (AIIoT) into PM monitoring, indoor air-quality management, and future advances. It summarizes existing PM monitoring approaches, including the use of sensors for data collecting. Light scattering theory is followed to characterize most low-cost PM AIIoT sensors.	AI for PM monitoring should combine current air control infrastructure with satellite data to produce an accurate estimate of air quality. AI in PM monitoring can identify air pollution from garbage burning, transportation, manufacturing, resource distribution, and population density. This needs an AI that can geospatially interpret air-quality data.

Table 6 Recent works on IoT and AI for robotics

	Authors and year	Methodology	Data set	Advantages	Limitations/future scope
[38]	Katy Borner et al., 2020	Many research articles have analyzed how AI, robotics, and IoT are utilized in research and how frequently they come together (be it two at a time or all three of them). Emergent ideas should display originality, persistence, and accelerated expansion, should establish a research community, according to the four-attribute model of technological emergence: burst detection and visualization, top organizations and funding, scientific mapping and classification systems, and coauthor networks.	WOS publications and NSF awards	Because AI, IoT, and robots have been converging, we should expect the border between hardware and software to blur much more rapidly than we had previously anticipated. The combined potential for innovation across all fields bodes well for the progress of humanity and the quality of its future.	Some of the current research in these areas is undertaken by military agencies, making it difficult to collect its results in public databases. Only 1998–2017 WOS publications and NSF grants were analyzed. As a part of future scope, authors may want to study more data sets, such as studying the federal business opportunities (FBO) database to further understand financing for sales and trading.

tion networks in vehicular models. Shaik and Malik [39] conducted a comprehensive survey on 5G wireless communication systems, addressing open issues, research challenges, channel estimation, multi-carrier modulation, and applications. Malik, Wadhwa, and Khinda [40] conducted a survey on device-to-device and cooperative communication for future cellular networks. Tiwari and Malik [41] focused on wideband microstrip antenna design for higher "X" band frequencies. Kaur and Malik [42] presented a study on multiband elliptical patch fractal and defected ground structures microstrip patch antennas for wireless applications. Shaik and Malik [43] retrospectively analyzed channel estimation techniques for 5G wireless communications. Finally, Malik and Singh [44] proposed a multiple bandwidth design of a microstrip antenna for future wireless communication. These papers collectively contribute to the understanding and application of IoT and AI in various domains.

3 Conclusion and Future Directions

IoT and AI have offered wonderful opportunities for organizations to enhance efficiency, visibility, and profitability. AI and IoT are altering our world. More companies than ever are making use of IoT. ML, AI, rapid feedback, and remote monitoring are here and not slowing down. Businesses that embrace the IoT revolution early avail themselves of several possibilities. The intention of this work is to present various new future directions in the field of AI and IoT by exploring their applications in various domains, such as healthcare, education, sustainability, robotics, pollution control, and much more. Every application of AI and IoT has unique advantages and disadvantages. But security and privacy concerns are the most common shortcomings emerging as open challenges. These will be stepping stones for future research. More real-time case studies are needed in order to measure the real-time effects of AI-assisted IoT on the environment.

References

1. Rahman, A., et al. (2022). Federated learning-based AI approaches in smart healthcare: Concepts, taxonomies, challenges and open issues. *Cluster Computing*, 1–41. https://doi.org/10.1007/s10586-022-03658-4
2. Shumba, A. T., Montanaro, T., Sergi, I., Fachechi, L., De Vittorio, M., & Patrono, L. (2022). Leveraging IoT-aware technologies and AI techniques for real-time critical healthcare applications. *Sensors, 22*(19). https://doi.org/10.3390/S22197675
3. Kamruzzaman, M. M., Alrashdi, I., & Alqazzaz, A. (2022). New opportunities, challenges, and applications of edge-AI for connected healthcare in internet of medical things for smart cities. *Journal of Healthcare Engineering, 2022*, 1. https://doi.org/10.1155/2022/2950699
4. Alghamdi, A., et al. (2020). Detection of myocardial infarction based on novel deep transfer learning methods for urban healthcare in smart cities. *Multimedia Tools and Applications*, 1–22. https://doi.org/10.1007/S11042-020-08769-X/TABLES/10

5. Lu, Z. X., et al. (2021). Application of AI and IoT in clinical medicine: Summary and challenges. *Current Medical Science, 41*(6), 1134–1150. https://doi.org/10.1007/S11596-021-2486-Z
6. Junaid, S. B., et al. (2022). Recent advancements in emerging Technologies for healthcare management systems: A survey. *Healthcare, 10*(10), 1940. https://doi.org/10.3390/HEALTHCARE10101940
7. Ramasamy, L. K., Khan, F., Shah, M., Prasad, B. V. V. S., Iwendi, C., & Biamba, C. (2022). Secure smart wearable computing through artificial intelligence-enabled internet of things and cyber-physical systems for health monitoring. *Sensors, 22*(3), 1076. https://doi.org/10.3390/s22031076
8. Kollu, P. K., et al. (2022). Development of advanced artificial intelligence and IoT automation in the crisis of COVID-19 detection. *Journal of Healthcare Engineering, 2022*, 1. https://doi.org/10.1155/2022/1987917
9. Sim, S., & Cho, M. (2021). Convergence model of AI and IoT for virus disease control system. *Personal and Ubiquitous Computing*, 1–11. https://doi.org/10.1007/S00779-021-01577-6
10. Huang, X. (2020). Application analysis of AI reasoning engine in microblog culture industry.*Personal and Ubiquitous Computing, 24*(3), 393–403. https://doi.org/10.1007/S00779-019-01338-6/FIGURES/13
11. Priyadarshini, I., Kumar, R., Sharma, R., Singh, P. K., & Satapathy, S. C. (2021). Identifying cyber insecurities in trustworthy space and energy sector for smart grids. *Computers & Electrical Engineering, 93*, 107204.
12. Fraga-Lamas, P., Lopes, S. I., & Fernández-Caramés, T. M. (2021). Green IoT and edge AI as key technological enablers for a sustainable digital transition towards a smart circular economy: An industry 5.0 use case. *Sensors, 21*(17). https://doi.org/10.3390/S21175745
13. Vo, M. T., Vo, A. H., Nguyen, T., Sharma, R., & Le, T. (2021). Dealing with the class imbalance problem in the detection of fake job descriptions. *Computers, Materials & Continua, 68*(1), 521–535.
14. MIT-BIH Arrhythmia Database v1.0.0. https://physionet.org/content/mitdb/1.0.0/ (accessed Nov. 18, 2022).
15. Sachan, S., Sharma, R., & Sehgal, A. (2021). Energy efficient scheme for better connectivity in sustainable mobile wireless sensor networks.*Sustainable Computing: Informatics and Systems, 30*, 100504.
16. Wagner, P., et al. (2020). PTB-XL, a large publicly available electrocardiography dataset. *Sci. Data, 7*(1). https://doi.org/10.1038/S41597-020-0495-6
17. Ghanem, S., Kanungo, P., Panda, G., et al. (2021). Lane detection under artificial colored light in tunnels and on highways: An IoT-based framework for smart city infrastructure. *Complex & Intelligent Systems*. https://doi.org/10.1007/s40747-021-00381-2
18. Time Series Classification Website. http://timeseriesclassification.com/description.php?Dataset=ECG5000 (accessed Nov. 18, 2022).
19. Sachan, S., Sharma, R., & Sehgal, A. (2021). SINR based energy optimization schemes for 5G vehicular sensor networks. *Wireless Personal Communications, 127*, 1023. https://doi.org/10.1007/s11277-021-08561-6
20. State of the world's nursing 2020: investing in education, jobs and leadership. https://www.who.int/publications/i/item/9789240003279 (accessed Nov. 18, 2022).
21. Priyadarshini, I., Mohanty, P., Kumar, R., et al. (2021). A study on the sentiments and psychology of twitter users during COVID-19 lockdown period.*Multimedia Tools and Applications, 81*, 27009. https://doi.org/10.1007/s11042-021-11004-w
22. World Population Prospects 2019: Highlights. Multimedia Library - United Nations Department of Economic and Social Affairs. https://www.un.org/development/desa/publications/world-population-prospects-2019-highlights.html. (accessed Nov. 18, 2022).
23. Azad, C., Bhushan, B., Sharma, R., et al. (2021). Prediction model using SMOTE, genetic algorithm and decision tree (PMSGD) for classification of diabetes mellitus. *Multimedia Systems, 28*, 1289. https://doi.org/10.1007/s00530-021-00817-2

24. Tang, X. (2019). The role of artificial intelligence in medical imaging research. *BJR open, 2*(1), 20190031. https://doi.org/10.1259/BJRO.20190031
25. Priyadarshini, I., Kumar, R., Tuan, L. M., et al. (2021). A new enhanced cyber security framework for medical cyber physical systems. *SICS Software-Intensive Cyber-Physical Systems, 35*, 159. https://doi.org/10.1007/s00450-021-00427-3
26. Coronavirus disease (COVID-19). https://www.who.int/emergencies/diseases/novel-coronavirus-2019 (accessed Nov. 19, 2022).
27. Singh, R., Sharma, R., Akram, S. V., Gehlot, A., Buddhi, D., Malik, P. K., & Arya, R. (2021). Highway 4.0: Digitalization of highways for vulnerable road safety development with intelligent IoT sensors and machine learning. *Safety Science, 143*, 105407. ISSN 0925-7535.
28. Sun, H., & Bai, S. (2022). Enterprise information security management using internet of things combined with artificial intelligence technology. *Computational Intelligence and Neuroscience, 2022*. https://doi.org/10.1155/2022/7138515
29. Bi, S., et al. (2022). A survey on artificial intelligence aided internet-of-things technologies in emerging smart libraries. *Sensors, 22*(8). https://doi.org/10.3390/S22082991
30. Sahu, L., Sharma, R., Sahu, I., Das, M., Sahu, B., & Kumar, R. (2021). Efficient detection of Parkinson's disease using deep learning techniques over medical data. *Expert Systems, 39*, e12787. https://doi.org/10.1111/exsy.12787
31. Bai, R., Zhao, J., Li, D., Lv, X., Wang, Q., & Zhu, B. (2020). RNN-based demand awareness in smart library using CRFID. *China Communications, 17*(5), 284–294. https://doi.org/10.23919/JCC.2020.05.021
32. Gültekin, Ö., Cinar, E., Özkan, K., & Yazıcı, A. (2022). Multisensory data fusion-based deep learning approach for fault diagnosis of an industrial autonomous transfer vehicle. *Expert Systems with Applications, 200*, 117055. https://doi.org/10.1016/J.ESWA.2022.117055
33. Gültekin, Ö., Cinar, E., Özkan, K., & Yazıcı, A. (2022). Real-time fault detection and condition monitoring for industrial autonomous transfer vehicles utilizing edge artificial intelligence. *Sensors, 22*(9), 3208. https://doi.org/10.3390/s22093208
34. Malik, P., Lu, J., Madhav, B. T. P., Kalkhambkar, G., & Amit, S. (Eds.). *Smart antennas: Latest trends in design and application*. Springer. ISBN 978-3-030-76636-8.https://doi.org/10.1007/978-3-030-76636-8
35. Roges, R., & Malik, P. K. (2021). Planar and printed antennas for internet of things-enabled environment: Opportunities and challenges.*International Journal of Communication Systems, 34*(15), e4940. https://doi.org/10.1002/dac.4940. (IF: 2.047) ISSN: 1099-1131.
36. Rahim, A., & Malik, P. K. (2021). Analysis and design of fractal antenna for efficient communication network in vehicular model. *Sustainable Computing: Informatics and Systems, Elsevier, 31*, 100586. https://doi.org/10.1016/j.suscom.2021.100586. ISSN 2210-5379.
37. Yang, C. T., Chen, H. W., Chang, E. J., Kristiani, E., Nguyen, K. L. P., & Chang, J. S. (2021). Current advances and future challenges of AIoT applications in particulate matter (PM) monitoring and control.*Journal of Hazardous Materials, 419*, 126442. https://doi.org/10.1016/j.jhazmat.2021.126442
38. Börner, K., et al. (2020). Mapping the co-evolution of artificial intelligence, robotics, and the internet of things over 20 years (1998-2017).*PLoS One, 15*(12), e0242984.https://doi.org/10.1371/JOURNAL.PONE.0242984
39. Shaik, N., & Malik, P. K. (2021). A comprehensive survey 5G wireless communication systems: Open issues, research challenges, channel estimation, multi carrier modulation and 5G applications. *Multimedia Tools and Applications, 80*, 28789. https://doi.org/10.1007/s11042-021-11128-z
40. Malik, P. K., Wadhwa, D. S., & Khinda, J. S. (2020). A survey of device to device and cooperative communication for the future cellular networks.*International Journal of Wireless Information Networks, Springer, 27*, 411–432.https://doi.org/10.1007/s10776-020-00482-8
41. Tiwari, P., & Malik, P. K. (2021). Wide band micro-strip antenna design for higher "X" band. *International Journal of e-Collaboration (IJeC), 17*(4), 60–74.https://doi.org/10.4018/IJeC.2021100105. (ISSN: 1548-3673).

42. Kaur, A., & Malik, P. K. (2021). Multiband elliptical patch fractal and defected ground structures microstrip patch antenna for wireless applications. *Progress In Electromagnetics Research B, 91*, 157–173. https://doi.org/10.2528/PIERB20102704. (ISSN: 1937-6472).
43. Shaik, N., & Malik, P. K. (2020). A Retrospection of Channel Estimation Techniques for 5G Wireless Communications: Opportunities and Challenges. *International Journal of Advanced Science and Technology, 29*(05), 8469–8479. ISSN: 2005-4238.
44. Malik, P. K., & Singh, M. (2019). Multiple bandwidth design of micro strip antenna for future wireless communication. *International Journal of Recent Technology and Engineering, 8*(2), 5135–5138. https://doi.org/10.35940/ijrte.B2871.078219. ISSN: 2277-3878.

Text Summarization for Big Data Analytics: A Comprehensive Review of GPT 2 and BERT Approaches

G. Bharathi Mohan, R. Prasanna Kumar, Srinivasan Parathasarathy, S. Aravind, K. B. Hanish, and G. Pavithria

1 Introduction

At present, where everything is fast and simple for the convenience of humans, a short version of the existing long version is necessary.

People do not have the time and patience to consume information as a whole and they have to be fed in a simple manner. Professionals who handle large reports also face such problems. This can be solved by summarization of text. Text summarization is defined as

> Automatic text summarization is the process of compressing and extracting information effectively from input documents while still retaining its key content. [1]

A basic understanding on what a summary is required before moving on to the text summarization. A summary is a minimized version of something that is created from one or more texts, delivers the key ideas from the original text, and is written in a concise manner. Automatic text summarization aims to show the source text as a short version form with semantics [2]. The goal of text summarization is to come up with methods to produce this summary efficiently and clearly. There are two types of summarizers used in our case. They are abstractive and extractive summarizers.

By choosing a portion of the entire sentence base, extractive summarization creates a summary of the text that has been provided. The text's most significant phrases and sentences are determined and chosen with a score that is calculated based on the words in the text. The approach of abstractive summarization begins with analyzing the text document to develop an interpretation. The computer makes

---

G. Bharathi Mohan (✉) · R. Prasanna Kumar · S. Aravind · K. B. Hanish · G. Pavithria
Department of CSE, Amrita School of Computing, Amrita Vishwa Vidyapeetham, Chennai, India
e-mail: g_bharathimohan@ch.amrita.edu

S. Parathasarathy
Oracle America, Lehi, UT, USA

R. Sharma et al. (eds.), *Data Analytics for Internet of Things Infrastructure*,
Internet of Things, https://doi.org/10.1007/978-3-031-33808-3_14

a prediction based on this interpretation: a summary. By paraphrasing parts of the actual text, it changes the essence of the text [3].

Text summarization is classified into two types: indicative and informative. Inductive summary just provides the core concept of the text to the consumer. This form of summary is typically 5–10% of the length of the main portion of the text. Informative summary methods, on the other hand, give exact information on the primary text. The useful summary should be 20–30% as long as the main material [2].

At this stage, most of the automatic text summarization techniques mainly use machine learning or deep learning methodologies and models. Deep learning techniques were used for the first time in abstractive text summarization in 2015, with the proposed methodology based on the encoder-decoder architecture. Deep learning models have produced amazing results in these applications and have been widely used recently [4].

Our work compares two models of Transformer, GPT2 and BERT, to find out which one is more efficient than the other. At the end of our analysis using ROUGE metrics, it has been confirmed that BERT gives a better performance compared to GPT2.

2 Related Works

Subha Shini and Ambeth Kumar developed and put into use recurrent neural network-based text summarization techniques. They have addressed different approaches and distinct datasets that are used to create text summaries, which decrease the time required to manually enter summaries for lengthy texts while maintaining the meaning produced by recurrent neural networks (RNNs) [5].

Two of the LSA-based summarizing techniques are described in [6]. The algorithm results are compared using their ROUGE scores after being assessed on Turkish and English documents. Both of our systems perform equally well on collections of Turkish and English documents; however, one of them yields the highest scores.

The most recent extractive text summarizing methods for several languages have been described in this study [7]. It is seen that excellent work has been done for several foreign languages, like Chinese and English. However, there is still no summary system for Bengali languages. Therefore, it is difficult to provide a summary method that makes use of several feature categories.

In [8], the authors have designed and implemented a text summarization that uses the BERT algorithm. They introduced a document-level encoder based on BERT which expresses the document into segments and obtains a representation of the given paragraph. They also used a two-staged fine-tuning approach to further improve the quality of the summaries that is generated.

In another study, text summarization was accomplished using multi-layered attentional peephole convolutional LSTM (Long Short-Term Memory). The goal is to create an input of a given lengthy text and construct an automatic text summation. They have improved the settings for MAPCol utilizing the central composite design (CCD) in conjunction with RSM, which produces summaries with good accuracy [9].

Jesse Vig introduced an open source tool as a way to visualize attention in the Transformer at various scales. The tool was demonstrated on GPT-2 and BERT, and three use cases were presented. To traverse the tool's three views, a uniform interface is intended to be created in future works [10].

A text summarization tool that uses the LSTM-CNN-based ATS framework (ATSDL) has been designed and implemented by some authors. The purpose is to construct new sentences which will explore more good-grained fragments than sentences generated. ATS is the task of forming summary sentences by combining facts from different sources and forming them into shorter content and preserving meaning [11].

In [12], the authors upgrade to a recurrent neural network architecture the most advanced model for abstractive sentence summarization. The model is a condensed form of the machine translation encoder-decoder structure. To produce headlines based on the first line of each news article, this model was trained on the Gigaword corpus. Gigaword data and the DUC-2004 challenge show that it performs far better than the prior state of the art, despite the fact that this model does not rely on any additional extraction properties.

The Improved Attention Layer-assisted-Recurrent Convolutional Neural network model is used for yet another text summarization (IA-RCNN). The Sequence-to-Sequence (S2S) paradigm was integrated with RCNN in the model, which was created for abstractive text summarization in a variety of text sources. The model that is suggested in this research is tested using various text sources. For real-time applications, its performance is adequate [13].

2.1 Text Summarization Using Deep Learning

In 2015, deep learning techniques were first used to summarize abstract text, and a model based on encoder/decoder architecture was proposed. Deep learning algorithms have become popular recently and have achieved excellent results in these applications.

Deep learning analyzes difficult situations as a decision aid. Deep learning uses feature extraction at different levels of abstraction to mimic the capabilities of the human brain. Deep learning is used in many NLP tasks because it uses multiple nonlinear layers of data processing to facilitate learning multi-level hierarchical representations of data. Various deep learning models were used for abstract summarization, including RNNs, convolutional neural networks (CNNs), and sequence-to-sequence models. In this section, deep learning models will be explored [27].

A large corpus is used to learn the contextual representations of the language. The BERT language representation is one of the new word embedding model extensions. BERT benefits from a fine-tuning and role-based approach (depending on specific job goals). In addition, transformers can learn the meaning of relationships between "word pairs" by using self-awareness to compute input and output representations [4].

2.2 Need for Text Summarization in Big Data Analytics

Big data is a technology that is used to manage massive amounts of data. It is a technique for storing, distributing, and processing massive amounts of data. Because of technology improvements and HTML 2.0, it is now possible to send data without specifying a tag. Specifically, social media tools such as WhatsApp, Facebook, Instagram, Google, and others are extremely effective for transmitting large amounts of organized and unstructured data [22]. When you view a video on YouTube, you will see advertising that are relevant to your interests at random intervals [23]. This advertising has been chosen for you based on your browsing history and preferences. The most current advancements in a certain issue may be found via Twitter trends.

Furthermore, Google text search and voice-based search have simplified searches nowadays. Today, most individuals use mobile phones to conduct various transactions, and these phones are also used to monitor user locations. Today, various mobile applications are available, and these programmers are installed based on the user's preferences and requirements [24]. These mobile applications are highly beneficial in understanding the user's interests and how much time they spend. These many scenarios suggest extensive data creation, with data being created in vast quantities at all times [25, 26]. As a result, a unique text summarizing approach is necessary to better grasp the hidden knowledge or information concealed behind this data.

3 BERT

BERT (Bidirectional Encoder Representations from Transformers) is used to get over the restriction of RNN (Recurrent Neural Networks), CNN (Convolution Neural Networks), and ANN (Artificial Neural Networks). BERT models are pre-trained on big data sets. So no additional training on the dataset is required. It uses an efficient uniform architecture along with sentence transformation in the resources. With this, the best results are observed in summarization. Unlike any context-free embedding model, BERT is a contextual embedding model [14]. BERT creates on top of the transformer architecture, but its purpose is distinct for pre-training data [15].

Example:
Consider these sentences:

Python is very dangerous. It can kill you.
I love Python. It's easy to understand.

If the word embedding is applied, it will consider the word python in two sentences together. This is not correct because one python is a snake and another is a programming language. So BERT helps us by keeping the contextual embedding of the model.

Advantages of the BERT model are as follows:

1. *Context* – It keeps the context of the words.
2. *Word Ordering* – It keeps the words ordering with the help of positional encoding or positional vectors.
3. *Embeddings* – It has token embedding, sentence embedding, and positional embedding. All these are BERT embedding.
4. *Out of Vocabulary* – It keeps out of vocabulary with the help of self-attention. Moreover, it takes care of those keywords, which are present or not present in the vocabulary.

Applications of the BERT model are as follows:

- Text Summarization
- NER – Name Entity Recognition
- Next Sentence Prediction
- Long Text Classification
- Sentiment Analysis
- Question Answering

3.1 BERT Architecture

The goal of Bidirectional Encoder Representations from Transformers (BERT), which pre-trains deep bidirectional representations from unlabeled text, is to simultaneously condition both left and right contexts across all layers. As a result, the pre-trained BERT model can be enhanced with just one additional output layer to create cutting-edge models for a variety of tasks, such as question answering and language inference, without requiring large changes to the architecture for each task [16].

A group of sentences from "s1,s2,s3,..., sn" have two possible outcomes, with xi = {0,1}indicating whether or not a given sentence will be chosen. The output vectors are tokenized instead of sentences thanks to a pre-trained MML (Masked Language Model). Instead of having multiple sentences, it only has two labels, Sa and Sb, and uses embedding to specify the various sentences. These embedding are appropriately changed to produce the necessary summaries (see Fig. 1) [17].

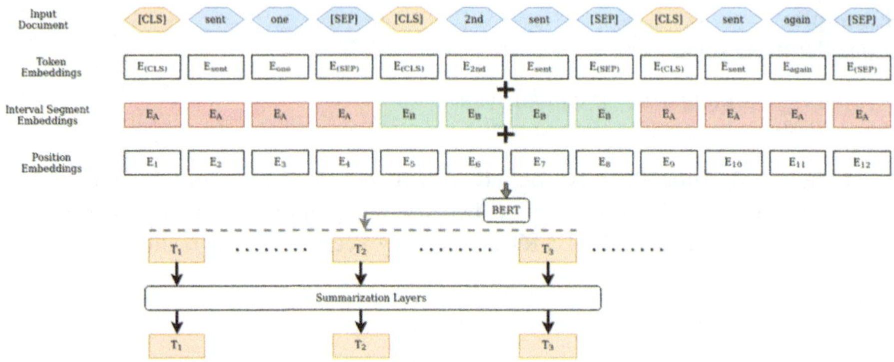

Fig. 1 Process flow of BERT

3.2 Phases in Generating the Summary

3.2.1 Input Document

Sentences from the input source are encoded during this phase. Each sentence has a CLS tag before it and a SEP tag at the end. To group the components of one or more sentences, use the CLS tag.

3.2.2 Interval Segment Embedding

Sentences in the input document will be differentiated during this phase. Each sentence is assigned to one of the labels mentioned above.

Depending on the value of i $\{s_i\}$ might be either$\{E_a$ or $E_b\}$. Essentially, the standard is E_a for even value i and E_b for the odd value i [18].

3.2.3 Embedding

The representation of words in their corresponding vector forms is referred to as embedding. Google makes use of this BERT model feature to help with understanding. Gaining access to various semantic functions, such as those for understanding the purpose of the input document and constructing word-related models.

Our input text is subjected to three different types of embedding before being fed to the BERT model layer:

Embedding tokens: The words are transformed into output vectors with fixed dimensions during this stage. At the beginning and conclusion of input sentences, the CLS and SEP tags are inserted.

3.2.4 Segment Embeddings

Using binary coding, the different inputs are distinguished in this phase.

For example,

Sentence1 – "I love books"

and Sentence2 – "I love sports."

Following the segment embedding operation: [CLS],I,Love,Books,[SEP],I, Love,Sports

Following the segment embedding operation: [0,0,0,0,0,1,1,1], Sentence1 = 0, Sentence2 = 1

3.2.5 Position Embeddings

The BERT model can handle input sequences of up to 512 bytes in length, and its output vector dimension is [512,768]. A word's placement in the input sentence in this phase changes the context of the sentence and should not have the same vectors.

3.2.6 Summarization

The self-attention layer is the primary distinction between recurrent neural networks and Bidirectional Encoder Representations from Transformers (BERT). The model aids in the representation of words and looks for connections between the words.

Simple Classifier: To predict the score Yi, a linear layer and sigmoid function are added to the BERT model. The sigmoid function sets a threshold that determines the range of probability that is mapped to a binary value Y^i [18].

$$Y\hat{}\mathrm{i} = \alpha\,(W_{\mathrm{o}}T_{\mathrm{i}} + b_{\mathrm{o}})$$

3.2.7 Inter Sentence Transformer

The simple classifier is not utilized in this stage. The BERT model is improved by the addition of additional transformer layers, which also help it to better identify the key ideas in the input document [18].

$$h\tilde{}1 = \mathrm{LN}\Big(h^{l-1} + \mathrm{MHAtt}\left(h^{l-1}\right)\Big)$$

$$hl = \mathrm{LN}\,(h\tilde{}1 + \mathrm{FFN}\,(h\tilde{}l))$$

$$h^0 = \mathrm{PosEmb}(T)$$

where

- T – Sentence output vector by BERT model.
- PosEmb – Positional Embeddings
- LN – Layer Normalization Function
- MHAtt – Multi-head Attention Function
- l – in-depth of the stacked layer.
- Sigmoid output layer – $Y\hat{}\mathrm{i} = \alpha(W_{\mathrm{o}}T_i + b_{\mathrm{o}})$.

4 GPT-2

The GPT tokenizer was used to tokenize the chosen files, although the context size of GPT models limited it to 512 or 1024 tokens (512 or 1024 tokens for GPT and GPT-2, respectively). Identical text files exist for BERT as well; 100 files were selected from the dataset containing the necessary tokens for training (Fig. 2).

Using the tokens that came before it, a language model makes a probabilistic prediction about the token that will come next in the sequence. It gains knowledge of the likelihood that a sentence or string of tokens will appear based on the text examples it has encountered during training. The following conditional probability can be used to represent it:

$$\mathrm{P}\left(w_1^T\right) = \mathrm{P}\left(w_t | w_1^{t-1}\right)$$

where w_{t} is the tth token, and writing sub-sequence $w_i^j = (w_i, w_{i+1}, \ldots, \mathrm{w}_{j-1}, \mathrm{w}_j)$.

The Transformer model variant known as GPT/GPT-2 only has the decoder portion of the Transformer network. They function like conventional uni-directional

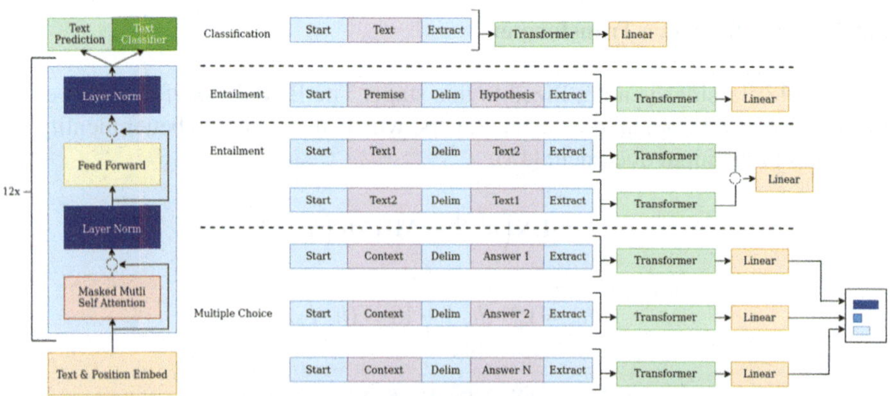

Fig. 2 Process flow of GPT 2

language models thanks to the multi-headed masked self-attention that enables them to focus on just the first I tokens at time step t. These models analyze tokens concurrently, as opposed to sequentially, like RNNs, by simultaneously predicting tokens for all time steps. These role models could be represented by:

$$h_0 = \mathrm{UW_e} + W_\mathrm{p},$$

$$h_i = \text{transformerblock}\,(h_{i-1})\,\forall i \,\epsilon\, [1, n]$$

$$P(u) = \text{softmax}\left(h_\mathrm{n} W_\mathrm{e}^\mathrm{T}\right)$$

where $U = (u_1, u_2, \ldots, u_k)$ are the tokens, n is the number of layers, W_e are the token embedding matrix, and W_p is the position embedding matrix and trained on the language model objective:

$$L(U) = \sum\nolimits_i^k P\left(u_i \mid u_{i-k}, \ldots\ldots, u_{i-1}; \beta\right)$$

where $U = (u_1, u_2, \ldots, u_k)$ are the tokens, k is the context size (maximum number of tokens which can be processed at once by model), and P is the conditional probability modeled by our language model with parameters β. Table 1 describes in the test data set with sample title and content [19].

Byte Pair Encoding is used by GPT2 to construct the vocabulary tokens. This implies that the tokens are typically word fragments. GPT-2 can predict the following token in a sequence because it was trained with the intention of causal language modeling (CLM). Using this feature, GPT-2 could produce text that is syntactically coherent. GPT-2 generates fabricated text samples in response to arbitrary input priming the model. The model has the capacity to alter how it looks to correspond with the style and information in the conditioning text [20].

Due to the GPU resource limitation, the abstractive summarization model is a pre-trained distil version of GPT-2. The DistilGPT2 can accept token lengths of up to 1024. It includes 6 transformer decoder layers and 12 attention heads. The Huggingface transformer package's Hugging face GPT-2 is utilized, which is implemented in Pytorch [14].

The GPT-2 performs comparably to supervise baselines on reading comprehension in a no-shot environment. In some activities, like summarizing, it performs the task qualitatively, but based on quantitative measurements, it still performs at a rudimentary level. The zero-shot performance of GPT-2, while suggestive as a research finding, is still a long way from being useful in terms of practical applications [19].

Table 1 Sample test data set

Title	Content
"Today, Macau is paying out again! Each permanent resident will earn 9000 patacas!"	"Macau's 2014 cash-sharing system will go into effect on July 2. MSAR permanent residents and non-permanent residents will get MOP 9000 and MOP 5400, respectively, at that time. The fiscal cost for this plan by the MSAR government is roughly MOP 5659 million. The cash-sharing plan was established in 2008 to distribute the benefits of economic progress to the general populace." [28]
"Cool! Jordan hosts the eighth international special forces tournament, in which Chinese soldiers compete."	"On May 3, local time, the eighth 'warrior race' international special forces competition was conducted in Amman, Jordan, at the King Abdullah Special Warfare Training Center, with 27 teams from various countries competing. China dispatched the armed police snow leopard assault team and the 14-player Hebei Yanshan team. Jordan hosted the seventh 'warrior competition,' an international special forces tournament in which Chinese soldiers participated."

5 Experiment Setup

5.1 About the Dataset

The non-anonymized BCC dataset that Kaggle provided was utilized, which is designed for condensing news items into two to three sentences. The dataset included thousands of text files with an average length of 15 lines and also included a predetermined summary. Using various measures, this summary is used to assess the precision of both models (GPT-2 and BERT). A few more model-specific pre-processing processes were carried out before feeding this data to the models. A few more pre-processing steps specific to the models were performed.

5.2 Training the Models

Some snippets that are relevant in training and summarization are provided here.

```
"from summarizer import Summarizer, TransformerSummarizer

    #this is the code snippet where the summariser is imported and transformer
summariser modules after installing transformers from hugging face#


GPT2_model                                                                    =
TransformerSummarizer(transformer_type="GPT2",transformer_model_key="gpt
2-medium")
    #this is the code snippet where the parameters are entered for transformer
summariser to train the gpt 2 model#


bert_model = Summarizer()


    #this is the code snippet where the summariser is trained for bert model#"
```

5.3 Evaluation Metrics

The term ROUGE, which stands for "Recall-Oriented Understudy of Gisting Evaluation," refers to a set of standards for rating texts that are produced automatically. It is typically employed to assess the effectiveness of a TS algorithm's summary. There are many metrics and ratings that have been proposed in the literature for the evaluation of text summarization results, but ROUGE is the most popular one. Since ROUGE operates similarly on both the abstractive and extractive algorithms, it does not produce outstanding results. In addition, numerous executions are typically preferable to a single one. It ignores the semantic and grammatical precision of the system and human summaries in favor of the overlap of n-grams (represented as a number value).

The process of writing a summary is easier to compare to an abstractive TS task than an extractive one because when writing a summary, a person can try to express his thoughts with new words and phrases after carefully reading and understanding one or more source texts, trying to cover as many topics as possible from the original text [21].

```
"from rouge_score import rouge_scorer

 # a list of the hypothesis documents

hyp = bert_summary

# a list of the references documents

ref = Summary_list

 # To create aa RougeScore object for rouge1

scorer = rouge_scorer.RougeScorer(['rouge1'])

 # a results contains precision recall and fmeasure

results = {'precision': [], 'recall': [], 'fmeasure': []}


 # for each of the document and hypothesis s pair

for (h, r) in zip(hyp, ref):

   # calculate the ROUGE

   score = scorer.score(h, r)

   # measurements are separated

   precision, recall, fmeasure = score['rouge1']

   # append them to list in the dictionary

   results['precision'].append(precision)

   results['recall'].append(recall)

   results['fmeasure'].append(fmeasure) "
```

5.4 Summary Snippets

Text given as input:
"Safety alert as GM recalls cars.

According to federal officials, the world's largest automaker, General Motors (GM), is recalling roughly 200,000 vehicles in the United States for safety reasons. According to the National Highway Traffic Safety Administration (NHTSA), the greatest recall comprises 155,465 trucks, vans, and SUVs (SUVs). This is due to potential brake system problems. In the year 2004 and 2005 recall of the product affects automobiles, according to GM. The Chevrolet Avalanche, Express, Kodiak, Silverado, and Suburban, as well as the GMC Savana, Sierra, and Yukon, have possible flaws. According to the NHTSA, if the hood was open while driving, pieces from a pressure accumulator in the braking system may shatter, endangering passengers.

It warned that doing so might enable hydraulic fluid to escape, making it more difficult to brake or steer and perhaps resulting in an accident.

Additionally, GM is recalling 19,924 Pontiac Grand Prix cars, SRX SUVs, and Cadillac XLR coupes from the 2004 model year.

This is due to the fact that in severely cold weather the accelerator pedal could not function effectively, necessitating greater braking.

In addition, the automaker is recalling 17,815 2005-model Buick Raniers, Chevrolet Trailblazers, GMC Envoys, and Isuzu Ascenders due to improper windshield installation that could result in a crash.

However, GM emphasized that it was unaware of any injuries connected to the issues.

The recall was made public after GM said last month that it anticipated weaker profitability this year than in 2004.

The largest automaker in the world is struggling with sluggish US sales, losses in its European division, and now a product recall.

In January, General Motors warned that increased healthcare expenditures in North America and decreased earnings at its financial services unit would negatively affect the company's performance in 2005."

Summary given as input
According to GM, the recalled products include cars from the 2004 and 2005 model years.

The NHTSA warned that if the hood was open while driving normally, a pressure accumulator in the braking system might fracture and cause injuries to individuals.

Federal officials have announced that General Motors (GM), the largest automaker in the world, is recalling roughly 200,000 vehicles in the US for safety reasons.

The recall comes after GM said last month that it anticipated weaker profitability this year than in 2004.

In addition, the automaker is recalling 17,815 2005 model-year Buick Raniers, Chevrolet Trailblazers, GMC Envoys, and Isuzu Ascenders because the windshield is improperly installed and could fly out in an accident.

The greatest recall, according to the National Highway Traffic Safety Administration (NHTSA), affects 155,465 pickups, vans, and sports utility vehicles (SUVs).

Summary by BERT model
"Safety alert as GM recalls cars.

According to federal laws, the United States is recalling around 200,000 cars manufactured by General Motors (GM), the world's largest carmaker.

According to the National Highway Traffic Safety Administration (NHTSA), 155,465 trucks, vans, and SUVs have been recalled (SUVs).

In addition, the company is recalling 17,815 Buick Raniers, Chevrolet Trailblazers, GMC Envoys, and Isuzu Ascenders from the 2005 model year because the windshield was mounted incorrectly and might fly out in a collision."

Summary by GPT2 model
"Safety alert as GM recalls cars.

Federal officials have announced that General Motors (GM), the largest automaker in the world, is recalling roughly 200,000 vehicles in the US for safety reasons.

The automaker is also recalling 17,815 2005 model-year Buick Raniers, Chevrolet Trailblazers, GMC Envoys, and Isuzu Ascenders because the windshield is improperly installed and could fly out in an accident.

The recall comes after GM said last month that it anticipated weaker profitability this year than in 2004."

6 Comparison of Results

The numerical values given by ROUGE metrics on precision, recall, and f-measure can be listed and the difference in values given by both the models can be seen in Tables 2 and 3.

Figures 3 and 4 represent the precision, recall, and f-measure given by the ROUGE Metrics. Comparing these bar graphs, it is found that BERT gives better

Table 2 Performance of GPT2

No.	Precision	Recall	F-measure
0	0.437908	0.807229	0.567797
1	0.323529	0.585106	0.416667
2	0.269231	0.555556	0.362694
3	0.270531	0.746667	0.397163
4	0.393162	0.567901	0.464646

Table 3 Performance of BERT

No.	Precision	Recall	F-measure
0	0.483660	0.649123	0.554307
1	0.458824	0.951220	0.619048
2	0.261538	0.548387	0.354167
3	0.415459	0.788991	0.544304
4	0.564103	0.750000	0.643902

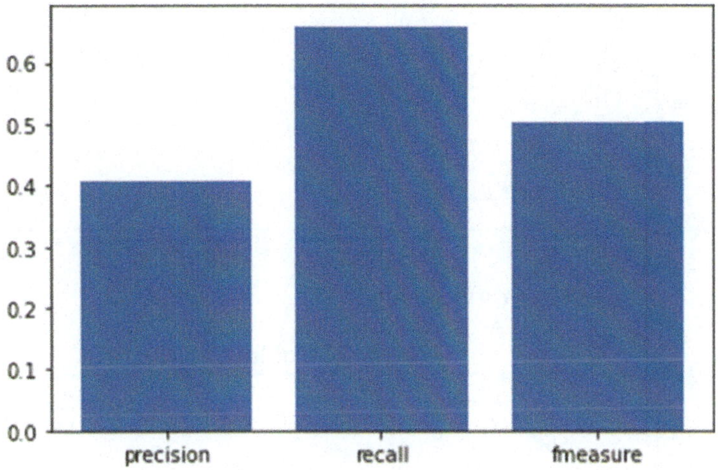

Fig. 3 Performance of GPT2

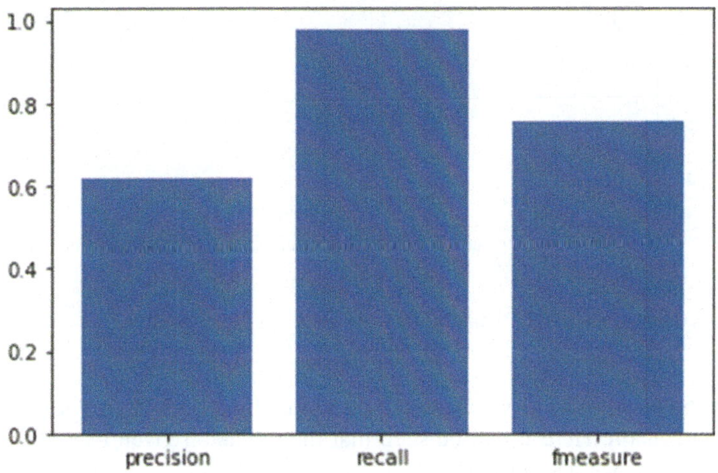

Fig. 4 Performance of BERT

results than GPT 2. These bar graphs are just representations of two random summaries that were generated by the respective Transformer models.

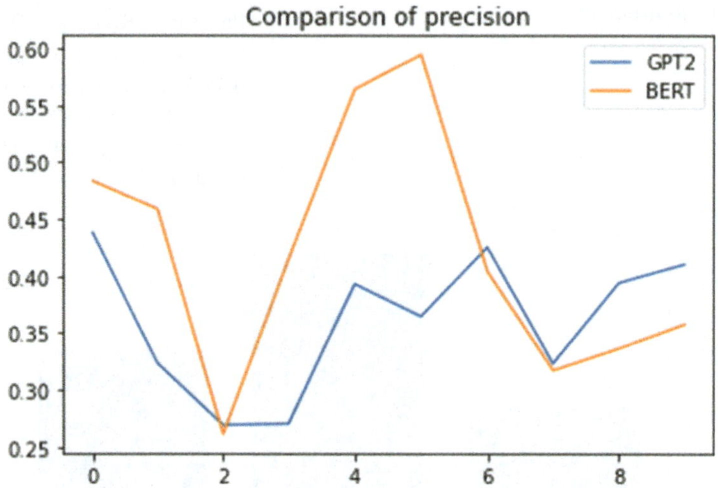

Fig. 5 Precision comparison of GPT2 and BERT

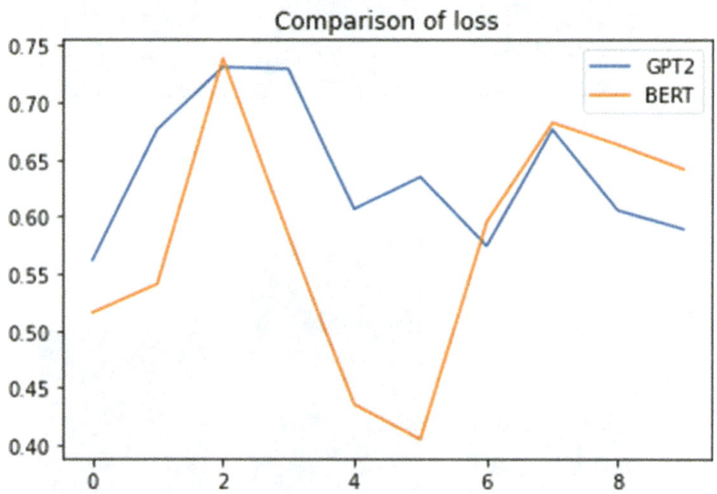

Fig. 6 Loss comparison of GPT2 and BERT

Figure 5 can be used to visualize the trend in which both the models perform in terms of precision. Here it can be seen that the precision given by GPT2 is lower compared to the one given by BERT.

From Fig. 6, the loss function of the BERT model seems to be comparatively better than GPT 2.

7 Conclusion

From all the visualizations and analysis done using ROUGE metrics on BERT and GPT 2 models, it is concluded that BERT gives better results than GPT 2. This conclusion has been reached by analyzing the numerical values obtained by ROUGE metrics and the results that the graph shows. BERT gives an average precision of 0.40 whereas GPT2 gives 0.38. Comparing this and the other factors mentioned by the metrics, it is concluded the BERT model for Big Data Analytics using summarization.

References

1. Ma, T., Pan, Q., Rong, H., Qian, Y., Tian, Y., & Al-Nabhan, N. (2022). T-BERTSum: Topic-aware text summarization based on BERT. *IEEE Transactions on Computational Social Systems, 9*(3), 879–890. https://doi.org/10.1109/TCSS.2021.3088506
2. Babar, S., Tech-Cse, M., & Rit (2013). Text summarization: An overview.
3. Gupta, A., Chugh, D., & Katarya, R. (2022). Automated news summarization using transformers. In *Sustainable advanced computing* (pp. 249–259). Springer.
4. Suleiman, D., & Awajan, A. (2020). Deep learning based abstractive text summarization: Approaches, datasets, evaluation measures, and challenges. *Mathematical Problems in Engineering, 2020.*
5. Shini, R. S., & Kumar, V. A. (2021). Recurrent neural network based text summarization techniques by word sequence generation. In *2021 6th international conference on inventive computation technologies (ICICT)* (pp. 1224–1229). IEEE.
6. Ozsoy, M. G., Alpaslan, F. N., & Cicekli, I. (2011). Text summarization using latent semantic analysis. *Journal of Information Science, 37*(4), 405–417.
7. Mahajani, A., Pandya, V., Maria, I., & Sharma, D. (2019). A comprehensive survey on extractive and abstractive techniques for text summarization. In *Ambient communications and computer systems* (pp. 339–351).
8. Liu, Y., & Lapata, M. (2019, August 22). *Text summarization with pretrained encoders*. arXiv preprint arXiv:1908.08345.
9. Rahman, M. M., & Siddiqui, F. H. (2019). An optimized abstractive text summarization model using peephole convolutional LSTM. *Symmetry, 11*(10), 1290.
10. Vig, J. (2019, June 12). *A multiscale visualization of attention in the transformer model*. arXiv preprint arXiv:1906.05714.
11. Song, S., Huang, H., & Ruan, T. (2019). Abstractive text summarization using LSTM-CNN based deep learning. *Multimedia Tools and Applications, 78*(1), 857–875.
12. Chopra, S., Auli, M., & Rush, A. M. (2016). Abstractive sentence summarization with attentive recurrent neural networks. In *Proceedings of the 2016 conference of the North American chapter of the association for computational linguistics: Human language technologies* (pp. 93–98).
13. Nagalavi, D., Hanumanthappa, M., & Ravikumar, K. (2019). An improved attention layer assisted recurrent convolutional neural network model for abstractive text summarization. *INFOCOMP Journal of Computer Science, 18*(2), 36–47.
14. Kieuvongngam, V., Tan, B., & Niu, Y. (2020, June 3). *Automatic text summarization of covid-19 medical research articles using bert and gpt-2*. arXiv preprint arXiv:2006.01997.
15. Miller, D. (2019, June 7). *Leveraging BERT for extractive text summarization on lectures*. arXiv preprint arXiv:1906.04165.

16. Devlin, J., Chang, M.-W., Lee, K., & Toutanova, K. (2019). BERT: Pre-training of deep bidirectional transformers for language understanding. In *Proc. Conf. North Amer.- Chapter Assoc. Comput. Linguistics, Hum. Lang. Technol* (Vol. 1, pp. 4171–4186). Association for Computational Linguistics. https://doi.org/10.18653/v1/n19-1423
17. Abdel-Salam, S., & Rafea, A. (2022). Performance study on extractive text summarization using BERT models. *Information, 13*(2), 67.
18. Liu, Y. (2019, March 25). *Fine-tune BERT for extractive summarization*. arXiv preprint arXiv:1903.10318.
19. Radford, A., Wu, J., Child, R., Luan, D., Amodei, D., & Sutskever, I. (2019). Language models are unsupervised multitask learners. *OpenAI blog, 1*(8), 9.
20. Montesinos, D. M. (2020, September 10). *Modern methods for text generation*. arXiv preprint arXiv:2009.04968.
21. Barbella, M., & Tortora, G. *Rouge metric evaluation for text summarization techniques*. Available at SSRN 4120317.
22. Bharathi Mohan, G., & Prasanna Kumar, R. Survey of text document summarization based on ensemble topic vector clustering model. In P. P. Joby, V. E. Balas, & R. Palanisamy (Eds.), *IoT based control networks and intelligent systems* (Lecture notes in networks and systems) (Vol. 528). Springer. https://doi.org/10.1007/978-981-19-5845-8_60
23. Mohan, G. B., & Kumar, R. P. (2022). A comprehensive survey on topic modeling in text summarization. In D. K. Sharma, S. L. Peng, R. Sharma, & D. A. Zaitsev (Eds.), *Microelectronics and telecommunication engineering . ICMETE 2021* (Lecture notes in networks and systems) (Vol. 373). Springer. https://doi.org/10.1007/978-981-16-8721-1_22
24. Kalpana, G., Kumar, R. P., & Ravi, T. (2010). Classifier based duplicate record elimination for query results from web databases. In *Trendz in Information Sciences & Computing (TISC2010)* (pp. 50–53). https://doi.org/10.1109/TISC.2010.5714607
25. Assegie, T. A., Rangarajan, P. K., Kumar, N. K., & Vigneswari, D. (2022). An empirical study on machine learning algorithms for heart disease prediction. *IAES International Journal of Artificial Intelligence (IJ-AI), 11*(3), 1066. 10.11591/ijai.v11.i3.pp1066-1073.
26. Mohan, G. B., & Kumar, R. P. (2022). Lattice abstraction-based content summarization using baseline abstractive lexical chaining progress. *International Journal of Information Technology*. https://doi.org/10.1007/s41870-022-01080-y
27. Yang, Z., Dong, Y., Deng, J., Sha, B., & Xu, T. (2021). Research on automatic news text summarization technology based on GPT2 model. In *2021 3rd international conference on artificial intelligence and advanced manufacture*. https://doi.org/10.1145/3495018.3495091

Leveraging Secured E-Voting Using Decentralized Blockchain Technology

Anushka Chaubey, Anubhav Kumar, Vikalp Pandey, Bharat Bhushan, and Priyambada Purohit

1 Introduction

Poor electoral systems have always been a major issue leading to election violence in many democratic countries, including India. Electronic voting was postulated as a potential replacement for the paper and ballot system. Over the years, e-voting has evolved as a viable alternative for many non-governmental elections. Various nations are now conducting a number of e-voting trials in the political sphere despite the numerous debates, disputes, and anomalies that have been brought up in relation to them. One of the well-known instances that sparked numerous concerns regarding vote tampering and miscounting was the Ohio e-voting incident in 2004 [1]. As a result, it is difficult to ensure that e-voting can be a precise and error-free practical approach for governmental elections yet. In this scenario, blockchain technology comes to the rescue.

Satoshi Nakamoto originally proposed the idea of blockchain [2]. It was the first time that a distributed system of network and cryptology were combined in order to execute peer-to-peer transactions of the cryptocurrency known as bitcoin in a secure and open fashion. Due to its intricacy and unpredictable nature, blockchain was initially used sparingly, but over time it has drawn attention on a global scale. A blockchain, which is a chain of blocks made up of cryptographically linked blocks, is essentially a data structure utilized throughout transactions that maintains data or information in the structure of blocks. Each block includes a unique hash value in addition to the preceding block's hash, which aids in creating the links between the

A. Chaubey (✉) · A. Kumar · V. Pandey · B. Bhushan · P. Purohit
School of Engineering and Technology, Sharda University, Greater Noida, UP, India

Faculty of Management Studies, SRM IST, Ghaziabad, UP, India
e-mail: 2019004401.anushka@ug.sharda.ac.in; 2019549001.anubhav@ug.sharda.ac.in; 2019641513.vikalp@ug.sharda.ac.in; bharat.bhushan@sharda.ac.in; priyambv@srmist.edu.in

R. Sharma et al. (eds.), *Data Analytics for Internet of Things Infrastructure*,
Internet of Things, https://doi.org/10.1007/978-3-031-33808-3_15

blocks. It is the responsibility of the nodes within the network to interconnect these blocks. To ensure the integrity and safety of the data, the blocks are encrypted, and the network itself validates the blocks. Basically, blockchain technology is a concept that makes use of a number of technologies, including decentralized networks, peer-to-peer transactions, smart contracts, and cryptographic algorithms.

Recently, blockchain has become a tool for boosting the effectiveness of technologies used in several industries. In order to get around some potential problems with e-voting, the voting systems that are based on blockchain technology have recently grown more and more significant. The immutable property of blockchain technology has rendered it a decentralized and distributed ballot box, leading to the proposal of blockchain-based voting systems as the forthcoming generation of contemporary electronic voting systems [3]. Governments are encouraged by blockchain technology to adopt intelligent sustainable voting machines and incorporate sustainability reports into voting systems. It guarantees that all parties have access to trustworthy information on sustainable assets. In spite of the fact that blockchain technology is rapidly being used to strengthen the security of the e-voting system, a number of problems still exist.

So as to address as many issues as possible that might emerge while implementing a blockchain-enabled e-voting system, it is very important to understand the concept of blockchain technology from its core. For this purpose, this chapter provides a very detailed understanding of blockchain technology and its integration and implementation with e-voting systems. The following contributions are made by this chapter: providing detailed information on the blockchain technology, its working and, integration with the e-voting system including the Elliptical Curve Digital Signature Algorithm; understanding how different types and features of blockchain technology help in improving the electronic voting mechanism; providing an overview on different consensus algorithms; and identifying a set of the previous paper and ballot and e-voting system's unfilled gaps. This chapter shows how blockchain technology, if integrated, can help overcome most of the major issues of an electronic voting system.

Further, this chapter has been organized in the following manner. Section 2 defines blockchain technology, its working, and, its features. Section 3 discusses the different types of blockchains and the associated consensus protocols. This section further outlines the unfilled gaps in the existing voting systems. Section 4 presents a literature survey on the various electronic voting systems in existence. Finally, this chapter concludes in Sect. 5 highlighting some open research directions to guide further research in the area.

2 Blockchain

Let us begin by understanding the meaning and significance of blockchain technology and how is it useful. Haber and Stornetta first put forth the idea of the blockchain in 1991. The major goal was to create a digital document timestamp that could not be altered. Satoshi Nakamoto is thought to have created the first blockchain-based system in 2008. It is also obvious that Bitcoin was the first widely used blockchain technology. The idea of a blockchain can be compared to a distributed, open, and secure data book. Most people believe that the idea will be a crucial part of industry 5.0 applications in the next years. While blockchain is well recognized in the cryptocurrency industry, one could easily make the case that its potential goes much beyond digital currency. Government agencies as well as private businesses have started experimenting with blockchain. In further sections, the definition of the blockchain (Sect. 2.1), its working (Sect. 2.2), and the features of the blockchain (Sect. 2.3) have been discussed.

2.1 *What Is Blockchain?*

Blockchain is defined as a digital database or a ledger that is shared among the nodes of a peer-to-peer system of a network. It is a database that electronically saves data or information in digital form [4–7]. It is a sort of distributed ledger technology (DLT) made up of an expanding list of data, known as blocks, that are safely connected to one another using cryptography. Every block includes the timestamp, a cryptographic hash value of the preceding block, and transaction information. The timestamp establishes the existence of the transaction data at the time the block was produced. They basically create a chain (same as the linked list), with each new block linked to the previous ones since each block carries information about the prior block [8]. Blockchain transactions are therefore irreversible in the sense that, once they have been stored, the contents of any specific block cannot be changed subsequently without changing all succeeding blocks.

The blockchain is a series of blocks that, like a traditional public ledger, contain an exhaustive list of transactional data [9]. Each block has a reference, known as the parent block, which is effectively a hash value of the block preceding it and points to it. The genesis block is the first block in a blockchain and is the only block without a parent block [10].

Blocks are made up of the block body and the block header. The block header consists of the following:

- The block version specifies the set of block verification guidelines that are to be followed.
- A 256-bit, cryptographic hash value called the parent block hash points to the block previous to it.
- The hash value of each transaction stored within the block is represented by the Merkle tree root hash.
- The timestamp is the demonstration of the current time in seconds since January 1, 1970, at 0:00 UTC.
- nBits is the shorthand for the current hashing goal.

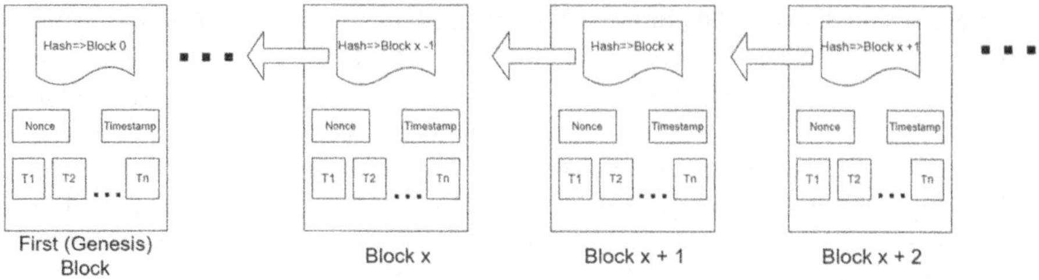

Fig. 1 Representation of a blockchain consisting of a sequence of blocks

- A 4-byte field known as a nonce is typically initialized with 0 and rises with each hash calculation.

Transactions and a transaction counter constitute the block body. A block's maximum capacity for transactions is determined by the size of every transaction and the block size. Asymmetric cryptography is used by blockchain to verify transaction authentication. The utilization of an asymmetric cryptographic digital signature in an unreliable environment is applied. Figure 1 represents an example of a blockchain consisting of a sequence of blocks.

Utilizing a peer-to-peer system of network and a decentralized timestamping server, a blockchain database can be unilaterally controlled. Via widespread cooperation driven by unit self-interest, they are verified. Since participants are minimally concerned about the security of the data, a design like this promotes a strong workflow. A digital asset loses its ability to be replicated indefinitely when a blockchain is used. The long-standing issue of double spending is resolved since it demonstrates that each unit of currency was transferred just once. A value-exchange protocol has been used to characterize a blockchain [11]. Since it produces a record that enforced offer and acceptance when the exchange agreement was correctly set up to document it, a blockchain can preserve title rights.

2.2 *Working of a Blockchain*

A blockchain can be defined as typically a continuous series of blocks of information that are cryptographically linked to each other and shared among the nodes of a network. Let us now understand how exactly this technology works.

By employing a digital signature that utilizes private key cryptography, a node begins a transaction among a decentralized blockchain network. On the blockchain platform, a transaction could be viewed as a data structure that reflects the exchange of digital assets among peers. Each transaction is kept in a pool of unconfirmed transactions, and the Gossip protocol, a flooding mechanism, is used to spread it throughout the network. Peers then need to select and authenticate these transactions in accordance with a set of predefined criteria. For instance, the nodes attempt to authenticate and confirm these transactions by determining if the initiator has enough balance to initiate a transaction or just by enforcing double spending in an attempt to deceive the system. Double spending is when you make a number of distinct transactions utilizing the very same amount of money as your input [12].

The miners confirm and validate the transaction, and then it is added to a block [13]. Peers who mine for blocks employing their computing resources are referred to as miners [14]. To publish a block, miner nodes must utilize enough of their computing power and resolve a computational problem. The miner who is able to figure out the riddle first wins and earns the opportunity to add a new block. Upon successfully establishing a new block, a tiny bonus is granted. The new block is then verified by every peer in the network using a consensus algorithm, which is a mechanism used in distributed systems to achieve consensus. A localized copy of every peer's immutable ledger is then created, and the new data block is added to the already existing chain. The transaction is finalized at this stage. A cryptographic hash reference is used by the following block to connect itself to the freshly formed block. The block now receives its first confirmation, and the transaction now receives its second confirmation. The transaction will also be reaffirmed every time a fresh block is incorporated into the chain. Typically, a transaction takes 6 network confirmations to be deemed complete [15]. A complete diagrammatic representation of the working of a blockchain is portrayed in Fig. 2.

A detailed overview of the process of transactions has been described further. Verifying the sender's identity is the initial stage in the transaction process, indicating that only the sender and not anybody else is requesting the transaction between both the sender and the recipient. Let us use a simple transaction between Robin and

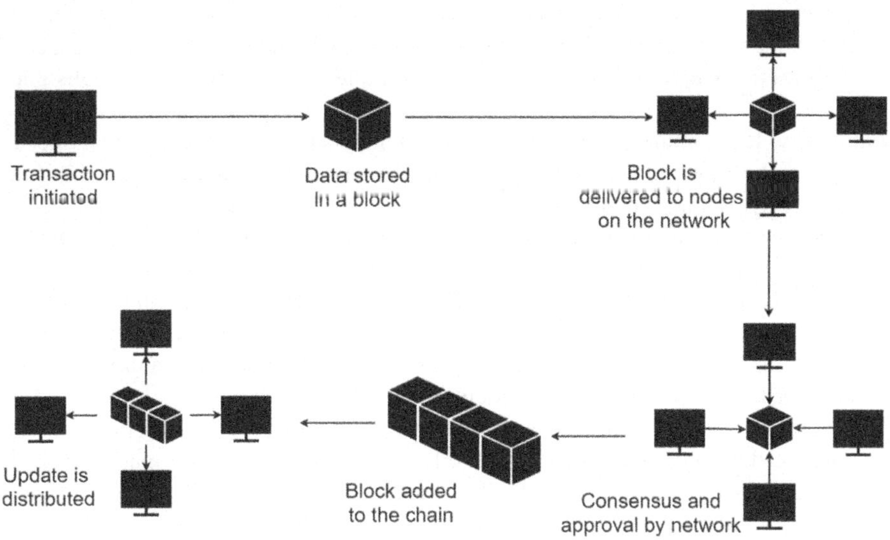

Fig. 2 Working of a blockchain

Ana as an example. Assume that Robin and Ana each have a balance in bitcoins and that Ana wants to send Robin 10 bitcoins. Ana will now broadcast a message in the blockchain network with the details of the transaction in order to send the money. Blockchain utilizes digital signatures (private and public keys) to accomplish this [16]. Robin's details, including his public address and the amount of transaction, as well as Ana's public key and digital signature, are provided for the broadcast. Ana created that digital signature using her private key. All miners individually perform transaction validation based on several standards. Blockchain employs the Elliptic Curve Digital Signature Technique (ECDSA) [17]. This algorithm makes sure that the funds can only be used by the people who actually own them.

2.2.1 Elliptic Curve Digital Signature Technique (ECDSA)

It is among the most challenging public key encryption techniques. Elliptic curve cryptography produces keys that are generally smaller than those produced by digital signing techniques. It is a type of public key cryptography that is based on the algebraic nature of elliptic curves on finite fields. Elliptic curve cryptography is mostly used to generate digital signatures and pseudo-random numbers, among other things [18]. Bitcoin uses the encryption technique ECDSA to make sure that only the rightful owners of currency may use it. It depends on the hash function and curve order chosen. These are, respectively, Secp256k1 and SHA256 (SHA256()) for bitcoin [19]. Some of the terms that describe the procedure of ECDSA are as follows:

Private key: A code that is only known to the person who created it. In essence, a private key can be described as a numeric code that was chosen at random. In Bitcoin, the funds can only be spent by someone who has a private key associated with them. A private key in Bitcoin is a unique 256-bit and 32 bytes unsigned integer.

Public key: It is a numeric value that is equivalent to a private key although it doesn't have to be hidden. A private key is used to calculate a public key, but not the other way around. Without disclosing the private key, a public key is utilized to check whether a signature can be deemed authentic. Public keys in Bitcoin are either compressed or not. A 256-bit integer, say x, and a prefix of either (0x03) or (0x02) constitute the 33 bytes compressed public keys. The earlier uncompressed keys include a prefix that is a constant (0x04), and two 256-bit integers that are known as x and y (2×32 bytes) and are 65 bytes long. A compressed key's prefix enables its y value to get calculated from its x value.

Signature: It is a code that demonstrates the completion of a signing process. The hash of the object that is to be signed and the private key are mathematically combined to create a signature. The signature is made up of the two integers r and s. Without requiring access to the private key, a mathematical procedure using the public key on the signature can be used to establish that it has been initially formed using the hash combined with the private key. The resulting signatures are 73, 72, or 71 bytes long, with probabilities of approximately 25%, 50%, and 25%, respectively. However, proportions that are considerably smaller are feasible along exponentially diminishing probabilities.

Let us take a look at some of the primitives used by ECDSA for obtaining a signature of a message and for its vice versa.

- s and r: Both these values together form the signature.
- z: It is the message's hash that is required to be signed.
- k: The nonce value used to determine the values of s and r.
- Q_A and d_A: They represent the public key point and private key number of the message, respectively. When an address inside the wallet is provided, a copy of this can be derived.

Signature Algorithm

The pair s and r (also known as the signature pair) are calculated using the signing algorithm from d_A to z.

- Identify the curve's group order n.
- Follow by generating a cryptographically secure value k (a random number) that ranges between 1 and $n - 1$.
- Calculate $(x, y) = k * G$, where G can be defined as secp256k1 curve's generator point.
- Determine $r = x \bmod n$. Create a new random k and restart if $r = 0$.
- Calculate $s = k^{-1}(z + r * d_A) \bmod n$. Generate a new random k and restart if $s = 0$.

Note: Reusing k after a signature has been created with it is not advised because of the defects that allow a perpetrator to obtain signed messages' private key if they are aware of k (the shared nonce) that was utilized in them.

Verification Algorithm

- The verification algorithm checks the consistency of the signature pairs z and Q_A,s and r.
- Make sure that r and s are both between 1 and n 1.
- Calculate $u_1 = z * s^{-1} \bmod n$ and $u_2 = r * s^{-1} \bmod n$.
- Make sure (x, y) is not the same as the infinitesimal point when you compute $(x, y) = u_1 * G + u_2 * Q_A$. When two points are added together that would not otherwise result in a point on the curve, for instance, two points with identical X values but reversed Y values, a peculiar point known as the point at infinity is created.
- The signature is valid if $r = x \bmod n$, it is invalid whenever a test fails or if anything else goes wrong.

  The complete process of ECDSA is represented in Fig. 3.

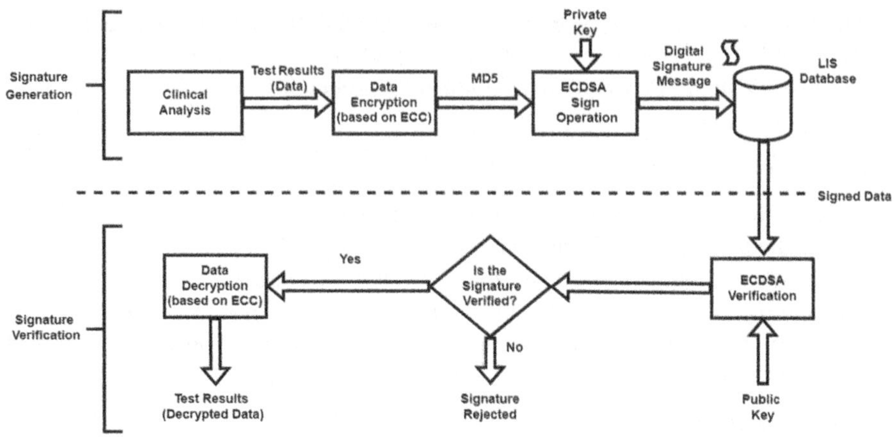

Fig. 3 Working of an elliptical curve digital signature technique

2.3 *Features of Blockchain Technology*

Blockchain is among the most exciting and popular technologies in today's era. Now that we already have a basic understanding of this technology, let us take a look at some of its primary features.

2.3.1 Immutability

The attribute of immutability is the inability to change or modify. One of the main characteristics of a blockchain that guarantees the network's permanence and immutability would be this. Since centralized information depends on the third-party gateway for its security, it is susceptible to theft and hacking. Blockchain systems are secure because of their decentralized organization; however, because of their immutability, it is almost impossible to alter or manipulate the data. These problems are easily resolved by enterprise blockchains [20]. Data can be easily modified with everyone's consent because they are little in size and connected as they have a common structure and purpose.

2.3.2 Auditability

A digital timestamp and distributed ledger are used as the verification and record, respectively, of every transaction that takes place within a blockchain system or network. As a consequence of this, through gaining authorization for any network node, it is probable to review and trace earlier records [21]. For instance, Bitcoin allows for the iterative tracing of all transactions, supporting the transparency and auditability of the data state of the blockchain. However, it becomes exceedingly challenging to track down the source of the money when it is spread across numerous accounts.

2.3.3 Persistency

Blockchain enables both producers and consumers to demonstrate that their data is true and undamaged by providing the framework by which veracity can be evaluated [22]. For instance, if a Blockchain has 10 blocks, block number 10 would have the prior block's hash, and the current block's information would be employed to build a new block. The result is, every block in the chain that already exists is linked and connected to every other block. Even transactions have connections to earlier transactions. Thus, a simple change to any transaction will drastically alter the block's hash. Anyone who wishes to edit any information must alter all the hash data from prior blocks, which is thought to be an enormously challenging task given the amount of effort required. Additionally, the block is validated by other members of the network once it is produced by a miner. Therefore, the network is able to identify any data tampering or fabrication. Due to this, blockchain is almost impervious to manipulation and is viewed as a distributed ledger that cannot be altered.

2.3.4 Decentralization

In conventional centralized transaction systems, the trusted centralized agency must verify every transaction (e.g., a bank). A wiser approach would be a distributed peer-to-peer blockchain structure to address the key issue of lift resilience, availability, and failover that decentralization requires, which is trust. In contrast to centralized systems, any two peers (P2P) can perform a transaction within the blockchain network without the need for central authentication. By utilizing several consensus techniques, blockchain can in this way lessen the trust issue. Additionally, it can alleviate performance bottlenecks at the central server and lower server expenditures (including development and operation costs).

2.3.5 Anonymity

A participant's genuine identity is concealed on a blockchain platform. Every member of the dispersed network is assigned an address. Instead of an actual identity, this address serves as that entity's identification. On the network, the addresses maintain the user's privacy [23]. An element of anonymity is provided by the trustless environment provided by blockchain technology.

3 Types, Consensus Protocols, and Unfilled Gaps

The meaning of blockchain, its working, and its different features and characteristics that make this technology a better alternative as compared to the traditional central ledger systems have been described so far. To get a deeper understanding of blockchain technology, understanding the different types that blockchain technology has been divided into, based on its characteristics and applications, is crucial. Blockchain is a technological mechanism that employs its own decentralized nodes to record, validate, transact, and exchange network data without the aid of a third party. Blockchain is based on intelligent distributed cryptographic and mathematical techniques. The participants come to an agreement and resolve the issue of cost-effective and dependable trust and value transmission. Blockchain networks could be utilized to adapt their functioning to the specific needs that an enterprise may have. Although the technology can initially seem perplexing, blockchain can be broadly divided into two broad categories and four different kinds of networks.

The different types of blockchains available have been described in Sect. 3.1. Further, in Sect. 3.2, the significance of the consensus protocol in creating a secure and reliable blockchain network and the different types of consensus protocols available for use have been studied. Following this, the various challenges faced by the existing voting systems and their unfilled gaps have been discussed later in Sect. 3.3.

3.1 Types of Blockchain

Permissioned networks and permissionless networks are the two primary forms of blockchain networks. Either of these sections can be used to group the four different types although some might use both. A permissionless or permissioned chain might depend on the needs of the company considering using blockchain technology [24]. Both the advantages and disadvantages of the networks within these two bigger divisions are distinct.

A deeper understanding of the four different forms of blockchains has been provided below.

3.1.1 Public Blockchain

Any system or node can participate in a public blockchain, making it a particularly inclusive sort of blockchain system. Anyone with access to a public blockchain can try to decrypt a block of data's hash and include it in the chain. A public blockchain network is still safe as an attack would take a lot of time and effort. Any potential attacker would be required to possess control over 51% of the chain's nodes. Permissionless blockchain networks are another name for this kind of blockchain technology. It should be no surprise that the exchanging and tracking of cryptocurrency transactions is the most popular use of public blockchain technology provided that the primary public blockchain was utilized for cryptocurrency. Although Bitcoin was the initial cryptocurrency, Litecoin, and several other cryptocurrencies soon followed, each running on its own blockchain. Notarizing papers and keeping a record of public property ownership data are two other potential uses of a public blockchain. The important thing to remember is that any information that requires to be both public and secure is best served by a public blockchain.

3.1.2 Private Blockchain

A private blockchain network is one that is either administered in kind of a closed environment or is managed by a single organization, like a company. In other terms, it often has a smaller size as compared to a public blockchain. The controlling entity among a private blockchain (also known as an enterprise/permissioned network of blockchain) determines who is allowed access to the chain and who shall validate all data before it is incorporated. Private blockchains can be used for a variety of tasks, including supply chain management, private voting, and safeguarding trade secrets. There are restrictions regarding who can obtain the information and incorporate information into the chain in a permissioned blockchain. The data is not accessible to outside parties. Due to the fact that there are lesser nodes, they are typically quicker as compared to public blockchains since data can be authenticated and uploaded to the blockchain much more quickly.

3.1.3 Consortium Blockchain

A consortium or federated blockchain is a partly private blockchain that is run by numerous entities rather than just one. Participants of a verified group can only access the blockchain, removing some of the hazards associated with letting a centralized authority supervise the network, as would be the case in one private blockchain. In contrast to a public blockchain, where any node wishing to join could reach consensus, a consortium blockchain's consensus process is managed by predetermined nodes, along with a validator station that can send, accept, and verify any transaction. Transactions can also be transmitted and received by member nodes; however, they would not be able to validate them. A consortium blockchain is mostly used in the payments and banking industries. Theoretically, banks might organize a consortium and choose which node should be in charge of authenticating all transactions. The technique could also be helpful for supply chain processes and research. The benefit of developing a consortium network is the fact that it provides access control similar to a private blockchain while being more robust and flexible than that of a public blockchain.

3.1.4 Hybrid Blockchain

A hybrid blockchain provides its developers with all of the features and benefits of both private and public blockchains. A hybrid blockchain often consists of a private, permission-based network alongside a public, permissionless network. The corporation has control over who may obtain and add information to the blockchain. The hybrid blockchain is owned by a private organization, but it cannot modify transactions. Given that it is more challenging to manipulate or influence 51% of the total nodes, it is more robust than from a private blockchain. In contrast to a public blockchain, a hybrid blockchain does not make the data and transactions it adds publicly available. They may, nevertheless, be publicly validated if necessary. A smart contract, which comprises a code fragment contained in a block and can start a consented event whenever an agreement milestone is achieved, is one example. Every participant's identity is hidden from other participants until they conduct a transaction, which is another essential feature of a hybrid blockchain. Hybrid blockchains can be used for a variety of purposes. Real estate, where only certain details should be kept hidden however listings and sales should be publicly disclosed, is among the areas on which there is a significant focus. It might also be used in the retail and financial sectors. In Table 1, all four types of blockchains have been compared.

3.2 Consensus Protocols

Blockchain is a distributed system of a network that attempts to provide data security and data integrity. Every transaction on a blockchain platform is regarded as being quite secure and authenticated because there is no central body to check and audit the transactions. The intricacy of the truth may exist because blockchain operates in

Table 1 Comparison between various blockchain networks

Public blockchain	Private blockchain	Consortium blockchain	Hybrid blockchain
Anyone is free to sign up and take part in the network	This system of the network is governed by a single corporation	The blockchain network is influenced by numerous organizations	Authenticated access; just a few components are private
Immutable, fully decentralized, and secure ledger system	Faster production, better energy efficiency, and privacy	A system that is scalable, decentralized, and lightning-fast	Flexible management of which data is made public and which is private
Transactions are transparent to all parties but remain anonymous	Data processing has been made simpler but is not accessible to everyone	Network security and privacy are maintained	A highly scalable, decentralized, and controlled system

a decentralized fashion and stores high-frequency transactions at the same time. In order to avoid harmful events like double-spending assaults, it is crucial to reach a consensus [25]. The consensus protocol enters the picture here. A consensus protocol is a technique used in computer science to achieve unanimity among distributed systems or components over a particular data item. A consensus protocol is a procedure that enables all participants throughout a decentralized computing system to reach a consensus about the ledger's current data state and also be capable of trusting unknown participants. Several consensus algorithms have been discussed below.

3.2.1 Proof of Work

A "Proof of Work" is really a fragment of information that is complex to produce (expensive, time-taking), yet simple to be authenticated by other participants. Blockchains that use the PoW algorithm make use of specialist nodes, or "miners," which use energy resources to provide security and accounting functions for the platform. Freshly minted coins are used to pay miners for their work. PoW aims to make it ridiculously expensive to hack the network. Since authentic resources must be employed to modify the ledger, it is impossible to "fake the work." From the viewpoint of behavioral economics, Bitcoin's PoW design strikes a very fine balance between incentives. Miners are motivated by rewards.

3.2.2 Proof of Burn

Iain Stewart's Proof of Burn algorithm attempts to address the problems with the proof-of-work approach. PoB protocol makes use of the concept of obliterating or burning the currencies, which reduces the requirement for high-energy resources during mining. It lessens the PoW's reliance on robust computational gear as a result. By transmitting the coins to a publicly recognized, untraceable, and authenticated address, they are burnt. The transmitted coins then lose their ability to be spent. Instead of having to wait months after burning the coins, the miner instantly gains the ability to contest for the generation of fresh blocks. The node has a greater likelihood of producing the subsequent block and earning rewards the more coins it burns. However, this approach does not guarantee that the node would be permitted to mine after burning a specific number of coins. Therefore, the node can experience significant financial loss before receiving its reward. Additionally, the likelihood of being awarded decreases as the count of miners rises [26].

3.2.3 Proof of Stake

The Proof of Stake-based database keeps a record of each validator (PoW's counterpart of a miner) and each party's individual investment in the blockchain network. In a PoS system, each validator makes an investment to increase their odds of mining the following block. The chances increase with increasing stakes. It does not, however, ensure that the validator having the biggest stake would be chosen. Comparable to a lottery, this mechanism selects the node for the role of a validator at random for block creation. Any person who tries to game the system forfeits his stake. Block construction is simple and does not require a lot of computer power, unlike PoW.

3.2.4 Delegated Proof of Stake

According to DPoS principle, only the nodes with stakes can elect block verifiers (also known as block producers) [27]. By using this voting method, the stakeholders grant the delegates they support the right to create blocks rather than doing so themselves, resulting in a computational power consumption of 0. The stakeholders will choose other nodes to take their place if the delegates are not able to produce blocks during their rounds. To arrive at a conclusion in a fair and democratic way, DPoS makes the best use of the shareholders' votes. DPoS is a low-cost and very efficient consensus protocol when compared to PoW and PoS. Other cryptocurrencies, including BitShares, EOS, are implementing DPoS.

3.2.5 Proof of Elapsed Time

While the Proof of Elapsed Time (PoET), a jackpot-based consensus technique of the blockchain network, has an equivalent activity flow to the PoW protocol, it uses significantly less computation power because it does away with the requirement for mining-intensive mechanisms, which reduces energy use and resource utilization [28]. In PoET, each node of the shared ledger has its own independent random timing that dictates whether or not it will incorporate a fresh block to the network and receive a reward. As a result, it primarily prioritizes efficiency and makes sure that every node is given an equal opportunity of becoming the following block generator.

3.2.6 Proof of Participation

Proof of Participation merges aspects from PoS and Federated consensus techniques to a novel algorithm aimed to demonstrate that a node is contributing valuable network work. The "Node Registry" is a consortium of nodes maintained by the PoP protocol. Anyone may submit a request to join the network, and the distributed protocol shall select who is accepted. This is a significant improvement above federated chains, such as Ripple, in which a central authority chooses who can join the network. Only nodes within that Node Registry are authorized to produce blocks, and every node has approximately the same probability of generating the subsequent block within the chain. To participate, nodes should wager tokens, and this technique serves to combat Sybil assaults. PoP is intended to compensate a broader range of participants with fresh issuance, hence reducing the centralizing effects of PoW and PoS.

3.2.7 Proof of Authority

Proof of Authority can be described as an image-based technique for consensus that capitalizes on the reputation and identification of block auditors. An organization of Ethereum engineers headed by Gavin Wood presented PoA as a response to spamming assaults against Ropstein Ethereum test net in March of 2017. PoA is comparable to PoS; however, auditors wager their reputation rather than coins. PoA is really only suitable for private blockchain systems due to the fact that network leaders choose trustworthy nodes/validators [29]. PoA networks typically depend on a limited group of validators, thus resulting in greater performance per layer. PoA is commonly regarded as a viable alternative for semi-trusted management of supply chain settings. Ideally, diverse businesses in such a supply network would be able to protect the confidentiality of their data while yet taking advantage of a shared network.

3.2.8 Proof of Importance

Proof of Importance, first established by NEM (XEM), chooses its miners based on specific criteria in a procedure known as "harvesting." The volume and quantity of transactions over the previous 30 days, the amount of invested currency, and network activity are typical determinants. These elements form the basis of the importance score given to nodes. The likelihood of being selected for harvesting a block and collecting the associated transaction fee increases with the score. Although comparable to PoS, PoI avoids the latter's propensity to automatically reward the wealthy by taking into consideration participants' total network support. As a result, simply placing a big POI bet does not ensure that you will win the block.

3.2.9 Proof of Capacity

Proof of Capacity, sometimes recognized as the proof of space, is a mining algorithm that bases mining rights on the amount of accessible space on a miner's hard drive, in contrast to the majority of its predecessors, which awards mining rights based on the computation power or coins staked [30]. In PoC, the process of "plotting" is used by miners to create a list of all feasible hashes beforehand. Then, a hard disc is used to store these plots. There are more potential solutions for the more storage space a miner has. The likelihood of having the right hash combination and receiving the reward rises as the proportion of solutions rises. PoC makes it possible for the common person to take part in the network because it does not call for expensive or specialized equipment. As a result, it is a more decentralized and less energy-intensive alternative to some of the more widely used consensus mechanisms. The system has not yet been adopted by many developers, and there are worries that it could fall victim to virus assaults.

3.2.10 Proof of History

Proof of History (PoH) offers evidence of historical occurrences, as the term implies. PoH technology enables "timestamps" to be incorporated directly into the blockchain, independently confirming the interval between transactions. This timestamping technique is made possible by a sequential-hashing verifiable delay function (VDF), like SHA-256. It functions by using the outcome of a transaction as the input for the subsequent hash, allowing everyone to plainly understand which event occurred in a specific order. PoH significantly minimizes the processing weight of the blockchain because the VDFs can only be solved by a single CPU score, making it faster and more energy-efficient than many of its peers. PoH has not yet undergone extensive testing because Solana is the only company that uses it [31].

Protocol architecture is replete with choices. Every architecture has its own advantages, disadvantages, and potential applications. From this perspective, it is logical to conclude that they are not genuinely in competition with one another. Bitcoin's application of PoW, for example, is really safe and open. PoW is an excellent consensus protocol for cryptocurrencies. On the other hand, it would have been a horrible option for blockchain-based social media. With blockchain being such a new sector, it is imperative that we maintain our research in order to find new blockchain architectures. New consensus techniques will enable further blockchain applications.

3.3 Challenges Faced by Existing Systems

Elections have always been a contentious issue. Paper ballots and electronic voting machines (EVMs) both have their share of issues and challenges. When elections were still decided by paper ballots, a party's victory was attributed to booth capturing; in modern times, it is linked to EVM tampering. EVMs were proposed as a better alternative to paper and ballot systems as they are time efficient and do not require huge manual labor for the counting of votes but due to their centralized nature, they are vulnerable to security threats. In the below-mentioned sub-sections, we will look at the several challenges faced by both the paper and ballot and digital e-voting systems.

3.3.1 Paper and Ballot Systems

The major issues of using the paper and ballot system for elections are as follows:

Wastage of Paper
It would be equivalent to going back to eating raw meat in the Stone Age after the discovery of fire to switch back to the paper from technological devices. Even the most polluting businesses are working to lower their carbon footprints, so using paper on such a large scale when there are other, more environmentally friendly solutions is a clear disregard for the environment.

Manual Ballot Counting Takes a Lot of Time
The manual counting of votes is laborious and prone to mistakes. It is quite tough to manually count crores of votes, and it takes far more man-days to tally the votes and determine the outcome. As a result, the votes are counted incorrectly, extending the already lengthy counting process.

Booth Capturing
Political parties have used force as a weapon to influence election results in their favor. In the media, a booth capture video became viral during the 2017 Odisha elections. In addition, CPM party booth capture incidents in West Bengal are public knowledge. The Chief Election Commissioner, Sunil Arora, stated about the return to paper ballots in an interview on February 1, 2019, "Political parties have a right to make their thoughts and concerns known because they're the biggest stakeholders after the voters. But we won't return to the era of paper ballots. We're not going back to the days when ballots were seized, force was employed to accomplish the task, and counting took an abnormally long time" [32].

Ballot Paper Manipulation Techniques
Other techniques of vote manipulation, besides booth capturing, have been used in previous elections. This includes the variations in the ink used when casting a ballot. A video from the Rajya Sabha elections in Haryana demonstrated how the voting periods for BJP and Congress MLAs differed, and how later, the votes made by Congress MLAs were disregarded due to a change in the voting pen. Even some voting papers may occasionally be altered voluntarily or forcibly, making the voting process dangerous for both voters and electoral commission members.

3.3.2 Digital E-Voting Systems

E-voting has improved the voting process's performance and credibility in contrast to traditional voting techniques. Electronic voting is frequently deployed in a variety of methods throughout elections due to its adaptability, ease of use, and competitive pricing as compared to traditional elections [33].

Despite these benefits, conventional electronic voting techniques have the danger of encountering excessive authority and manipulated data, so reducing the voting process' essential impartiality, confidentiality, opacity, untraceability, and verifiability [3]. Due to the fact that e-voting techniques are consolidated and licensed by the crucial organization that regulates, analyzes, and supervises the procedure of an e-voting platform, this is a potential obstacle to a fair voting procedure. Recent issues in democratic countries like the United States and India bolster this thesis and demonstrate its validity. It is crucial to prevent the erosion of voter confidence [34].

It is recommended that e-voting systems have the following qualities:

- Receipt-freeness prohibits the production of any receipts as evidence of a voter's support for a certain candidate [35].
- Fairness, preliminary outcomes that could have an impact on other voters' choices are not acceptable [36, 37].
- Data integrity guarantees that every vote is recorded accurately and that it cannot be altered in any way after it has been logged [38].
- Voter anonymity and privacy: Voters' names and the candidates they support should not be made public [39]. Only eligible democrats should be allowed to cast ballots [40].
- Dependability and robustness; voting platforms must function error-free. Software and procedures ought to be created without any malicious code or mistakes [41].
- Individuality should not let voters cast multiple ballots [42].
- Verifiability: Voters should be able to ascertain that their ballots were appropriately counted [43].

4 Recent Advances

Habib et al. [44] discussed applications of an electronic voting platform utilizing the Ethereum technology and smart contract technology. Their describes the old approaches and the drawbacks of using them and how untrustworthy it is in a democratic country so it suggests an alternative approach using the technologies of this new era to move to a technique that is cheaper, faster, and easier to implement as compared to old approaches which are costly, easy to manipulate, and always questionable. New approaches ensure data integrity, increased voter count, and transparency. The proposed system in the forehand mentioned work ensures the security measures using fingerprint authentication and artificial intelligence used for face recognition. Zhang et al. [45] proposed a message authentication and transmission system that verifies authority while hiding the identity. The work suggests that this system can be used in various scenarios like complaint boxes, questionnaires, outcome assessments, opinion collection, and many more. This verifies the organizer and the voters with a design that implies that authentication can be carried out while preserving anonymity which is implemented using the blind signature. It solves the problem of trust between the organizer and the voter and defines the true meaning of democratic voting.

Sayyad et al. [46] described blockchain to be a "cryptographically secure transactional singleton machine with shared-state." This work explains the working based on a network that is decentralized and records transaction history and no single machine can alter the record on the decentralized network. Further, the work explains that the data is recorded in the blocks and every block is associated with two blocks, the previous and the subsequent one, through the hash algorithm. In case someone attempts to manipulate the information in any of the blocks, the chain becomes invalid. Thus, the data once stored, becomes immutable. The employed hash algorithm is SHA-256 which is a function that converts the raw data into hash data using a mathematical expression. It explains some limitations including that 51% of computational power can modify the transaction data. Another limitation specifies that on updating the system divides the network nodes into two types: new nodes, and old nodes, and after the change of version, old nodes cannot connect with the new nodes. Mohammedali et al. [47] presented a framework that is effective and stronger; it has a wallet that is used for the creation of a key pair. An administrator first adds the voter data to the voting system after checking its private key with the parent pair and, therefore, confirms that information. All blocks are associated to one another via the previous hash field. All the blocks are arranged chronologically. This makes it immutable. The framework has an administrator who verifies the outcome with the previously stored data. Authorities can see the results but still cannot change them. The framework is created using the Java programming language to make a portable application, therefore, motivating the voters and increasing the voter count and hence, making the framework less expensive and more secure. SHA 512 increases security even more as compared to SHA 256.

Rezvani et al. [48] proposed security policies of CIA (Confidentiality, Integrity, and Availability) model and their parameters which are relevant to the application of the E-voting system. It explains the CIA model into four categorical Voting Governmental Policies such as pre-registration, multi-candidate, and multi-casting. Confidentiality is a set of rules which stops unauthorized access to the data. Some majors of confidentiality are eligibility, no impersonation, ballot secrecy, vote casting secrecy, voting data secrecy, and privacy. Integrity ensures the trustworthiness of data without any forgery and it constitutes No Double voting, Ballot Immutability, self-tallying, and no ballot forging. Availability means the security of hardware, software, and database. It should resist online hacking attacks like DoS attacks, keys should be kept secret, and prevent loss of voting keys. Sukheja et al. [49] pointed out the issue that software developers who built the web application can manipulate the data, or can add extra voters and remove voters which grants them authorization that can be misused. Therefore, he suggested that a different committee should be appointed (except software developers who built it) which should ensure DAOs contracts. The committee should be able to add or remove members and grant the member the type of authorization according to the system or it should be autonomous. An institution should be answerable for any type of fraud or theft that takes place. This would help in the transparency of blockchain and make it more secure.

Rathee et al. [50] discussed that the e-voting system's privacy and security threats create a significant issue that could result in hackers committing a variety of scams to rig the election. Therefore, a possible challenge is establishing a trustworthy channel of communication by separating genuine devices from untrustworthy ones by calculating their level of trust using an optimizer. In order to identify and address the numerous hazards brought on by an intrusion at multiple levels, they developed a safe method using IoT devices utilizing blockchain. Benabdallah et al. [51] presented that due to its ability to eliminate TTP, decentralize transactions, provide transparency, and completely secure data storage. Blockchain is already being offered as a new technological foundation for a variety of applications. Additionally, it enables the usage of smart contract that automates and executes licensing terms, inside their context. Li et al. [52] proposed that the e-voting protocol allows multiple options and self-tallying options. The work demonstrates that these protocols fulfill complete tracing, likability, privacy, and secrecy. Furthermore, the work suggests that e-voting is realistic and can be used for practical applications after examining the time and gas expense of activities.

Gao et al. [53] discussed an e-voting mechanism based on blockchain that offers a transparent process. By utilizing certificate-less and software encryption, this approach can also audit voters who practice malfunctioning and withstand quantum assaults. Following an evaluation of the performance, it was determined that the proposed work is appropriate for local elections and offers certain benefits in regard to effectiveness and safety when there are few voters. Farooq et al. [54] presented that without using actual polling sites, their proposed technology offers an architecture that could be used to perform voting activities online. They suggested a design that uses adaptive consensus algorithms. Voting is safer because of the chain security methodology used in the electoral platform. When a transaction is being carried out, smart contracts could offer a safe link connecting the network and the client. Shahzad et al. [55] proposed that the architecture outlined in the proposed study considers the utility of hashing methods, the building and securing of blocks, the collection of information, and declaring outcomes by utilizing a flexible blockchain strategy. This work makes the argument that the proposed framework understands data and security management issues with blockchain technology and offers a better representation of e-voting. Huang et al. [56] aimed to make data anonymous and verifiable, and hence they used proof of knowledge encryption which is homomorphic. The result of evaluation and testing and the result of evaluation and checking, as well as compared with comparable ideas, demonstrate that our work has benefits in terms of robustness and security, which demonstrates that our plan is both scalable and practical.

Tran et al. [57] discussed that considering the developments in the field of blockchain, it will soon open new possibilities for applications such as smart contract systems, agriculture, smart cities, and e-voting. As a consequence, there will be a growing need for enhanced security that preserves transparent and decentralized technology that is expected to support blockchain developments. Zaghloul et al. [58] presented an idea that is safe and protects voter's anonymity by using secure multiparty calculations that are carried out by parties with various allegiances. Additionally, the presented work uses a blockchain powered by smart contracts as a tamper-proof bulletin board that is open to the public to save votes forever and avoid duplicate voting. Their examination of confidentiality and safety demonstrates that the suggested system offers voter identification and is safe against vulnerability management threats. In this suggested approach, the findings of the research study and cell phone modeling demonstrate the viability of the suggested strategy for large-scale elections. Li et al. [59] aimed for a system that meets all the security criteria, maximum voting secrecy, conflict-free, and fair voting system. This work examines the processing time and assesses the running expenditure of every method on the blockchain platform by simulating them on a computer, Raspberry Pi, and a smartphone. The outcome of this experiment shows how useful it is.

Panja et al. [60] presented security evidence to demonstrate that their proposed system gives the highest level of ballot data secrecy. They put their concept into practice by leveraging Ethereum blockchain as an open forum to register election processes as openly auditable transactions. The testing results and their studies demonstrate the protocol's capability to be used in real-life applications. Zaghloul et al. [61] presented a comprehensive assessment and study of the main difficulties, possibilities, confidentiality, and safety concerns relating to Cryptocurrency and blockchain-based technology. Their discussion on confidentiality and security characteristics concludes with a summary of the most recent technological developments. Their findings could help miners create effective tactics for engaging in mining and maximizing income. Gao et al. [62] presented the results of their tests that demonstrate the effectiveness of the suggested MASE-based classification and method in achieving the best possible exchange between computing performance and interactions. The result of this escapes the FHE-based paradigm's inefficiencies. Table 2 summarizes the literature survey of the abovementioned papers.

5 Conclusion

In this chapter, the importance and advantages of an e-voting platform based on blockchain are highlighted. This chapter uses blockchain technology to present a systematic analysis that compiles the most recent e-voting research. The concept of blockchain and its characteristics and working are presented prior to details on existing electronic voting techniques. Following this, a number of flaws in the current electronic voting architecture, opportunities presented by the blockchain framework to enhance electronic voting, and existing e-voting system solutions integrated with blockchain are noted and explored.

The blockchain can serve as an appropriate arrangement for decentralized electronic voting, according to many researchers. Moreover, all electors and outside spectators can view the electoral records stored in the above-proposed systems. This chapter focuses on the five major features and different consensus algorithms of blockchain that make this technology highly suitable to be incorporated in

Table 2 A brief summary of literature survey

References	Year	Contribution
Habib et al. [44]	2021	Highlighted the applications of an electronic voting platform using Ethereum technology and smart contract technology
Zhang et al. [45]	2019	Proposed a message authentication and transmission system that verifies authority while hiding the identity
Sayyad et al. [46]	2019	Explained the working of a decentralized network and recording transaction history on the decentralized network
Mohammedali et al. [47]	2019	Presented a framework with an administrator verifying the outcome of stored data
Rezvani et al. [48]	2019	Proposed a security policy of confidentiality, integrity, and availability
Sukheja et al. [49]	2019	Pointed out the possibility of data manipulation without any authorization
Rathee et al. [50]	2021	Discussed the challenges in establishing a trustworthy channel of communication
Benabdallah et al. [51]	2022	Presented the ability to eliminate TTP and provide transparent and secure data
Li et al. [52]	2021	Proposed the protocols of e-voting which allows multiple options
Gao et al. [53]	2019	Discussed a mechanism that offers a transparent process of blockchain-based e-voting
Farooq et al. [54]	2022	Presented an adaptive framework that can be used for e-voting purposes
Shahzad et al. [55]	2019	Proposed the use of hashing algorithms, construction, and sealing of blocks
Huang et al. [56]	2022	Demonstrated the use of proof of knowledge for better security and efficiency
Tran et al. [57]	2021	Discussed the possibility of applications of smart contract systems and e-voting
Zaghloul et al. [58]	2021	Presented an idea that is safe and protects voters' anonymity
Li et al. [59]	2022	Proposed a system that meets all the security criteria, including maximum voting secrecy
Pania et al. [60]	2020	Presented evidence that their system gives the highest level of ballot data secrecy
Zaghloul et al. [61]	2020	Examined all the difficulties, possibilities, confidentiality, and safety concerns
Gao et al. [62]	2022	Demonstrated the effectiveness of MASE-based classification in achieving the best computing performance and interaction

e-voting systems making them completely immutable, secure, highly persistent, and maintaining user anonymity.

A decentralized voting system using blockchain could be the future of electronic voting. This framework would be capable of overcoming most of the significant flaws of the conventional electoral systems. Although blockchain technology holds out a lot of promise, its existing limitations may prevent it from realizing its full potential. Research on the fundamentals of blockchain technology needs to be intensified in order to enhance its features and support for complicated apps that can operate on the blockchain architecture.

References

1. Cetinkaya, O., & Cetinkaya, D. (2007). Verification and validation issues in electronic voting. *Electronic Journal of e-Government, 5*(2).
2. Idrees, S. M., Aijaz, I., Jameel, R., & Nowostawski, M. (2021). Exploring the blockchain technology: Issues, applications and research potential. *International Journal of Online & Biomedical Engineering, 17*(7).
3. Taş, R., & Tanrıöver, Ö. Ö. (2020). A systematic review of challenges and opportunities of blockchain for E-voting. *Symmetry, 12*(8), 1328.
4. Bhushan, B., Sinha, P., Sagayam, K. M., & J, A. (2021). Untangling blockchain technology: A survey on state of the art, security threats, privacy services, applications and future research directions. *Computers & Electrical Engineering, 90*, 106897. https://doi.org/10.1016/j.compeleceng.2020.106897
5. Goyal, S., Sharma, N., Kaushik, I., & Bhushan, B. (2021). Blockchain as a solution for security attacks in named data networking of things. *Security and Privacy Issues in IoT Devices and Sensor Networks*, 211–243. https://doi.org/10.1016/b978-0-12-821255-4.00010-9
6. Staff, E. (2016). Blockchains: The great chain of being sure about things. *The Economist, 18*(7).
7. Narayanan, A., Bonneau, J., Felten, E., Miller, A., & Goldfeder, S. (2016). *Bitcoin and cryptocurrency technologies: A comprehensive introduction*. Princeton University Press.
8. Haque, A. K. M. B., Bhushan, B., Hasan, M., & Zihad, M. M. (2022). Revolutionizing the industrial internet of things using blockchain: An unified approach. In V. E. Balas, V. K. Solanki, & R. Kumar (Eds.), *Recent advances in internet of things and machine learning. Intelligent systems reference library* (Vol. 215). Springer. https://doi.org/10.1007/978-3-030-90119-6_5
9. Saxena, S., Bhushan, B., & Ahad, M. A. (2021). Blockchain based solutions to secure Iot: Background, integration trends and a way forward. *Journal of Network and Computer Applications*, 103050. https://doi.org/10.1016/j.jnca.2021.103050
10. Zheng, Z., Xie, S., Dai, H. N., Chen, X., & Wang, H. (2018). Blockchain challenges and opportunities: A survey. *International Journal of Web and Grid Services, 14*(4), 352–375.
11. Bheemaiah, K. (2015). Block chain 2.0: The renaissance of money. *Wired*. Wired.com. Accessed 10 Apr.
12. Karame, G. O., Androulaki, E., & Capkun, S. (2012). Two bitcoins at the price of one? Double-spending attacks on fast payments in bitcoin. *Cryptology EPrint Archive*.
13. Monrat, A. A., Schelén, O., & Andersson, K. (2019). A survey of blockchain from the perspectives of applications, challenges, and opportunities. *IEEE Access, 7*, 117134–117151.
14. Kroll, J. A., Davey, I. C., & Felten, E. W. (2013, June). The economics of Bitcoin mining, or Bitcoin in the presence of adversaries. In *Proceedings of WEIS* (Vol. 2013, No. 11).
15. del Castillo, M. (2017). *Chain is now working on six 'Citi-Sized'blockchain networks*.
16. Karame, G. O., & Androulaki, E. (2016). *Bitcoin and blockchain security*. Artech House.
17. Yuan, Y., & Wang, F. Y. (2018). Blockchain and cryptocurrencies: Model, techniques, and applications. *IEEE Transactions on Systems, Man, and Cybernetics: Systems, 48*(9), 1421–1428.
18. Soni, D. K., Sharma, H., Bhushan, B., Sharma, N., & Kaushik, I. (2020). Security issues & seclusion in bitcoin system. In *2020 IEEE 9th international conference on communication systems and network technologies (CSNT)*. https://doi.org/10.1109/csnt48778.2020.9115744

19. https://en.bitcoin.it/wiki/Elliptic_Curve_Digital_Signature_Algorithm. Accessed on 24 Sept 2022.
20. https://data-flair.training/blogs/features-of-blockchain/. Accessed on 26 Sept 2022.
21. Yu, H., Yang, Z., & Sinnott, R. O. (2018). Decentralized big data auditing for smart city environments leveraging blockchain technology. *IEEE Access, 7*, 6288–6296.
22. Shrier, D., Wu, W., & Pentland, A. (2016). Blockchain & infrastructure (identity, data security). *Massachusetts Institute of Technology-Connection Science, 1*(3), 1–19.
23. Wang, Q., Li, X., & Yu, Y. (2017). Anonymity for bitcoin from secure escrow address. *IEEE Access, 6*, 12336–12341.
24. https://data-flair.training/blogs/types-of-blockchain/. Accessed on 7 Nov 2022.
25. Bhushan, B., Sahoo, C., Sinha, P., & Khamparia, A. (2020). Unification of Blockchain and Internet of Things (BIoT): Requirements, working model, challenges and future directions. *Wireless Networks*. https://doi.org/10.1007/s11276-020-02445-6
26. Bansal, N., Singhal, M., Rastogi, M., & Arora, L. (2021). Understanding and analyzing consensus algorithms for blockchain. *NVEO-Natural Volatiles & Essential Oils Journal | NVEO*, 1794–1812.
27. Zhang, S., & Lee, J. H. (2020). Analysis of the main consensus protocols of blockchain. *ICT Express, 6*(2), 93–97.
28. Xiao, Y., Zhang, N., Li, J., Lou, W., & Hou, Y. T. (2019). Distributed consensus protocols and algorithms. *Blockchain for Distributed Systems Security, 25*, 40.
29. Kaur, S., Chaturvedi, S., Sharma, A., & Kar, J. (2021). A research survey on applications of consensus protocols in blockchain. *Security and Communication Networks, 2021*.
30. Sankar, L. S., Sindhu, M., & Sethumadhavan, M. (2017, January). Survey of consensus protocols on blockchain applications. In *2017 4th international conference on advanced computing and communication systems (ICACCS)* (pp. 1–5). IEEE.
31. https://101blockchains.com/what-is-consensus-algorithm/. Accessed on 12 Nov 2022.
32. https://en.wikipedia.org/wiki/Electronic_voting_in_India#:~:text=EVMs%20are%20easier%20to%20transport,easier%20than%20ballot%20paper%20system. Accessed on 18 Nov 2022.
33. Jardí-Cedó, R., Pujol-Ahulló, J., Castella-Roca, J., & Viejo, A. (2012). Study on poll-site voting and verification systems. *Computers & Security, 31*(8), 989–1010.
34. Sharma, R., Gupta, D., Maseleno, A., & Peng, S.-L. (2022). Introduction to the special issue on big data analytics with internet of things-oriented infrastructures for future smart cities. *Expert Systems, 39*, e12969. https://doi.org/10.1111/exsy.12969
35. Sharma, R., Gavalas, D., & Peng, S.-L. (2022). Smart and future applications of internet of multimedia things (IoMT) using big data analytics. *Sensors, 22*, 4146. https://doi.org/10.3390/s22114146
36. Sharma, R., & Arya, R. (2022). Security threats and measures in the internet of things for smart city infrastructure: A state of art. *Transactions on Emerging Telecommunications Technologies*, e4571. https://doi.org/10.1002/ett.4571
37. Zheng, J., Wu, Z., Sharma, R., & Lv, H. (2022). Adaptive decision model of product team organization pattern for extracting new energy from agricultural waste, 102352. *Sustainable Energy Technologies and Assessments, 53*(Part A). https://doi.org/10.1016/j.seta.2022.102352. ISSN 2213-1388.
38. Mou, J., Gao, K., Duan, P., Li, J., Garg, A., & Sharma, R. (2022). A machine learning approach for energy-efficient intelligent transportation scheduling problem in a real-world dynamic circumstances. *IEEE Transactions on Intelligent Transportation Systems*. https://doi.org/10.1109/TITS.2022.3183215
39. Priyadarshini, I., Sharma, R., Bhatt, D., et al. (2022). Human activity recognition in cyber-physical systems using optimized machine learning techniques. *Cluster Computing*. https://doi.org/10.1007/s10586-022-03662-8

40. Priyadarshini, I., Alkhayyat, A., Obaid, A. J., & Sharma, R. (2022). Water pollution reduction for sustainable urban development using machine learning techniques. *Cities, 130*, 103970., ISSN 0264-2751. https://doi.org/10.1016/j.cities.2022.103970
41. Pandya, S., Gadekallu, T. R., Maddikunta, P. K. R., & Sharma, R. (2022). A study of the impacts of air pollution on the agricultural community and yield crops (Indian context). *Sustainability, 14*, 13098. https://doi.org/10.3390/su142013098
42. Bhola, B., Kumar, R., Rani, P., Sharma, R., Mohammed, M. A., Yadav, K., Alotaibi, S. D., & Alkwai, L. M. (2022). Quality-enabled decentralized dynamic IoT platform with scalable resources integration. *IET Communications, 00*, 1–10. https://doi.org/10.1049/cmu2.12514
43. Deepanshi, I. B., Garg, D., Kumar, N., & Sharma, R. (2022). A comprehensive review on variants of SARS-CoVs-2: Challenges, solutions and open issues. *Computer Communications*. https://doi.org/10.1016/j.comcom.2022.10.013. ISSN 0140-3664.
44. Ahasan Habib, A. K. M., Hasan, M. K., Islam, S., Sharma, R., Hassan, R., Nafi, N., Yadav, K., & Alotaibi, S. D. (2022). Energy-efficient system and charge balancing topology for electric vehicle application. *Sustainable Energy Technologies and Assessments, 53*(Part B), 102516. https://doi.org/10.1016/j.seta.2022.102516. ISSN 2213-1388.
45. Zhang, Q., Xu, B., Jing, H., & Zheng, Z. (2019). Ques-chain: An ethereum based e-voting system. *arXiv preprint arXiv:1905.05041*.
46. Sayyad, S. F., Pawar, M., Patil, A., Pathare, V., Poduval, P., Sayyad, S., et al. (2019). Features of blockchain voting: A survey. *International Journal, 5*, 12–14.
47. Mohammedali, N., & Al-Sherbaz, A. (2019). Election system based on Blockchain technology. *International Journal of Computer Science and Information Technology (IJCSIT), 11*(5), 13–31.
48. Rezvani, M., & Khani, H. (2019). E-voting over blockchain platforms: A survey. *Journal of Network Security and Data Mining, 2*(3), 1–14.
49. Sukheja, D., Indira, L., Sharma, P., & Chirgaiya, S. (2019). Blockchain technology: A comprehensive survey. *Journal of Advanced Research in Dynamical and Control Systems, 11*(9), 1187–1203.
50. Rathee, G., Iqbal, R., Waqar, O., & Bashir, A. K. (2021). On the design and implementation of a blockchain enabled e-voting application within IoT-oriented smart cities. *IEEE Access, 9*, 34165–34176.
51. Benabdallah, A., Audras, A., Coudert, L., El Madhoun, N., & Badra, M. (2022). Analysis of blockchain solutions for E-voting: A systematic literature review. *IEEE Access*.
52. Li, H., Li, Y., Yu, Y., Wang, B., & Chen, K. (2020). A blockchain-based traceable self-tallying E-voting protocol in AI era. *IEEE Transactions on Network Science and Engineering, 8*(2), 1019–1032.
53. Gao, S., Zheng, D., Guo, R., Jing, C., & Hu, C. (2019). An anti-quantum e-voting protocol in blockchain with audit function. *IEEE Access, 7*, 115304–115316.
54. Farooq, M. S., Iftikhar, U., & Khelifi, A. (2022). A framework to make voting system transparent using blockchain technology. *IEEE Access, 10*, 59959–59969.
55. Shahzad, B., & Crowcroft, J. (2019). Trustworthy electronic voting using adjusted blockchain technology. *IEEE Access, 7*, 24477–24488.
56. Huang, J., He, D., Chen, Y., Khan, M. K., & Luo, M. (2022). A blockchain-based self-tallying voting protocol with maximum voter privacy. *IEEE Transactions on Network Science and Engineering, 9*(5), 3808–3820.
57. Tran, Q. N., Turnbull, B. P., Wu, H. T., De Silva, A. J. S., Kormusheva, K., & Hu, J. (2021). A survey on privacy-preserving blockchain systems (PPBS) and a novel PPBS-based framework for smart agriculture. *IEEE Open Journal of the Computer Society, 2*, 72–84.
58. Zaghloul, E., Li, T., & Ren, J. (2021). d-BAME: Distributed blockchain-based anonymous mobile electronic voting. *IEEE Internet of Things Journal, 8*(22), 16585–16597.
59. Li, Y., Susilo, W., Yang, G., Yu, Y., Liu, D., Du, X., & Guizani, M. (2020). A blockchain-based self-tallying voting protocol in decentralized IoT. *IEEE Transactions on Dependable and Secure Computing, 19*(1), 119–130.

60. Panja, S., Bag, S., Hao, F., & Roy, B. (2020). A smart contract system for decentralized borda count voting. *IEEE Transactions on Engineering Management, 67*(4), 1323–1339.
61. Zaghloul, E., Li, T., Mutka, M. W., & Ren, J. (2020). Bitcoin and blockchain: Security and privacy. *IEEE Internet of Things Journal, 7*(10), 10288–10313.
62. Gao, C., Li, J., Xia, S., Choo, K. K. R., Lou, W., & Dong, C. (2020). Mas-encryption and its applications in privacy-preserving classifiers. *IEEE Transactions on Knowledge and Data Engineering, 34*(5), 2306–2323.

Multilayer Security and Privacy Provision in Internet of Things Networks: Challenges and Future Trends

Kashif Naseer Qureshi, Thomas Newe, Rosheen Qazi, and Gwanggil Jeon

1 The Internet of Things

In today's world, technology has changed society theatrically and vividly. Innovative advanced technologies and technical assistance are making users' lives easier by providing various opportunities, such as reducing labor, saving time and money, and providing fast and convenient data communication services [1]. The research in the field of the Internet of Things (IoT) has focused on many technical facets and various application scenarios, from a simple smart dust bin to smart homes, smart healthcare, smart industries, and smart transportation services. Smart homes, or home automation, is a new concept, in which home appliances are controlled with the help of IoT, such as heating and cooling options on smartphones to control air conditioners [2]. In smart transportation, traffic systems make vehicle parking easier with the help of this technology. IoT is also enabling smart agricultural systems, where with the help of different sensors and smart devices, we can now control the various features necessary for productive agriculture. Ozone detection, air-quality control, and temperature monitoring are some additional examples of these networks [3]. IoT has had many impacts on healthcare too, with the help of sensing devices and many other kinds of smart equipment; for example, doctors can now remotely examine their patients. There are smart wearable gadgets available,

K. N. Qureshi (✉) · T. Newe
Department of Electronic & Computer Engineering, University of Limerick, Limerick, Ireland
e-mail: kashifnaseer.qureshi@ul.ie; thomas.newe@ul.ie

R. Qazi
Department of Computer Science, Bahria University Islamabad, Islamabad, Pakistan

G. Jeon
Department of Embedded Systems Engineering, College of Information Technology, Incheon National University, Incheon, South Korea
e-mail: gjeon@inu.ac.kr

R. Sharma et al. (eds.), *Data Analytics for Internet of Things Infrastructure*,
Internet of Things, https://doi.org/10.1007/978-3-031-33808-3_16

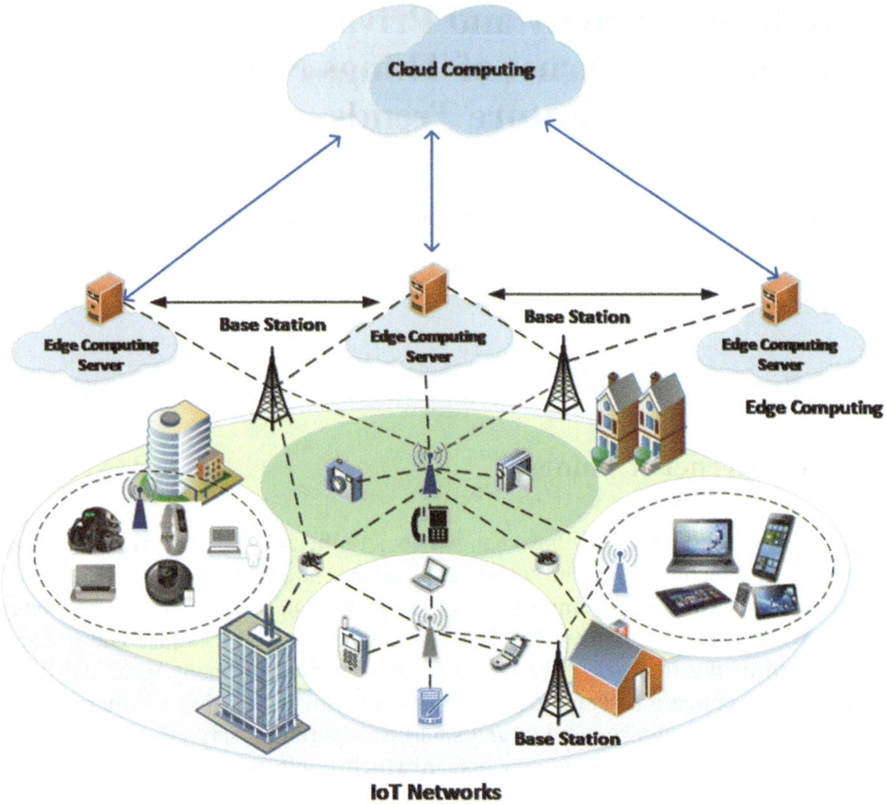

Fig. 1 IoT networks and data communication

such as smart glasses, smart watches, and smart sensors. The life cycle of an IoT device follows four major steps: amass, link, investigate, and act. The "things" in IoT are categorized into three types: feeling things (sensors), shrinking things (nanotechnologies), and thinking things (embedded intelligent systems or devices). Figure 1 shows several IoT networks and how data are communicated.

IoT is being used to make communication easier and more advanced between different entities or facilitate human interactions in various simulated atmospheres. However, IoT is not only about these different entities, networks, and patrons in the connected system but also about different surface areas. IoT has several essential elements, namely entity devices or things, the cloud, mobile applications, encryption algorithms, network interfaces, authentication mechanisms, and physical security. IoT has been committed to providing widespread interconnection, inclusive intellectual services, and rigorous data perceptions. At the same time, as IoT is making our lives easier and more flexible, some challenges need to be considered, such as privacy and confidentiality.

2 Architecture and Technologies of IoT

The layered architecture of IoT includes three crucially different types of layers: the application layer, the perception layer, and the network layer [4]. All these layers use different technologies, depending on their functionalities. In the application layer, which communicates with the system's firmware, various service platforms are used to provide efficient services. The perception layer is used to sense the data or information for further processing. The network layer contains several types of microcontrollers that enable the sharing or transferring of data between different objects (things) within an IoT-enabled network [5].

IoT networks have five basic characteristics: event-driven, ambient intelligence; flexible structures; semantic sharing; and complex access technologies [6]. With these characteristics and features, IoT devices provide efficient and inoperative integration between human society and physical modern systems [7]. The communication technologies of IoT can be divided into two categories: noncellular technologies and cellular technologies. Different technologies are used in IoT networks, such as electronic product codes (EPCs) and radio-frequency identification technology (RFID).

Table 1 shows the layer-based technologies in IoT networks.

3 Security Requirements in Distributed IoT Applications

Many studies and official documents have mentioned that distributed IoT is an efficient option to consider when deploying secure and efficient IoT-enabled systems or networks in the real world [8]. The only important purpose of all this progress and research is to provide secure, confidential, and authorized IoT-enabled systems and networks. In distributed IoT networks, all the devices or things are connected and can retrieve, process, and combine information and then provide it to all the

Table 1 Layer-wise IoT technologies

IoT layers	Technologies
Application layer	Message queues telemetry transport (MQTT) Advanced message-queuing protocol (AMQP) Extensible messaging and presence protocol (XMPP)
Network layer	IPV6 Routing protocol for low-power and lousy networks IPV6 low-power wireless personal area network
Physical/perception layer	Cellular devices Low-energy Bluetooth devices Low-power wide-area network Radio-frequency identification Wi-Fi

Table 2 Security challenges and solutions in distributed IoT

Security challenges	Distributed IoT	Solutions
Identity and authentication	Node to node	Authentication and authorization mechanism
Access controls	Heterogeneous policies	Access-control lists and roles based on access-control lists
Protocol and network security	Unknown peers	Datagram transport layer security protocols and symmetric key-based protocols
Privacy concerns	Flexible	Privacy-by-design approach
Trust and governance	Detailed information	Untrusted framework
Fault tolerance	Detailed point of view	Middleware layer

entities in the system. Different IoT services are reliable and offer scalability, viability, interoperability, and data management. Securing IoT-enabled devices and networks is a major concern nowadays, and the security threats to these devices need to be removed [9]. There are several security challenges in IoT, such as network attacks, heterogeneous attacks, identity attacks, privacy attacks, trust attacks, and fault tolerances. A distributed network may pose fewer chances for attackers to control the network because an attacker can take control of some parts of the system but not the whole system. Table 2 shows the security challenges and their solutions for distributed IoT-enabled systems or networks.

Attackers are always interested in stealing sensitive information such as passwords, credit or debit card numbers, and locations. This is why most attackers try to target at least one of the components of an IoT network. The attacker intends to gain access to and steal important information. Smart home devices connect various other devices. Smart home devices are made by different manufacturers and can have different security levels. The security level of a smart home device depends on the attackers' objectives.

Smart healthcare systems are vulnerable to multiple threats because of their versatility. These smart devices are used to monitor patients' vital signs, such as body temperature, body movement, blood pressure, and heartbeat. As the number of devices increases, the risk of security breaches likewise increases. For example, several devices are being used by different users in a smart health environment—i.e., by doctors, patients, nurses, admin staff, etc.—and each device has its own programming logic. Thus, keeping up on updates to its firmware and other functionalities is not easy.

Smart wearable gadgets are a new trend nowadays. These gadgets are used to measure metrics such as walking steps, heart rate, and running steps. Stumpy operator risks and high-viability attacks are common threats to wearable devices [10]. The attacker targets the data on the cloud by using techniques such as keylogging, trojans, and malware. High operator risks, low-viability attacks, medium operator risks, and medium-viability attacks are some of the categories of risky attacks. Smart wearable devices are being compromised because of these attacks. Smart cities also face many threats, such as advanced tenacious threats, a lack of security aptitude, recurrent breaches, defective processes, and the rise of outsourcing.

4 Existing Challenges and Issues in IoT

Recent advancements in IoT and the rapid trend toward adopting IoT have made IoT an ideal vulnerable target for attackers to breach data privacy, confidentiality, and data integrity [11]. The following are some of the many vulnerabilities in IoT-enabled systems:

- Effortlessness and comfort of use
- Uncertain default login credentials
- Ignoring software updates
- Communications that haven't been encoded
- Uncertain user interfaces
- Ignoring confidentiality defenses

There are different security issues at the application layer of the IoT model. Some of these issues are the verification and confirmation of individuality and illegal access to the system [12]. Figure 2 shows the architecture of IoT security.

To prevent illegal access to the system, appropriate schemes and mechanisms need to be implemented to verify and confirm users' identities. In the application layer, the data is transferred from one location to another via different communication links, which move data from one place to another. This can cause serious security issues, such as the confidentiality, integrity, and secrecy of data. Thus, to avoid these types of issues, strong mechanisms for storing and retrieving data should be implemented [13]. Managing big data is another challenge for IoT applications that aim to prevent systems from having vulnerabilities. Better smart systems need to be designed to manage these big data. Application layer software susceptibilities need clear and understandable code because weak code and vulnerable code can lead to serious issues. The attacker can gain an advantage to compromise these systems. Lastly, a proper mechanism needs to be implemented to avoid buffer flow.

4.1 Security Issues in the Network Layer

Interconnectivity, network infrastructure, and communication protocols are the three basic services being provided by the network layer in IoT-enabled devices. Eavesdropping and forging are two commonly known security issues that are being faced by the network layer in the IoT architecture [14]. However, processing in IoT-enabled devices generates massive numbers of data; therefore, securing these data in such huge numbers is a challenging issue in IoT. TLS/SSL, IPsec, and several other mechanisms have been proposed to solve security issues in the network

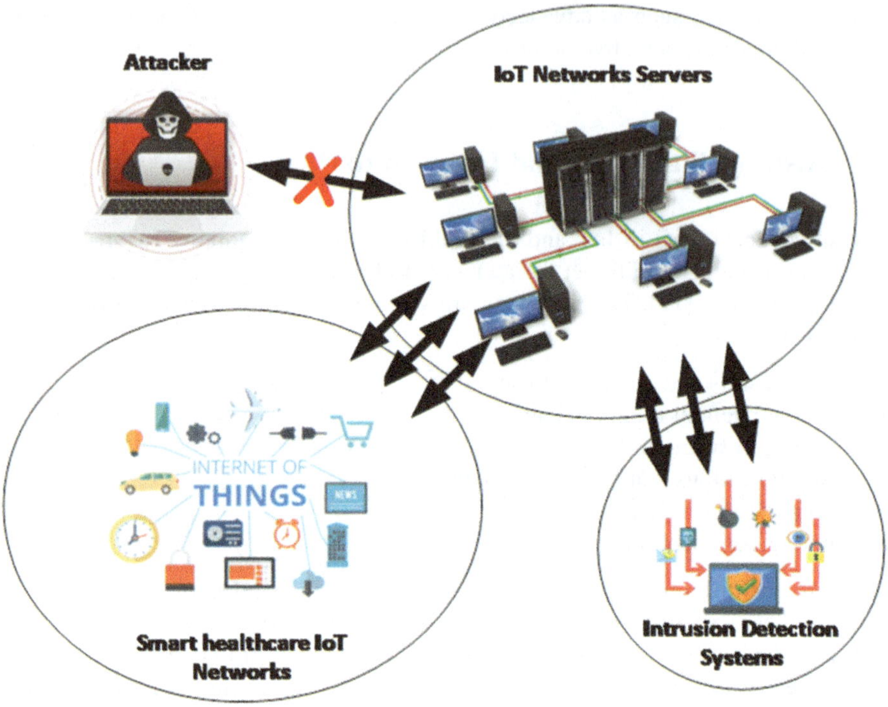

Fig. 2 IoT security architecture

Table 3 Possible attacks on network layers and solutions to them

Issue	Solution
Eavesdropping	The encrypted algorithm on the sender and receiver ends
DoS	Intrusion detection and protection systems
Damaging confidentiality	Certificates, data encryptions, and access controls
Damaging integrity	Encrypted communications and end-to-end authentications

layer. Integrity, availability, and confidentiality are some of the major concerns in this layer. To provide security for the network layer, security certificates and data-encrypting algorithms need to be implemented. Some of the attacks that can occur in the network layer are denial-of-service (DoS) and routing attacks [15]. The issues in the network layer can be divided into two categories: customary security problems and compatibility issues. Illegally accessing networks; eavesdropping; damaging the confidentiality, integrity, and availability (CIA) of the network; DoS attacks; and other exploits fall under the customary security problems of network layers. Table 3 shows some of the possible exploits and some solutions to these security issues in network layers.

4.2 *Security Issues at Physical Layer*

The physical layer contains low-power devices such as physical devices, sensors, and actuators; as the name implies, their computational powers are low, so applying security mechanisms to this layer is a difficult challenge for developers. Damaging the integrity and damaging the authenticity of the network are some of the major issues facing IoT devices in the physical layer. However, threats from the outer environment and security threats from hackers and viruses are also common in this layer. On the other hand, sensor network security issues are also being faced by the physical layer of the IoT architecture. The nodes in the network are responsible for transmitting data from one end to another end of the communication channel [16]. Introducing malicious code, fabrication, and illegally reading content are some of the security issues for IoT-enabled systems. Applying an authentication mechanism and ensuring confidentiality with the help of an access-control mechanism are solutions that can secure the physical layer. Table 4 shows the various security challenges and security requirements of the physical layer.

5 Countermeasures for Security in IoT

This section lists several countermeasures against various attacks on the different layers of IoT-enabled systems. Application layer security measures for IoT-enabled systems are shown in Table 5 [17].

Table 4 Security challenges and requirements of the physical layer

Security challenges	Security requirements
Limited resources	Lightweight encryption techniques
DoS vulnerability	Protecting the data on sensor nodes
Confidentiality of data	Security certificates
Integrity of data	Authentication and key agreement mechanisms

Table 5 Security measures for the application layer

Security measures	Description
Substantiation and validation	This precaution will help restrict and limit access to a system. Only authorized or confirmed users will be allowed to access the system.
Invasion discovery	This will help notify the system of various attacks on it and can also help provide various solutions to these threats.
Risk valuation	Risk valuation can help secure the application layer in IoT applications by comparing different standards of the threats in order to examine the risk level of the threats.
Data security	Updating firewall settings and encrypting data can help secure the data in the application layer of IoT devices.

Table 6 Countermeasures that secure network layers in IoT devices

Security measures	Description
IPsec and PPSK	Internet protocol security (IPsec) should be applied to authenticate network access, and a private preshared key (PPSK) mechanism can be used to enhance the confidentiality, integrity, and authenticity of a layer.
Disabling default passwords	The default passwords should be disabled after installing new network devices to maintain strong password policies and the periodic changing of passwords.
Intrusion prevention system (IPS)	If a network has an IPS, then it can keep track of all the packets headed toward their respective destinations; this provides security at the network layer of wired IoT applications.

Countermeasures for the network layer improve the security of the IoT architecture. Researchers are currently working on deploying a stumpy resource–starving firewall in IoT devices to provide packet inspection to the layer. Table 6 shows the countermeasures that secure the network layer.

The physical layer's security measures are categorized into mainly two parts. The first part of this layer contains nodes, controllers, and managers. The second part is the perception network, which connects the physical layer to the network layer [18]. Several technologies are in the perception layer, such as RFID, ZigBee, sensor nodes, and sensor gateways. There are different types of security concerns at the perception layer, such as encryption, authentication, authorization, and privacy. Different solutions are available to control the threats to the perception layer, such as authentication mechanisms, authorization mechanisms, cryptography algorithms, stenography algorithms, and image-processing mechanisms.

5.1 Attacks on and Threats to IoT

Various types of attacks are launched against IoT devices, such as vulnerability exploitation, malware, password, sniffing, spoofing, botnet, remote-access, and data leakage attack usually occur through web and emails, via mobile data storage devices such as USB, optical media, and laptops [19]. In vulnerability exploitation attacks, the three types of vulnerabilities are buffer overflow, code injection, and cross-site scripting. By exploiting a vulnerability, a hacker can gain access to the system and disrupt the system's processes. Malware is another type of attack, can it can be divided into subtypes, such as ransomware and spyware. A password attack is used to steal the credentials of a user by bombarding it with countless passwords until the attack finds the right password. Sniffing attacks are used to intercept the Internet's incoming and outgoing data traffic toward a smart device. In spoofing attacks, a malicious node impersonates another device or user in a network to launch attacks against a network host, steal data, spread malware, or bypass access controls. This attack is also known as the monkey-in-the-middle (MITM) attack. Botnet attacks are used for a wide variety of cybercrimes, such as DoS attacks. A remote-access attack is another malicious action that targets computer networks.

In application layer attacks, an attacker/hacker uses malicious code to compromise the system. The code is injected into this layer to make the attack possible. In some cases, the attacker can also interrupt the system as an authorized user and harm the system [19]. Normally, a DoS attack is an example of an application layer attack where the hacker pretends to be an authenticated user to interrupt the system. In phishing attacks, the attacker uses emails to attack the system by collecting the victim's credentials and then damaging the data.

The network layer performs data routing, determines the path of data, and forwards data to establish a connection. This layer transfers information from a particular source to its destination [20]. Neglect selective-forwarding attacks and greed selective-forwarding attacks are two different types of attacks in the context of wireless communication networks, specifically aimed at selective forwarding attacks. These attacks involve a malicious node selectively forwarding or dropping packets in order to disrupt or manipulate network communication. The main difference between neglect selective-forwarding attacks and greed selective-forwarding attacks lies in the behavior of the attacking nodes. Sybil attacks usually occur in peer-to-peer networks, where an attacker generates some strained identities to behave like a diverse node in the network. This behavior reduces network proficiency and disrupts data veracity, the application of resources, and the overall performance of the network. The entity in a peer-to-peer network is software that can access all the local resources in the network and broadcast itself in the network with the help of an identity, and the attacker may present multiple identities to a network to act as separate nodes. This strategy helps the attacker to reach an uneven level of control in the network.

The sinkhole attack is also known as the black hole attack, which creates a conflict between various resources. In the hello flood attack, a node sends several odd messages that are unrelated to communication and cause obstruction over the channel, which may slow down the communication process in the network. Acknowledgment flooding attacks fall under the category of DoS attacks, where acknowledgments (acks) are used to send wrong information to the intended adjacent nodes and thus cause congestion in the network. In the physical layer, attacks focus on forging collected data and destroying perception devices. In a tag-cloning attack, nodes or tags are easily available to read and modify, or copies of tags can be generated. This attack makes it difficult to distinguish between the original tag and the copy tag [19]. In an eavesdropping attack, the attacker can find important data, such as passwords, PIN codes, or any other information being transferred from a tag to a reader or from a reader to a tag. The attacker can then use this information for their own purposes [19].

A radio-frequency (RF)-jamming attack can interrupt the signal coming from a radio frequency and can introduce noise signals to disrupt the normal data communication over the channel. In freshness attacks, the attacker uses some malicious code or an object (device/thing) to transfer data to the endpoint with some genuine identification information to gain trust from the endpoint host. This usually aims to abolish the cogency of certification [18]. In side-channel attacks, encrypted or plain text is used to determine the encryption key that is being used for data encryption. Moreover, the attacker deploys different techniques on encryption devices to obtain the key. One of these attacks is known as a timing attack—where an attacker identifies the key by investigating the time required to execute the algorithm that encrypts or decrypts the data [18].

5.2 *Defenses Against IoT Attacks on Each Layer*

Various types of defense mechanisms have been developed against several attacks on each layer of the IoT architecture. The application layer's defense mechanisms are access-control lists and firewalls. In the network layer, the devices are connected to a wired or wireless medium, where the attackers can easily gain control of the communication over the channel. The defense mechanisms against network layer attacks include data privacy, secure routing, and authentication. The information in the perception layer can be rather sensitive; therefore, it needs a secure physical design to prevent changes to the data. On the other hand, the communication can be blocked by using jamming node attacks on the layer, in which case a secure IP channel is needed to ensure the confidentiality of the data. In addition to this, the signals can be dropped, or any distortion in the communication can stop the data transmission. Authentication is then required for devices to continue the data communication over the channel.

6 Privacy Issues in IoT

In an IoT-enabled environment, sensor nodes or actuators are connected with the help of several types of hardware and software. This interconnection between devices leads to the probability that information will leak through some unsanctioned and data-manipulated methods. Moreover, the devices are transferring different types of personal information; therefore, ensuring the privacy of data in an IoT-enabled environment is one of the major tasks for the developers and researchers [21]. Eavesdropping and data tracking are used to steal user data and violate user privacy [22].

Privacy in an IoT-enabled environment can be guaranteed by enhancing user access controls, improving transparency, and keeping apace of the latest IoT technologies. The confidentiality, integrity, and availability (CIA) traits of network security are considered as some of the most important security models [15]. Confidentiality in an IoT-enabled environment must be contained to only authorized users and entities in order to ensure their data privacy. However, the insufficient authentication of users and devices degrades user privacy [23]. Moreover, in an

IoT-enabled environment, it is necessary to keep user data secret during data transmission. Therefore, sensors in IoT networks are equipped to hide messages from attackers and thus remain confidential. Integrity means the original data are maintained throughout data transmission in IoT networks. Tempering with or modifying the data can cause a data breach, which violates data integrity. Integrity in an IoT-enabled environment ensures that no data or information has been modified during the data communication between sender and receiver [24]. Ensuring integrity in IoT-enabled networks is not an easy task; therefore, many encryption algorithms are being used to ensure the integrity of messages. The process of enabling access to the authorized user at any time and any place, as is required, is known as availability, one of the CIA security traits. Availability in an IoT-enabled environment guarantees that the services provided by IoT devices are not being interrupted and are available for usage in real time [25]. This security service is prone to serious threats, such as DoS attacks.

6.1 Existing Security Models for IoT Networks

The authors in [13] presented many of the security measures for IoT networks, namely authentication, intrusion detection, risk assessment, and data security. The study further suggested that security threats can be controlled to provide better services for IoT-enabled devices. In another study [26], several security models were proposed to provide security and privacy. That study further explained IoT objectives such as communication, security and privacy, device discovery, network discovery, content access, and service access. Securing privacy and providing the best security protection to users are two of the major concerns in IoT networks. Mosenia and Jha measured various threats at different layers in the IoT architecture [27]. Oracevic et al. mentioned several key challenges in IoT, such as security, privacy, identity management, standardization in interoperability, data deluges, and controlling access [28]. Their paper also explained more about the various threats at different layers in the IoT architecture.

Efficiently designed media access control (MAC) protocols are used to provide quality of service (QoS) in an IoT environment, better energy efficiency, and better data throughput; they are also used to reduce delay. Moreover, data-fusion and data-aggregation techniques are used to remove redundant data from devices [29]. The main requirements of IoT networks are scalability, interoperability, security, and proper QoS. A review of the existing studies affirms that future technologies should be more efficient so that they can better face numerous challenges and risks [30].

IoT devices and networks are ideal targets for an extensive array of security attacks, which can lead to data manipulation and data leakage. Different researchers have proposed different security models to secure IoT-enabled systems or networks. In [31], Shaikh et al. proposed a solution to secure an IoT network and proposed using a secure compiler to secure coding and virtual machines to prevent stack overflow attacks on IoT servers. There are different vulnerabilities and weaknesses in the source codes of different programs. Therefore, secure compilers are needed to help analyze the code and remove weaknesses from the input program. Secure compilers can be designed with six general parts: a scanner, a parser, a syntax-directed translator, a semantic analyzer, an intermediate code generator, and an optimizer. These all parts produce high-quality and trustworthy coding environments for IoT services with the help of protected coding rules.

Stack overflow is one of the famous and common vulnerabilities in IoT devices because it helps hackers to attack these systems. To avoid this type of vulnerability, researchers have presented virtual machine–based stack protection. This strategy separates stack frames, where memory is shared between different stack frames, and whenever a new function call is made, a new stack frame is generated against it. In this technique, an attacker hacks a stack frame because the virtual machine–based stack is controlled to block the sensitive information and then passes the exception to the system to notify it of the attack. Thus, the secure compiler can help analyze the code to provide better programming to the enabled systems. This also reduces the cost of the development of IoT systems, whereas virtual machine–based stack monitoring reduces the cost of foiling glitches in IoT systems. Gafurov and Chung proposed different solutions for securing different layers in IoT-enabled systems [32]. They proposed that two-way authentication mechanisms must be used to secure IoT-enabled devices.

Datagram transport layer security (DTLS) is a protocol that operates on the basis of using the rivest shamir adleman (RSA) algorithm and is specially designed for IPV6 to secure communication between the transport layer and the application layer in the IoT model. This protocol should be placed between these two layers to ensure the message's confidentiality, integrity, and authenticity and ensure end-to-end latency while consuming little energy [32]. The attribute-based signature scheme secures privacy in IoT-enabled devices, uses an attribute tree, and lays out all the unforgeable assumptions in the Diffie–Hellman algorithm. The users cannot copy any data for authentication or authorization in IoT-enabled systems.

The fuzzy trust-based access-control (FTBAC) approach [32] has been used to develop trust-based IoT-enabled systems or networks. There are three layers in an FTBAC framework, and each layer is different and provides a different functionality. The first layer of this framework is the device layer, which is responsible for all communication between the different devices in a system. The next layer is the request layer, which collects information and calculates its fuzzy trust value. The last layer is the access-control layer, which maps the trust value onto access permission. The results of this approach are remarkable in that it provides suppleness and scalability in IoT-enabled systems thanks to its cryptographic protection. The secure middleware layer, as proposed in [33], can help exchange data in a protected and secure environment. vulnerability information risk technical understanding and sharing (VIRTUS) eXtensible messaging and presence protocol (XMPP) are two types of middleware used in IoT-enabled devices to provide secure communication between devices.

7 The Future of IoT

Setting consistent standards is one of the major challenges for IoT-enabled networks. There are not enough standards to ensure consistency in IoT-enabled devices or systems. Connectivity issues still need to be accounted for in IoT networks. For example, one company is working on wearable gadgets, while another is working on smart medical appliances. In this case, the likelihood that these products will be able to communicate is minimized because they use different protocols, which likewise minimizes interoperability. If they both used same standard for communication, they would have more interoperability.

IoT is opening new paths toward an automated future, but at the same time, it comes with new risks. Several organizations are using smart grids, smart devices, and intelligent or smart manufacturing environments. All these fields are facing different challenges, such as data privacy and data security. Thus, global standards need to be built to increase the interoperability of IoT-enabled devices. The business industry is facing several challenges, such as the cost of operators, maintenance, and support, as well as some discovery issues. However, IoT has solved these issues by providing easy solutions to implement promotional strategies. IoT is the future of technology and the Internet because it automates almost every technological application in the world, such as smart wearable devices, smart homes, smart cities, smart health devices, smart transportation, smart vehicles, smart artificial intelligence–enabled devices, and many more.

7.1 Top-Ten IoT Developments

This section explains some of the major worldwide ventures of IoT networks that are listed by the Open Web Application Security Project (OWASP). They claim that IoT is not all about only devices or networks; rather, several areas are involved in IoT, and each of these areas needs to be evaluated. The elements of IoT include devices, the cloud, mobile applications, network interfaces, software, encryption algorithms, authentication mechanisms, and physical security measures. Table 7 shows various methods to control several threats to IoT networks. When assessing the overall security of an IoT-enabled system, the OWASP categorizes these areas as follows:

- Insecure web interfaces
- Insufficient authentication and authorization
- Insecure network services
- Lack of transport encryption
- Privacy concerns
- Insecure cloud interfaces
- Insecure mobile interfaces
- Insufficient security configurability

Table 7 Methods to control threats to IoT networks

Threat	Control method
Insecure web interfaces Insufficient authentication	Strong passwords and passcode-recovery mechanisms Use two-factor authentication
Insecure network services	Remove buffer overflow and provide network ports
Lack of transport encryption	Use SSL, TLS, and other encryption mechanisms
Privacy concerns	Collect only protected or encrypted data
Insecure cloud interfaces	Set session timeouts, and use two-factor authentication
Insecure mobile interfaces	Use two-factor authentication
Insufficient security	Ensure data are encrypted during transit
Insecure software/firmware	Keep devices updated
Poor physical security	Secure data-storage medium, and limit admin capabilities

- Insecure software or firmware
- Poor physical security

8 Summary

In this chapter, the different IoT security issues at different layers were discussed. The countermeasures and the developments in IoT security were deemed essential and necessary. Some of the countermeasures for these security threats were also discussed. The architecture and different technologies used in IoT were identified, and an overall view of different layers was laid out. IoT, as a complex system, integrates many layers, and the security problems come mostly from how the system is integrated, which is why the different layers are vulnerable to so many security problems; for instance, privacy protection is required for *every* layer. Finally, the number of IoT devices distributed in various systems and the future of IoT applications were discussed.

References

1. Al-Fuqaha, A., Guizani, M., Mohammadi, M., Aledhari, M., & Ayyash, M. (2015). Internet of things: A survey on enabling technologies, protocols, and applications. *IEEE Communications Surveys & Tutorials, 17*(4), 2347–2376.
2. Abbasnejad, E., Abdelrasoul, M., Abdou, A., Acquaviva, A., Ahmadinia, A., Ali, S., Ban, Y., Barnawi, A., Beckmann, K., & Bellizia, D. (2018). 2017 Index IEEE transactions on emerging topics in computing Vol. 5. *IEEE Transactions on Emerging Topics in Computing, 6*(1), 1.
3. Qureshi, K. N., Kaiwartya, O., Jeon, G., & Piccialli, F. (2021). Neurocomputing for Internet of Things: Object Recognition and Detection Strategy. *Neurocomputing, 485*, 263.

4. Tewari, A., & Gupta, B. (2018). Security, privacy and trust of different layers in Internet-of-Things (IoTs) framework. *Future Generation Computer Systems*.
5. Vashi, S., Ram, J., Modi, J., Verma, S., & Prakash, C. (2017). Internet of Things (IoT): A vision, architectural elements, and security issues. In *2017 International Conference on I-SMAC (IoT in Social, Mobile, Analytics and Cloud)(I-SMAC)* (pp. 492–496). IEEE.
6. Silva, B. N., Khan, M., & Han, K. (2018). Internet of things: A comprehensive review of enabling technologies, architecture, and challenges. *IETE Technical Review, 35*(2), 205–220.
7. Silva, D. R., Nogueira, M. B., Rodrigues, M. C., Costa, J. S., Silveira, D. V., & Oliveira, G. M. (2019). A concrete architecture for smart solutions based on IoT technologies. *IEEE Instrumentation & Measurement Magazine, 22*(2), 52–59.
8. Alnoman, A., Sharma, S. K., Ejaz, W., & Anpalagan, A. (2019). Emerging edge computing technologies for distributed IoT systems. *IEEE Network, 33*, 140.
9. Qureshi, K. N., Sikandar, A., & Dhawankar, P. (2022). Next-generation connected traffic using UAVs/drones. In *Secure and digitalized future mobility: Shaping the ground and air vehicles cooperation,* p. 65.
10. Qureshi, K. N., Alhudhaif, A., Haider, S. W., Majeed, S., & Jeon, G. (2022). Secure data communication for wireless mobile nodes in intelligent transportation systems. *Microprocessors and Microsystems*, 104501. https://doi.org/10.1016/j.micpro.2022.104501
11. Roman, R., Zhou, J., & Lopez, J. (2013). On the features and challenges of security and privacy in distributed internet of things. *Computer Networks, 57*(10), 2266–2279.
12. Ahmad, M., Younis, T., Habib, M. A., Ashraf, R., & Ahmed, S. H. (2019). A review of current security issues in Internet of Things. In *Recent trends and advances in wireless and IoT-enabled networks* (pp. 11–23). Springer.
13. Kalita, A. *Internet of Things*, 2017.
14. Yashaswini, J. (2017). A review on IoT security issues and countermeasures. *Oriental Journal of Computer Science and Technology, 10*, 454.
15. Khalaf, R. H., & Mohammed, A. H. (2018). Confidentiality and integrity of sensing data transmission in IoT application. *International Journal of Engineering & Technology, 7*(4.25), 240–245.
16. Borgohain, T., Kumar, U., & Sanyal, S. (2015). Survey of security and privacy issues of internet of things, *arXiv preprint arXiv:1501.02211*.
17. Swamy, S. N., Jadhav, D., & Kulkarni, N. (2017). Security threats in the application layer in IOT applications. In *2017 International Conference on I-SMAC (IoT in Social, Mobile, Analytics and Cloud)(I-SMAC)* (pp. 477–480). IEEE.
18. Xiao, W., Miao, Y., Fortino, G., Wu, D., Chen, M., & Hwang, K. (2022). Collaborative cloud-edge service cognition framework for DNN configuration toward smart IIoT. *IEEE Transactions on Industrial Informatics, 18*(10), 7038–7047. https://doi.org/10.1109/TII.2021.3105399
19. Erhan, L., Ndubuaku, M.U., Mauro, M.D., Song, W., Chen, M., Fortino, G., Bagdasar, O., & Liotta, A. (2020). Smart anomaly detection in sensor systems: A multi-perspective review. *arXiv: Learning*.
20. Fortino, G., Messina, F., Rosaci, D., & Sarné, G. M. L. (2020). Using blockchain in a reputation-based model for grouping agents in the Internet of Things. *IEEE Transactions on Engineering Management, 67*(4), 1231–1243. https://doi.org/10.1109/TEM.2019.2918162
21. Morshed, G., & Yuan, L. (2017). Big data in cloud computing: An analysis of issues and challenges. *International Journal of Advanced Studies in Computer Science And Engineering, 6*(4), 345–350.
22. Das, M. L. (2015). Privacy and security challenges in Internet of Things. In *International conference on distributed computing and internet technology* (pp. 33–48). Springer.
23. Valea, E., Da Silva, M., Flottes, M.-L., Di Natale, G., Dupuis, S., & Rouzeyre, B. (2019). Providing confidentiality and integrity in ultra low power IoT devices. In *2019 14th international conference on Design & Technology of Integrated Systems in Nanoscale Era (DTIS)* (pp. 1–6). IEEE.
24. Aman, M. N., Sikdar, B., Chua, K. C., & Ali, A. (2018). Low Power Data Integrity in IoT Systems. *IEEE Internet of Things Journal, 5*(4), 3102–3113.

25. Xiong, J., Ren, J., Chen, L., Yao, Z., Lin, M., Wu, D., & Niu, B. (2018). Enhancing privacy and availability for data clustering in intelligent electrical service of IoT. *IEEE Internet of Things Journal, 6*(2), 1530–1540.
26. Mohammed, A. (2017). *Security issues in IoT* (pp. 933–940). IJSRSET.
27. Mosenia, A., & Jha, N. K. (2016). A comprehensive study of security of internet-of-things. *IEEE Transactions on Emerging Topics in Computing, 5*(4), 586–602.
28. Oracevic, A., Dilek, S., & Ozdemir, S. (2017). Security in internet of things: A survey. In *2017 International Symposium on Networks, Computers and Communications (ISNCC)* (pp. 1–6). IEEE.
29. Parashar, A., & Rishishwar, S. (2017). Security challanges in IoT. In *2017 Third International Conference on Advances in Electrical, Electronics, Information, Communication and Bio-Informatics (AEEICB)* (pp. 446–449). IEEE.
30. Rajan, A., Jithish, J., & Sankaran, S. (2017). Sybil attack in IOT: Modelling and defenses. In *2017 International Conference on Advances in Computing, Communications and Informatics (ICACCI)* (pp. 2323–2327). IEEE.
31. Shaikh, F., Bou-Harb, E., Neshenko, N., Wright, A. P., & Ghani, N. (2018). Internet of malicious things: correlating active and passive measurements for inferring and characterizing internet-scale unsolicited IoT devices. *IEEE Communications Magazine, 56*(9), 170–177.
32. Gafurov, K., & Chung, T.-M. (2019). Comprehensive survey on internet of things, architecture, security aspects, applications, related technologies, economic perspective, and future directions. *Journal of Information Processing Systems, 15*(4), 797–819.
33. Hosenkhan, M. R., & Pattanayak, B. K. (2020). Security issues in Internet of Things (IoT): A comprehensive review. In *New paradigm in decision science and management* (pp. 359–369). Springer.

A Methodology for the Development of Soft Sensors with Kafka-ML

Antonio Jesús Chaves, Cristian Martín, Luis Llopis Torres, Enrique Soler, and Manuel Díaz

1 Introduction

The Internet of Things (IoT) has definitely contributed to the application of data science due to data growth in recent years [1]. Any kind of device is able to send data periodically, even with high rates, to servers, cloud, or fog platforms. This has made it possible to have large amounts of data on the platforms and has given them the possibility of applying different processes to improve decision-making. For example, analytic techniques can be applied in order to estimate, predict, or monitor values and actions. Machine learning (ML) and artificial intelligence (AI) are the main fields where data are treated to carry out these predictions and recommendations. In this sense, the data availability and ML techniques can contribute to improving the decision-makings and the automated response of the IoT systems.

However, there are two challenges that are difficult for data management. Sensors, which are mainly the data generators, are very expensive in some cases (expensive sensors can make product development unprofitable) or, simply, there is no sensor to produce these data. In this case, ML techniques can also help to infer these data from the information provided by other sensors that can have some correlation with the necessary data.

A soft sensor is the combination of hardware sensors and software (mostly through machine learning) to estimate parameters that cannot be calculated directly or it is expensive to obtain [2]. The development of soft sensors has been greatly facilitated by achievements in data science, computing and communication technologies, statistical tools, and machine learning techniques [3]. Therefore, soft-

A. J. Chaves · C. Martín (✉) · L. L. Torres · E. Soler · M. Díaz
ITIS Software Institute, University of Málaga, Málaga, Spain
e-mail: chaves@uma.es; cristian@uma.es; lmllopis@uma.es; esoler@uma.es; mdiaz@uma.es; http://itis.uma.es/

R. Sharma et al. (eds.), *Data Analytics for Internet of Things Infrastructure*,
Internet of Things, https://doi.org/10.1007/978-3-031-33808-3_17

sensor development is a low-cost alternative to manage variables that can be inferred when there is no data available to use in a formula or no such formula exists. Soft sensors can be designed as a source of variable estimation.

Kafka-ML [4] is an open-source framework to manage the pipeline of TensorFlow and PyTorch machine learning (ML) models on Kubernetes [5]. This tool allows the design, training, and inference of ML models by using data streams (with Apache Kafka as the data provider), enabling scalability, fault tolerance, and distributed models [6]. Since Kafka-ML is designed to work with data streams (for both training and inference), the framework can be used in an IoT ecosystem to manage real-time data pipelines, typical of sensorization in IoT, and to integrate ML models to take decisions and predict values. In this sense, Kafka-ML arises as a powerful tool for the design and development of soft sensors.

There is a large literature about soft sensors, each applying different tools but, from our point of view, there is a lack of a generic platform to support soft-sensor development. In this chapter, we present a methodology based on Kafka-ML for soft-sensor design and development. In order to show the suitability of the platform, we present the design, implementation, and prototype of a soft sensor to calculate harmful nitrate levels from underground water.

The main contributions of this chapter are:

- Presenting Kafka-ML as an open-source framework for the design and development of soft sensors
- Providing a streaming framework for the continuous prediction and visualization of soft sensors
- Using Kafka-ML to obtain a fast evaluation and validation of different models and scenarios, given the complexity of soft-sensor design
- Demonstrating through a use case the prediction of harmful nitrate levels from underground water

The rest of the chapter is structured as follows. Section 2 presents the related work regarding Kafka-ML and soft-sensor design. Section 3 presents the proposed methodology of soft-sensor design using Kafka-ML. Section 4 describes the proposed methodology by using a water pollution monitoring use case and evaluates several key elements of the proposed approach. Finally, Sect. 5 summarizes the work, with a discussion on further potential extensions.

2 Related Work

The IoT has contributed to increasing the development of hardware sensors. Almost any IoT device is equipped with sensors, and it is able to send information using wireless communication technology such as Bluetooth or WiFi. However, when the necessary knowledge is unavailable, or the modeling process is too complicated, the construction of soft sensors is a promising complementary technology [3].

The design of a soft sensor can be carried out by applying different techniques, namely model-driven and data-driven [2]. Model-driven soft sensors are based on first principle models (FPM), extended Kalman filter [7], and adaptive observer [8]. FPMs describe the physical and chemical background of the process, but it is difficult to model malfunctions or changing conditions. Data-driven soft sensors have been presented as a design alternative.

The design of data-driven soft sensors can apply statistical methods and ML techniques. Some examples are principal component and partial least squares regression, support vector machine, artificial neural network, and deep learning [3, 9, 10]. There are many different application fields for data-driven soft sensors such as:

- On-line prediction. Combining real-time data and estimated variables to carry out predictions about the processes
- Monitoring. Using estimation to monitor variables without sensor devices
- Fault detection. Applying the on-line prediction to detect malfunctions in the monitored system

Neural network techniques are broadly used to estimate values by minimizing the distance between the network and target outputs. When the model is trained, it is possible to predict the output simulating a real sensor. Soft sensors using this technique are designed in many different fields. For example, in [11], a control of distillation column soft sensor is designed. Kadlec et al. [2] present data-driven soft sensors for the industry. The chemical industry is applying techniques for variable estimations to substitute expensive sensors or data that cannot be directly sensorized. One example, in this context, is the estimation of the nitrate concentration as described in [12, 13].

Many works have researched the forecasting of nitrate concentration based on ML techniques. In [14], the ability of artificial neural networks to model groundwater nitrate of the Arak Aquifer is presented. Nitrate monitoring is very important in cities. In [15], an array of soft sensors to improve nitrate concentration accuracy is proposed. Soft sensors for nitrate estimation are also designed for treatment water plants as shown in [16]. Another complementary work [17] applies soft computing models to simulate nitrate contamination.

All these proposals apply the traditional stages for the application of machine learning techniques, i.e., feature extraction, model training, and inference. Additionally, data can be used in an offline or real-time mode, but it is necessary to adapt the framework to the particular characteristics of the design. We propose a methodology based on the Kafka-ML framework to simplify soft-sensor design.

Apache Kafka [18] is an open-source distributed event streaming platform used by thousands of companies for high-performance data pipelines, streaming analytics, data integration, and mission-critical applications. In particular, Kafka can be used as a real-time data provider (fault-tolerant and scalable streaming data) to the machine learning models in order to take decisions or forecasting. Many works deal with Kafka and ML, such as [19, 20] or [21].

Kafka-ML [4] is a framework to manage the pipeline of TensorFlow/Keras and PyTorch (Ignite) ML models on scalable containerized platforms (Kubernetes). The Kafka-ML pipeline allows the design, training, validation, and inference of ML models. The training and inference datasets for the ML models can be fed through Apache Kafka; thus they can be directly connected to data streams such as the ones provided by the IoT. Other approaches such as Kubeflow [22], NVIDIA DIGITS [23], Amazon SageMaker [24], Algorithmia [25], and Valohai [26] have similar goals or have provided some of the functionalities, but Kafka-ML is the first open-source framework to provide an ML/AI pipeline solution to integrate ML/AI and data streams. This chapter proposes a methodology to use this framework to design and implement soft sensors.

3 Proposed Methodology for Soft-Sensor Development with Kafka-ML

Kafka-ML provides an open-source framework that enables the management of the pipeline of machine learning techniques with data streams including steps such as training and inference. Soft sensors are precisely, in most cases, data stream generators that infer a unique property where machine learning and deep learning techniques are adopted. In this section, a methodology to integrate these two paradigms is presented. The proposed methodology (Fig. 1) for the design of soft sensors through Kafka-ML includes:

1. Soft-sensor design and data ingestion into Kafka-ML for the training of ML models

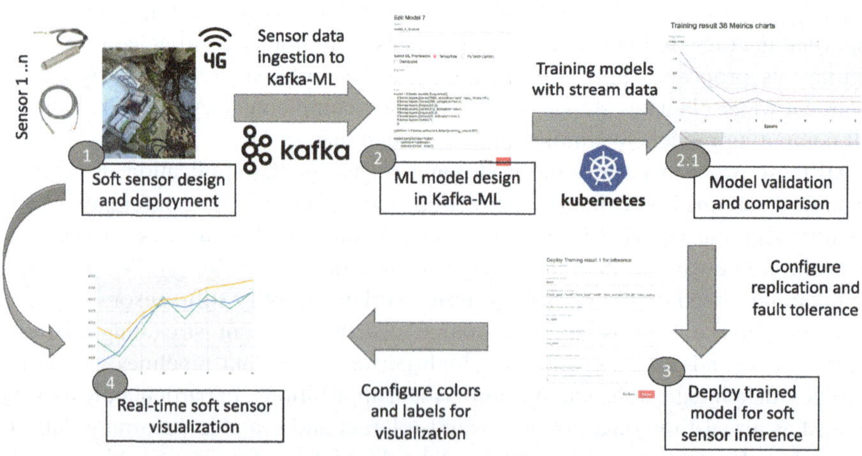

Fig. 1 Proposed soft-sensor development methodology in Kafka-ML

2. Design, training, and validation of ML models (even multiple in parallel) with the data streams received
3. Trained model deployment for soft-sensor inference, where real-time sensor data will be ingested into the model as input to obtain predictions of the desired property
4. Soft-sensor prediction visualization through a user-friendly interface in Kafka-ML

As you can see, you only need the Kafka-ML framework for the design of soft sensors with this methodology. Next, each step of the methodology is described in detail.

3.1 Soft-Sensor Design and Data Ingestion to Kafka-ML

The first decision to take into account in this methodology is the selection of the physical sensors to design the soft sensor. This decision can be influenced by several factors, such as the sensors available to predict the desired feature, their cost, real-time response, and the combination of all of them. Although in the inference phase it will only be necessary to send the measurements to predict the target variable, in the training phase it is also necessary to send the real value of the measurement to be predicted to carry out supervised learning. This can be done in a laboratory or by using a sensor (probably very expensive) that will later be replaced by the developed soft sensor. Once the sensors have been selected and properly installed in the study area, the sensor data can be sent to the Kafka-ML framework for further analysis and training. Two of the libraries developed in Kafka-ML can be used for this purpose: RAW (binary format) and Apache Avro [27].

On the one hand, the RAW library can be used to send and encode measurements directly in binary. Apache Kafka works with a binary format for high performance. The RAW library handles the data types and encodes them transparently to the user. The user only needs to code an array or matrix with a list of measurements to obtain a prediction from the soft sensor, and the library handles the encoding, shape, and dispatching.

On the other hand, users can also use the most popular library for encoding data streaming pipelines, Apache Avro. Using a defined Avro schema for the studied dataset, users only have to send a simple JSON with the Avro library to Kafka-ML. The advantage of this data format is that most big data and data stream platforms support Apache Avro, and therefore, this facilitates hypothetical integration with other systems.

Unlike many streaming data platforms, sending measurements for feature prediction does not mean that these measurements are consumed and discarded. Kafka-ML exploits the Apache Kafka feature for data retention, and stream data can be reused as many times as necessary. Thanks to this, Kafka-ML not only allows the reuse of the generated datasets for the training of multiple models in parallel, thus optimizing

Datasources received

Filter

Description	Deployment	Input format	Input configuration	Kafka topic	Validation rate	Test rate	Total msg	Time	Sent again
UK_Data dataset	10	RAW	{"data_typ ...	UK_Data:0:0:44319	0.15	0.15	44319	2022-11-09T11:46:45Z	
RondaData dataset	1	RAW	{"data_typ ...	RondaData:0:3531:4	0.15	0.15	1177	2022-11-09T08:53:48Z	

Fig. 2 Kafka-ML data stream management form

time and resources, but also for training new models later on. Figure 2 shows the Kafka form for data stream management. Once data streams are sent to Kafka-ML, they are registered so they can be easily dispatched to other training models in this form.

3.2 *Model Design, Training, and Validation*

The second step of the methodology covers the definition and model training through the data stream sent to Kafka-ML in the previous step. In Kafka-ML, users can easily define multiple models in popular machine learning and deep learning frameworks. Currently, Kafka-ML supports TensorFlow and PyTorch, allowing developers to use the one that best suits their needs. Once users have defined a set of models to evaluate, they can make them available for training. One of the common practices in machine learning is to continuously evaluate different models, hyperparameters, and architectures until reaching the one that best fits the studied dataset. Kafka-ML facilitates this thanks to its data stream management and simultaneous model deployment. In the soft-sensor design, users can create several models to compare their performance with data streams previously received from physical sensors.

Kafka-ML also supports model deployment on infrastructures with both CPUs and GPUs. In the case of computationally intensive models such as deep learning models, Kafka-ML exploits GPU acceleration for fast training. For advanced users who want to exploit Kafka-ML's features to their fullest, another alternative is to deploy partitioned models in a distributed way [6, 28]. This would enable the deployment of deep learning models in the cloud-to-things continuum for optimizing response times.

The definition of models in Kafka-ML is a very easy task. Figure 3 shows how to define a new machine learning model. In the creation form, a name for the model, optional imports for required libraries, and the source code of the model itself in the selected framework have to be specified. For instance, Listing 1 shows a Kafka-ML TensorFlow code example for the definition of a soft-sensor model with 4 input variables and 3 neural layers of 64, 32, and 16 neurons, respectively. This code also

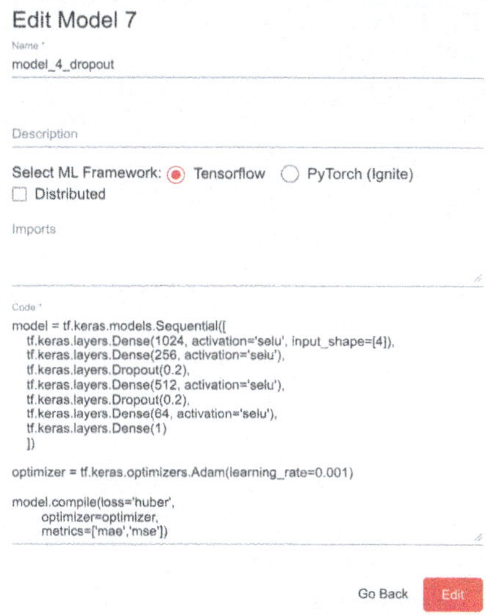

Fig. 3 Create model form in Kafka-ML

defines the optimizer (Adam), the loss function (mean squared error), and finally the metrics that will be displayed when the model is trained. In this case, the mean absolute error (MAE) and mean squared error (MSE) are used. In the soft-sensor design, usually regression models are involved so the output layer of the model is a continuous value.

Listing 1 Example soft-sensor TensorFlow/Keras ML code with 4 sensor inputs.

```
model = Sequential([
Dense(64, activation='relu', input_shape=[4]),
Dense(32, activation='relu'),
Dense(16, activation='relu'),
Dense(1)
])
model.compile(loss='mse',
optimizer='adam', metrics=['mae', 'mse'])
```

Once a set of models for the soft sensor has been defined in Kafka-ML, it is time to create a configuration. A configuration is a logical union of one or more models that will be trained together. The configuration is then deployed with the corresponding training parameters such as the batch size and the number of epochs. Figure 4 shows the form where the user can set those parameters. Once deployed, a Docker container will be created in Kafka-ML (Kafka-ML is a microservice architecture) and will start reading the data stream sent previously from Apache Kafka. It is not a problem if the data have not been sent yet because the models will wait until the data stream is available.

Users can check the training metrics and compare the results obtained by the defined models in real time since the first epoch. This allows users not only to

Fig. 4 Kafka-ML model deployment form

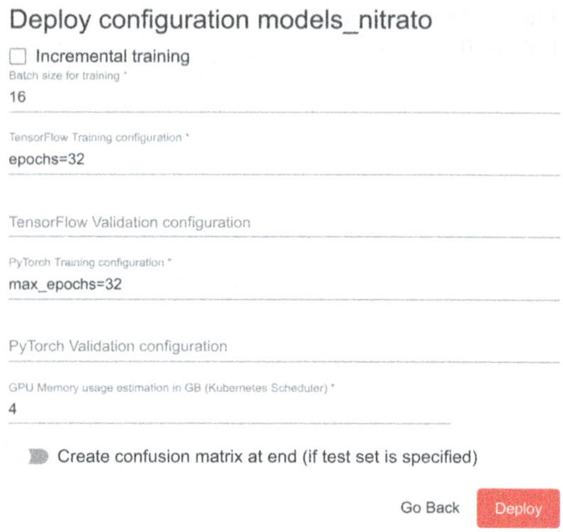

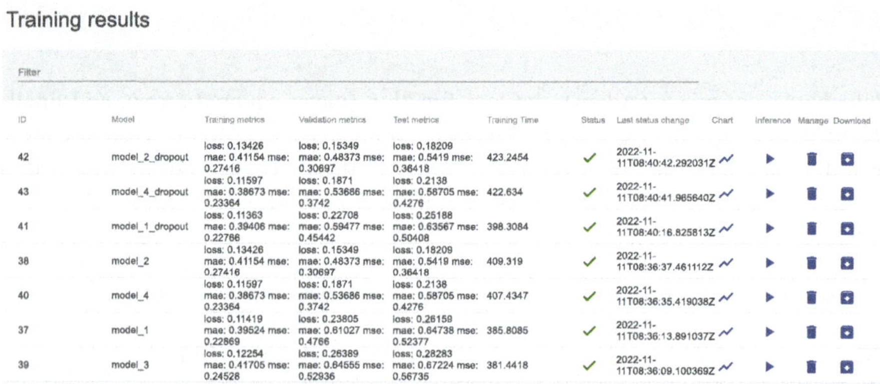

Training results

Filter

ID	Model	Training metrics	Validation metrics	Test metrics	Training Time	Status	Last status change	Chart	Inference	Manage	Download
42	model_2_dropout	loss: 0.13426 mae: 0.41154 mse: 0.27416	loss: 0.15349 mae: 0.48373 mse: 0.30697	loss: 0.18209 mae: 0.5419 mse: 0.36418	423.2454	✓	2022-11-11T08:40:42.292031Z				
43	model_4_dropout	loss: 0.11597 mae: 0.38673 mse: 0.23364	loss: 0.1871 mae: 0.53686 mse: 0.3742	loss: 0.2138 mae: 0.58705 mse: 0.4276	422.634	✓	2022-11-11T08:40:41.965640Z				
41	model_1_dropout	loss: 0.11363 mae: 0.39406 mse: 0.22766	loss: 0.22708 mae: 0.59477 mse: 0.45442	loss: 0.25188 mae: 0.63567 mse: 0.50408	398.3084	✓	2022-11-11T08:40:16.825813Z				
38	model_2	loss: 0.13426 mae: 0.41154 mse: 0.27416	loss: 0.15349 mae: 0.48373 mse: 0.30697	loss: 0.18209 mae: 0.5419 mse: 0.36418	409.319	✓	2022-11-11T08:36:37.461112Z				
40	model_4	loss: 0.11597 mae: 0.38673 mse: 0.23364	loss: 0.1871 mae: 0.53686 mse: 0.3742	loss: 0.2138 mae: 0.58705 mse: 0.4276	407.4347	✓	2022-11-11T08:36:35.419038Z				
37	model_1	loss: 0.11419 mae: 0.39524 mse: 0.22869	loss: 0.23805 mae: 0.61027 mse: 0.4766	loss: 0.26159 mae: 0.64738 mse: 0.52377	385.8085	✓	2022-11-11T08:36:13.891037Z				
39	model_3	loss: 0.12254 mae: 0.41705 mse: 0.24528	loss: 0.26389 mae: 0.64555 mse: 0.52936	loss: 0.28283 mae: 0.67224 mse: 0.56735	381.4418	✓	2022-11-11T08:36:09.100369Z				

Fig. 5 ML models and training metric result table in Kafka-ML

visualize and compare the performance of their models but also to stop the training of a model as soon as they realize that it is not getting acceptable results and not to wait until the end of the training. Once all the models have been trained, users evaluate the best-performing models, download them in h5 format, and prepare the selected ones for soft-sensor inference (next step).

Figure 5 shows the resulting metrics form where the trained models can be compared. These metrics are the ones defined when designing the model. Furthermore, metrics can be studied in more detail (e.g., learning curves) by selecting the chart option. In this case, a chart with the metrics such as the one displayed in Fig. 6 shows the progression of the metrics with respect to the epochs. In this interactive graph, users can select which metrics are displayed and compare the results of training and validation, for example, to identify whether overfitting or underfitting has occurred.

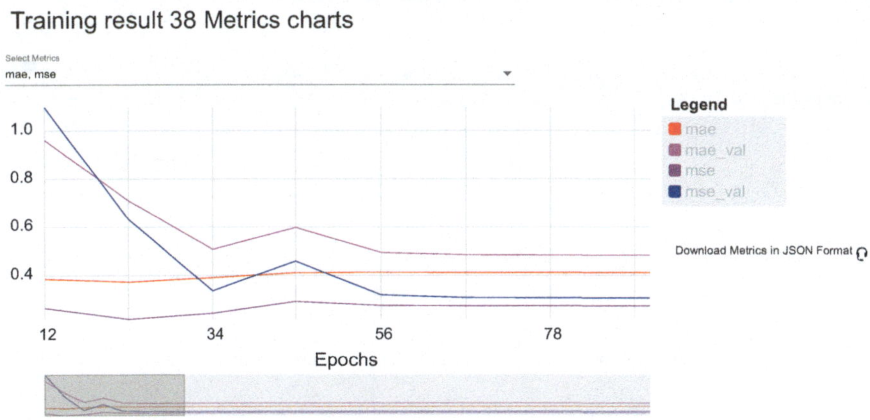

Fig. 6 Metric visualization chart in Kafka-ML

3.3 Model Deployment for Soft-Sensor Inference

Once the model or models have been trained, validated, and selected, users can deploy them to make data-driven inferences and obtain real-time predictions for the soft sensor. In this way, they do not have to use another tool for this purpose (e.g., TensorFlow Serving [29], BentoML [30], etc.). Kafka-ML offers the possibility to deploy these models directly from the application and through Kafka topics, receive, and return the output as appropriate. To deploy the models, users can configure the number of replicas of the model, the Kafka topic where the measurement data will arrive, and the output topic where the soft-sensor predictions will be sent, among other parameters. From the moment of deployment, the deployed models expect sensor data in an input topic, and they will send the soft-sensor prediction result to the configured output topic.

This closes a loop where the data reach Kafka-ML and is used in the different phases of the pipeline using data streams. In addition, the application allows the deployment of these models with the possibility of being fault-tolerant if necessary, deploying several replicas of the necessary components, or partitioning the input and output topics for load balancing.

3.4 Final Application: Kafka-ML Visualization

The last step when users have a trained and deployed model is to visualize the soft-sensor data over time. Normally, in these situations, an ad hoc API is developed to visualize the soft-sensor predictions via a web form (e.g., a Python Flask API that imports the trained model and serves the predictions through a graphical interface).

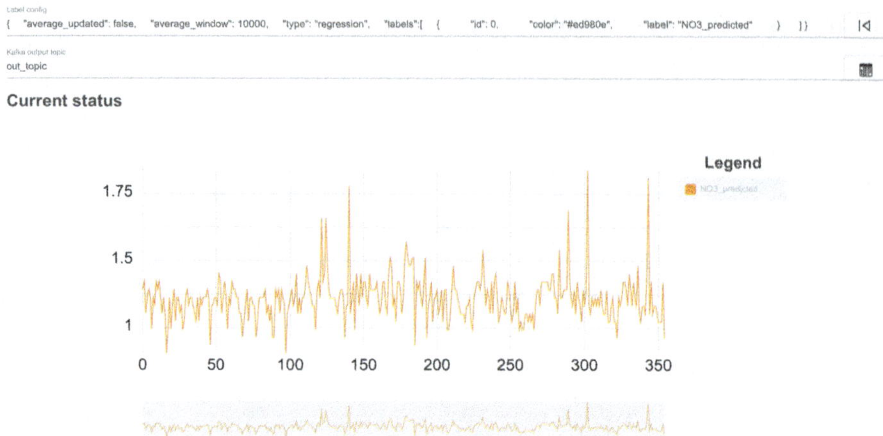

Fig. 7 Real-time soft-sensor prediction visualization in Kafka-ML

Kafka-ML has also considered this case, and it is prepared for fast application delivery. Finally, users can visualize the soft-sensor inferences in real time.

Figure 7 shows an example of the visualization of predictions on one of the models that we will later observe in the evaluation. To use this tool, users only need to configure the output topic they configured in the inference, and design aspects such as the color of the graph, the number of outputs, and the type of model (regression in this case). Once connected, the soft-sensor results will be displayed as shown in the previous example. Note that the data displayed are real-time data, and there is no system to display historical data. This API is currently intended for fast data visualizations as presented in this case.

Using this visualization means that users do not have to leave Kafka-ML and move everything previously developed to other tools such as Grafana [31] to visualize predictions in real time, accelerating the development and usage of soft sensors with data streams.

4 Water Pollution Monitoring Use Case

To demonstrate how the methodology works, we introduce the use of this methodology for the development of a soft sensor for water pollution monitoring. In particular, we present a soft sensor for the prediction of nitrate (NO_3^-) concentration in river watersheds. To do so, we will carry out this development with a public dataset taken from the United Kingdom (UK) rivers in an attempt to implement a soft sensor capable of detecting the amount of nitrate in water, a parameter that is difficult and costly to measure continuously.

In the following subsections, we will discuss the dataset used, as well as the models to be evaluated in order to compare the results and choose the best one to deploy in a soft sensor.

4.1 Methods and Datasets

Consider the challenge of measuring nitrate levels in water due to the lack of data and sensors; the first step is to look for a public dataset to address the problem and study whether it is viable to do so or not. In this first step, we have made use of the UK government's Open Data on river water quality monitoring [32]. This dataset is public and open, with a large amount of data available.

The dataset contains multiple information about the water quality monitoring stations, such as their geographical coordinates, station identifier name, as well as different variables measured over time. The variables change depending on the station, although the main ones include the following:

- Alkalinity
- Biochemical oxygen demand
- Conductivity
- Dissolved copper
- Dissolved oxygen
- Dissolved iron
- Nitrate
- Nitrite
- Ammonium
- pH
- Dissolved phosphor
- Suspended solids
- Dissolved zinc

Among these measured characteristics is the nitrate, which is difficult to measure directly, and the sensors to do so are very expensive. A clear candidate for a soft sensor. From the rest of the variables, we must select which ones that, when assembling a real device with attached sensors, are, as a whole, easier and cheaper to obtain than the nitrate sensor we are trying to simulate.

Therefore, from the previously mentioned variables, we selected alkalinity, conductivity, pH, and suspended solids, as these are the ones that are easier to measure with low-cost sensors.

After selecting the variables, we slightly preprocess the dataset by cleaning possible outliers, normalizing the data for better training, and sending it to Kafka-ML simulating that the data have been sent by an IoT device.

Regarding machine learning models, various model architectures have been evaluated in order to determine which of these best fit the data and the problem

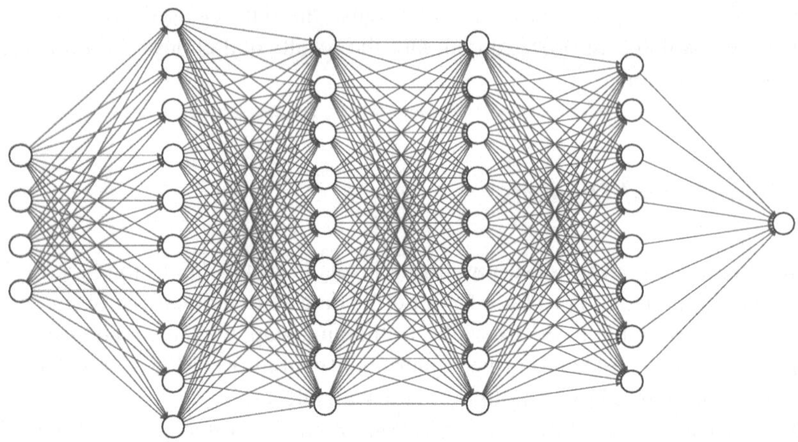

Fig. 8 Proposed soft-sensor ML model architecture

to be solved. Given the data, the architecture that suits us best is a fully connected (known as dense) one. One of these architectures can be seen in Fig. 8.

4.2 Evaluation

For the evaluation of the soft-sensor performance in Kafka-ML, the tool has been deployed on a cluster architecture with the following computation capacities:

- **Hardware configuration**. Kafka-ML has been deployed on our own cluster of 7 state-of-the-art servers. Each server has an Intel(R) Xeon(R) Gold 6230R CPU with two NVIDIA(R) Tesla(R) V100 GPUs as well as 384 GB of RAM. The client that sent the data and where the results were measured has an Intel(R) Core(TM) i9-10900K CPU and 64 GB of RAM.
- **Software configuration**. Each node runs Kubernetes v1.21.6 and Docker 20.10.8 on top of Ubuntu 20.04.3 LTS. A Kubernetes master was deployed in one of the nodes, while the remaining are Kubernetes workers. The client PC runs Ubuntu 21.04.

We evaluated a total of 4 different model architectures, each of which different number of layers and neurons per layer. In addition, the previous ones have been slightly modified by adding Dropout layers [33]. All models have been trained using Huber loss as loss function, Adam optimizer, and scaled exponential linear units (SELU) as activation function in hidden layers. Table 1 shows the different model architectures evaluated.

Table 1 Model architectures used during evaluation

#	Description	Architecture
1	Model with 3 layers	512 × 128 × 64
2	Model 1 with a dropout layer	512 × 128 × Dropout(0.2) × 64
3	Model with 4 layers	1024 × 512 × 512 × 256
4	Model 3 with dropout layer	1024 × 512 × Dropout(0.2) × 512 × 256
5	Similar to Model 1, changing neurons distribution	64 × 1024 × 64
6	Model with 4 layers with fewer neurons	1024 × 256 × 512 × 64
7	Model 6 with dropout layers	1024 × 256 × Dropout(0.2) × 512 × Dropout(0.2) × 64

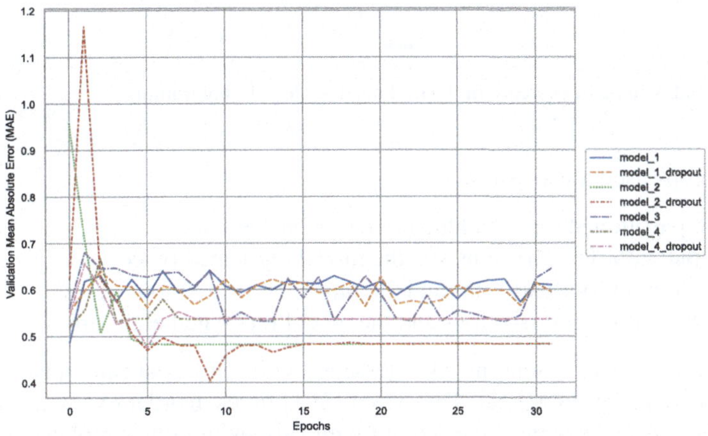

Fig. 9 Validation MAE decrease of the evaluated models during training

For this evaluation, these models have been defined in Kafka-ML, grouped, and deployed with the same training setup. Specifically, this consists of a training of 32 epochs with a batch size of 16 samples. MAE and Mean metrics have been used to compare the precision of the models. Figure 9 shows the MAE decrease for the different models and Fig. 10, the corresponding for the MSE.

As can be seen in the MAE and MSE validation, we found that `model_2` as well as the modified model with Dropout layer obtained the best performance. Its architecture can be seen in Fig. 8. Once selected `model_2` as the baseline model, we move on to the inference experiments using this model.

For the inference evaluation, we measure the response time of the inference service. In order to measure this, we have carried out different experiments by deploying the service in different setups. In these setups, we have varied the number of clients sending sensor data to be inferred. In addition, the number of replicas and the number of Kafka partitions were also modified.

These benchmarks are important since it is possible that in some use cases the user may require high availability of the predictive models due to the need of having multiple sensors deployed across a measurement area.

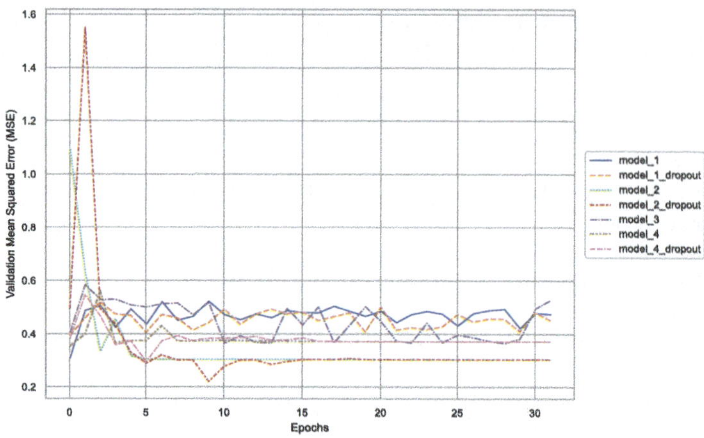

Fig. 10 Validation MSE decrease of the evaluated models during training

The setups used are as follows:

- One un-partitioned topic and the model replicated once
- One topic with two partitions and the model replicated twice
- One topic with four partitions and the model replicated four times
- One topic with eight partitions and the model replicated eight times

Each test consists of sending 512 messages with sensor data and measuring how long the model takes to predict. This was done 32 times in order to obtain an average result. These tests have been carried out with various inference replicas setups and various partitions in the Kafka topics. Furthermore, each test has been replicated with different numbers of clients (specifically with 1, 2, 4, 8, 16, and 32 clients, respectively), making requests, seeking to evaluate how the system behaves when faced with a multitude of requests.

The most basic use case of Kafka-ML is the setup where we use one topic in a single partition and no replication of the inference module. It can be observed in Fig. 11 that this setup works fine with a few clients, but as soon as we scale the number it gets overloaded and the latency increases.

Having a greater number of replicas of the model and a greater distribution of the data in Kafka improves this result. This can be seen in cases requiring higher availability, where the latency is clearly lower (except when we have already reached a limit that cannot be improved by replication and partitioning). Therefore, it can be concluded that having more replication and partitioning of data can reduce latency in cases where higher availability is required.

Another test was carried out to measure the response capacity of Kafka-ML at different data transmission frequencies. In this test, clients (simulating IoT devices) generated data at certain time intervals, and we seek to observe the responsiveness of Kafka-ML as well as the point where there is no longer a bottleneck in the prediction.

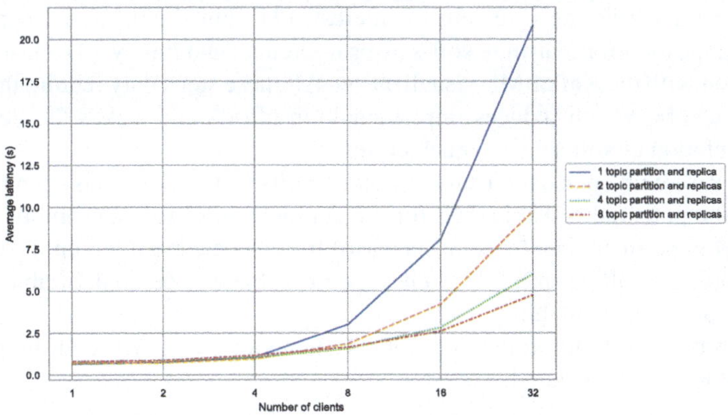

Fig. 11 Average inference latency response with a different number of clients, topic partitions, and replicas

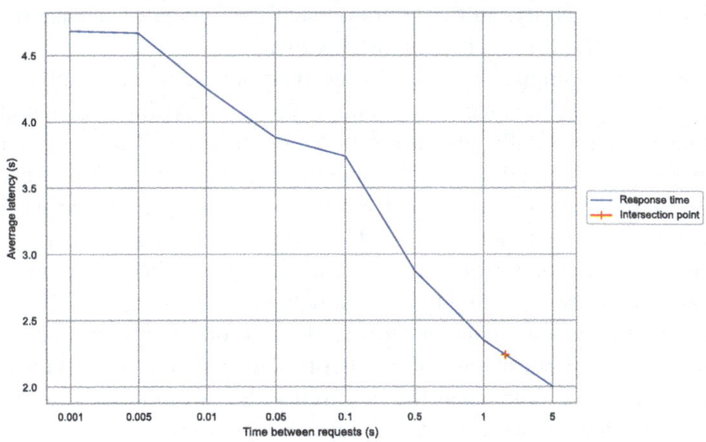

Fig. 12 Responsiveness of Kafka-ML toward data transmission at different time intervals

As a result (Fig. 12), it was found that from approximately 2.25 s onward, Kafka-ML is able to respond to data sent at this frequency without any bottleneck.

5 Conclusions and Future Work

In this chapter, we have presented a methodology for the definition and deployment of soft sensors using Kafka-ML, a framework for the management of ML/AI pipelines with data streams. In this methodology, we have presented step by step how to design a soft sensor including: (1) soft-sensor design and IoT data dispatching and integration into Kafka-ML; (2) ML model design, training, and validation for predicting a feature of interest; (3) trained ML model deployment and data ingestion for real-time soft-sensor inference; and finally, (4) a fast-delivery application with the Kafka-ML visualization. All these steps only require the Kafka-ML framework, which reduces implementation efforts and makes the design and implementation of soft sensors much easier.

A use case with open data from the water quality of UK rivers has been presented to show the potential of Kafka-ML for the deployment of soft sensors. In addition, a series of experiments in different scenarios have been carried out on the proposed methodology, resulting in a low-latency product that can be configurable for fault tolerance and high availability.

As a future work, the following improvements to Kafka-ML and the proposed methodology are considered:

- User experience improvement. From a user's point of view, we have observed some possible improvements that could enhance Kafka-ML for the proposed methodology. For example, the model creation view could be redefined to simplify it, if necessary, so that users outside the computer science domain are able to define ML/AI models in an intuitive way.
- Automatic hyperparameter optimization. The automatic selection of the best hyperparameters for the model would improve the evaluation of different models, allowing to automatically obtain the best combination of training parameters. Furthermore, and related to the aforementioned, this would allow users external to the domain to create fine-grained solutions in a fully automated way. In order to integrate this functionality, one of our choices is to adapt certain features of frameworks related to this area, such as Ray Tune [34] or Optuna [35], and integrate it into a new module to use at any time.
- Pretrained ML models inclusion. Researchers in other areas may already have ML models that perform the job of a soft sensor thanks to previous research efforts. Kafka-ML could open the possibility that these models could be integrated and usable for real-time data reception from sensors.
- Data preprocessing and postprocessing. In most ML application use cases, data preprocessing or postprocessing is necessary for better model performance. While this is achievable before sending the data to the model, either for training or for inference, a potential functionality for Kafka-ML and the proposed methodology would be the integration of data processing activities as well as statistical analysis for this.

References

1. Díaz, M., Martín, C., & Rubio, B., (2016). State-of-the-art, challenges, and open issues in the integration of Internet of Things and cloud computing. *Journal of Network and Computer applications* ,*67*, 99–117.
2. Kadlec, P., Gabrys, B., & Strandt, S., (2009). Data-driven soft sensors in the process industry. *Computers & Chemical Engineering, 33*(4), 795–814.

3. Jiang, Q., Yan, X., Yi, H., & Gao, F., (2020). Data-driven batch-end quality modeling and monitoring based on optimized sparse partial least squares. *IEEE Transactions on Industrial Electronics, 67*(5), 4098–4107.
4. Martín, C., Langendoerfer, P., Zarrin, P. S., Díaz, M., & Rubio, B., (2022) Kafka-ML: Connecting the data stream with ML/AI frameworks. *Future Generation Computer Systems, 126*, 15–33.
5. Kubernetes. https://kubernetes.io/es/docs/concepts/overview/what-is-kubernetes/, Retrieved October 9, 2022.
6. Carnero, A., Martín, C., Torres, D. R.D. R., Garrido, D., Díaz, M., & Rubio, B., (2021). Managing and deploying distributed and deep neural models through Kafka-ML in the cloud-to-things continuum. *IEEE Access, 9*, 125,478–125,495.
7. Welch, G., & Bishop, G., (2001). *An introduction to the Kalman filter*. Chapel Hill: University of North Carolina, Department of Computer Science.
8. Bastin, G., & Dochain, D., (1990). *On-line estimation and adaptive control of bioreactors* (Vol. 1). Amsterdam: Elsevier.
9. Sun, Q., & Ge, Z., (2021). A survey on deep learning for data-driven soft sensors. *IEEE Transactions on Industrial Informatics, 17*(9), 5853–5866.
10. Yan, W., Shao, H., & Wang, X., (2004). Soft sensing modeling based on support vector machine and Bayesian model selection. *Computers & Chemical Engineering, 28*(8), 1489–1498.
11. Rani, A., Singh, V., & Gupta, J., (2013). Development of soft sensor for neural network based control of distillation column. *ISA Transactions, 52*(3), 438–449.
12. Singh, S., Anil, A. G., Kumar, V., Kapoor, D., Subramanian, S., Singh, J., Ramamurthy, P. C., (2022). Nitrates in the environment: A critical review of their distribution, sensing techniques, ecological effects and remediation. *Chemosphere, 287*, 131996.
13. Haimi, H., Corona, F., Mulas, M., Sundell, L., Heinonen, M., Vahala, R. (2015). Shall we use hardware sensor measurements or soft-sensor estimates? Case study in a full-scale WWTP. *Environmental Modelling & Software, 72*, 215–229.
14. Zare, A. H., Bayat, V., Daneshkare, A. P., (2011). Forecasting nitrate concentration in groundwater using artificial neural network and linear regression models. *International Agrophysics, 25*, 187–192.
15. Corona, F., Mulas, M., Haimi, H., Sundell, L., Heinonen, M., & Vahala, R., (2013). Monitoring nitrate concentrations in the denitrifying post-filtration unit of a municipal wastewater treatment plant. *Journal of Process Control, 23*(2), 158–170. IFAC World Congress Special Issue.
16. Haimi, H., Mulas, M., Corona, F., & Vahala, R., (2013). Data-derived soft-sensors for biological wastewater treatment plants: An overview. *Environmental Modelling & Software, 47*, 88–107.
17. Haghbin, M., Sharafati, A., & Dixon, B. e. a., (2021). Application of soft computing models for simulating nitrate contamination in groundwater: Comprehensive review, Assessment and future opportunities. *Computat Methods Eng, 28*, 3569–3591.
18. Kreps, J., Narkhede, N., Rao, J., et al., (2011). Kafka: A distributed messaging system for log processing. In *Proceedings of the NetDB* (Vol. 11, pp. 1–7).
19. Wan, Z., Zhang, Z., Yin, R., & Yu, G., (2022). KFIML: Kubernetes-based fog computing IoT platform for online machine learning. *IEEE Internet of Things Journal, 9*(19), 19463–19476.
20. Kumar, K., Sharma, N. A., & Ali, A. B. M. S., (2021). Machine learning solutions for investigating streams data using distributed frameworks: Literature review. In *2021 IEEE Asia-Pacific Conference on Computer Science and Data Engineering (CSDE)* (pp. 1–6).
21. Ariyaluran Habeeb, R. A., Nasaruddin, F., Gani, A., Amanullah, M. A., Abaker Targio Hashem, I., Ahmed, E., & Imran, M., (2022). Clustering-based real-time anomaly detection—a breakthrough in big data technologies. *Transactions on Emerging Telecommunications Technologies, 33*(8), e3647. e3647 ett.3647.
22. Kubeflow, (2022). https://www.kubeflow.org/.
23. Yeager, L., Bernauer, J., Gray, A., & Houston, M., (2015). DIGITS: The deep learning GPU training system. In *ICML 2015 AutoML Workshop* (pp. 1–4).

24. Liberty, E., Karnin, Z., Xiang, B., Rouesnel, L., Coskun, B., Nallapati, R., Delgado, J., Sadoughi, A., Astashonok, Y., Das, P., et al., (2020). Elastic machine learning algorithms in Amazon SageMaker. In *Proceedings of the 2020 ACM SIGMOD International Conference on Management of Data* (pp. 731–737).
25. Algorithmia - the enterprise MLOps platform, (2022). https://algorithmia.com/.
26. Valohai, (2022). https://www.valohai.com/.
27. Vohra, D., (2016). Apache Avro. In *Practical Hadoop Ecosystem* (pp. 303–323). Berlin: Springer.
28. Torres, D. R., Martín, C., Rubio, B., & Díaz, M., (2021). An open source framework based on Kafka-ML for distributed DNN inference over the cloud-to-things continuum. *Journal of Systems Architecture, 118*, 102214.
29. Olston, C., Fiedel, N., Gorovoy, K., Harmsen, J., Lao, L., Li, F., Rajashekhar, V., Ramesh, S., & Soyke, J., (2017). TensorFlow-serving: Flexible, high-performance ML serving. arXiv preprint arXiv:1712.06139.
30. BentoML: A faster way to ship your models to production, (2022). https://www.bentoml.com/.
31. Grafana: The open observability platform, (2022). https://grafana.com/.
32. OpenDataNI, River water quality monitoring 1990 to 2018... http://data.europa.eu/88u/dataset/river-water-quality-monitoring-1990-to-2018.
33. Srivastava, N., Hinton, G., Krizhevsky, A., Sutskever, I., & Salakhutdinov, R., (2014). Dropout: A simple way to prevent neural networks from overfitting. The Journal of Machine Learning Research, 15(1), 1929–1958.
34. Liaw, R., Liang, E., Nishihara, R., Moritz, P., Gonzalez, J. E., & Stoica, I., (2018). Tune: A research platform for distributed model selection and training. arXiv preprint arXiv:1807.05118.
35. Akiba, T., Sano, S., Yanase, T., Ohta, T., & Koyama, M., (2019). Optuna: A next-generation hyperparameter optimization framework. In *Proceedings of the 25th ACM SIGKDD International Conference on Knowledge Discovery and data mining* (pp. 2623–2631).

Index

R. Sharma et al. (eds.), *Data Analytics for Internet of Things Infrastructure*, Internet of Things, https://doi.org/10.1007/978-3-031-33808-3

The manufacturer's authorised representative in the EU is Springer Nature Customer Service Centre GmbH, Europaplatz 3, 69115 Heidelberg, Germany. If you have any concerns regarding our products, please contact ProductSafety@springernature.com

Printed and bound by CPI Group (UK) Ltd, Croydon, CR0 4YY

28/09/2026

02224075-0014